Lecture Notes in Mathematics 2103

For further volumes:
http://www.springer.com/series/304

Manfred Knebusch • Tobias Kaiser

Manis Valuations and Prüfer Extensions II

Springer

Manfred Knebusch
Fakultät f.Mathematik
Universität Regensburg
Regensburg, Germany

Tobias Kaiser
Fakultät f.Informatik u.Mathematik
Universität Passau
Passau, Germany

ISBN 978-3-319-03211-5 ISBN 978-3-319-03212-2 (eBook)
DOI 10.1007/978-3-319-03212-2
Springer Cham Heidelberg New York Dordrecht London

Lecture Notes in Mathematics ISSN print edition: 0075-8434
ISSN electronic edition: 1617-9692

Library of Congress Control Number: 2014930747

Mathematics Subject Classification (2010): 13A18, 13A15, 13F05, 13F30, 13B30, 11J61

Printed on acid-free paper

Springer is part of Springer Science+Business Media (www.springer.com)

Preface

This is the second volume of a book devoted to Prüfer extensions of rings (here always commutative and with 1) and to valuations on rings related to Prüfer extensions. Following the three Chaps. I–III of Volume I [KZ], the present volume contains three more Chaps. 1–3, and nevertheless by no means exhausts what is known nowadays about Prüfer extensions even by us.

While Chaps. I–III of Volume I are strongly intertwined, Chaps. 1–3 of Volume II are nearly independent of each other.

The reader should follow Sects. 1–3 of Chap. 1, but then may read the rest of Chaps. 1–3 in any order and also with long time intervals in between.

Also by no means the whole content of Volume I plays a role in Volume II. Needed are from Chap. I: §1–§5 and §6 up to Corollary 6.11; from Chap. II: §1, §2 up to Theorem 2.6, §3 up to Theorem 3.3, §5 up to Proposition 5.2, major parts of §6–§8, and §10 up to Theorem 10.2. In Chap. III only §1–§3 are relevant to Volume II, the one exception being the "PM-hull" $\mathrm{PM}(A, R, \mathfrak{p})$ of a pair $(A, \mathfrak{p})$ in a ring extension $A \subset R$, with $\mathfrak{p}$ a prime ideal, better a maximal ideal, of A (cf. [Vol. I, Theorem III.5.3][1]), which shows up in Chap. 1.

Nowadays there exists an enormously extended literature on Prüfer domains and a still large body of papers about the more general "Prüfer rings with zero divisors." In our framework Prüfer rings with zero divisors are the same objects as the Prüfer extensions $A \subset \mathrm{Quot}(A)$, or sometimes $A \subset Q(A)$, with A a commutative ring, $\mathrm{Quot}(A)$ its total ring of quotients and $Q(A)$ its complete ring of quotients (cf. [Vol. I, p. 39]). If we assume that A is an integral domain, we fall back on Prüfer domains. Then $\mathrm{Quot}(A) = Q(A)$.

On the other hand the literature on Prüfer extensions is meager. But Prüfer extensions seem to be mandatory in particular for large parts of real and p-adic algebra. All this is discussed in the Introduction to the book in Volume I. We urge the uninitiated reader to look first into the Introduction, also for understanding our goals and motivation in writing this book.

---

[1] This means Theorem 5.3 in Chap. III of Volume I.

For the expert we mention one item, which explicitly has nearly no presence in the literature on Prüfer extensions (but see [Huc, pp. 30–31] for Prüfer rings with zero divisors). This is the theory of *PM-valuations* and *PM-extensions* ("PM" is an acronym for "Prüfer–Manis.") A valuation v on a ring R is called PM, if v is Manis and the extension $A_v \subset R$ is Prüfer, and then $A_v \subset R$ is called a PM-extension. PM-extensions can be characterized in various ways without mentioning valuations. Perhaps the most striking one runs as follows: if A is a subring of a ring R, then $A \subset R$ is PM iff A is integrally closed in R and the set of subrings B of R with $A \subset B \subset R$ is totally ordered by inclusion ([Vol. I, Theorem III.3.1]) If $A \subset R$ is PM, then the Manis valuation v on R with $A_v = A$ is uniquely determined up to equivalence, cf. [Vol. I, Theorem III.3.12]. Manis valuations in general still show some pathologies, which are absent for PM-valuations. All these are explained in [Vol. I, Chap. III §1–§3].

All three chapters of Volume II deal with families of valuations instead of studying properties of just one valuation. PM-valuations play here center stage.

In Chap. 1 we aim at analyzing a Prüfer extension $A \subset R$ in terms of the set $S(R/A)$ of nontrivial PM-valuations v on R over A (i.e., with $A \subset A_v$), called the *restricted PM-spectrum* of R over A, in order to understand the lattice of overrings of A in R. We engage $S(R/A)$ as poset[2] ($v \le w$ iff $A_v \subset A_w$), although it would be more comprehensive to view $S(R/A)$ as a topological space, namely as subspace of the valuation spectrum Spv(R), equipped with one of its well-established spectral topologies (cf. [HK]), whose specialization relation restricts to the partial ordering above. We exhibit several types of Prüfer extensions, where the poset viewpoint has seizable success.

Chapter 2 is devoted to approximation in a ring with respect to finite—and then also suitable infinite—systems of Manis valuations. Primordial approximation theorems already show up in Manis' seminal paper [M]. In Chap. 2 we embed the deep results of Gräter on approximation by a finite system of Manis valuations ([Gr], [Gr$_1$], [Gr$_2$], cf. also [Al$_1$]) in our framework. Gräter's theorems may be viewed as a grand generalization of the weak approximation theorems for finitely many valuations on a field, as presented, e.g., in the books [E] and [Rib]. They relate to arbitrary Prüfer extensions,[3] a novum in the literature on approximations in the beginning eighties of last century. It fits well with our thinking that all Manis valuations relevant in Gräter's papers turn out to be PM.

A second source for our Chap. 2 is a paper by Griffin from the late sixties [G$_3$], where—in succession of work by Ribenboim—a "reinforced approximation theorem" has been proved and analyzed for certain infinite systems of valuations on the quotient fields of suitable Prüfer domains, there called "generalized Krull rings." This theorem generalizes the classical strong approximation theorem over a

[2]= partially ordered set.

[3]Gräter uses the equivalent term "R-Prüferring," which we also use alternatively to "Prüfer extension," cf. [Vol. I, Definition 1 in I §5].

Dedekind domain and then Krull domain (cf. [Bo, Chap. 7], [E]). In [G_2] Griffin has then established approximation theorems for a special class of Prüfer extensions.

In [Gr_2] Gräter shows the "reinforced approximation theorem" for arbitrary Prüfer extensions given by finitely many valuations and expands this also to Prüfer extensions "of finite type." Moreover, he elaborates deep connections between his various approximation theorems in [Gr]–[Gr_2]. His results have been incorporated only once in a book, namely the book [Al-M] by Alajbegović and Močkoř. There approximation theorems are investigated also in other systems, not only rings. We give a new presentation of Gräter's results, stressing thereby the case of possibly infinite families "with finite avoidance."[4]

We hope that Chap. 2 provides easier access to Gräter's approximation theorems and their enlargements than [Al-M], due to the fact that we can refer to a much more complete general theory of Prüfer extensions than present in [Al-M]. Nevertheless the way from classical weak and strong approximation to the "reinforced approximation theorem" at the end of Chap. 2 remains long and needs patience by the reader. Among all chapters in the book Chap. 2 is the most demanding.

Chapter 3 aims at applications of the Prüfer theory to arbitrary (commutative) ring extensions. For a given ring extension $A \subset R$ we produce a Prüfer extension $B \subset T$ in various ways, together with a homomorphism $j : R \to T$ mapping A into B, called a *Kronecker extension* of $A \subset R$. Kronecker extensions generalize the Kronecker function rings (e.g., [Gi, §32], [Ha-K]) in the classical literature.

The restricted PM-spectrum $S(T/B)$ of a Kronecker extension $B \subset T$ gives us a family $(w \circ j \mid w \in S(T/B))$ of valuations on R over A, which can serve for a valuation theoretic description of various A-modules in R, since $S(T/B)$ does this so well for B-modules in T.

For constructing Kronecker extensions we use—suitably defined—*star operations* $I \mapsto I^*$ on the semiring $J(A, R)$ of all A-submodules of R (cf. Chap. 3, Sect. 3) and then also "partial star operations" defined on certain subsets of $J(A, R)$, cf. Chap. 3, Sect. 7.

In the classical setting, where A is a domain and R its quotient field, such star operations have been defined on the set of fractional ideals of A, leading to a *star multiplicative ideal theory* with a very extended literature. In Chap. 3 we display, beside Kronecker extensions, the basics of such a theory in arbitrary ring extensions. Everything is more complex than in the classical theory, due to the fact that for a ring extension $A \subset R$ a principal ideal of A usually is not R-invertible. Nevertheless the pattern is natural and manageable. But we go in the book only so far, that a reader with experience in the classical star multiplicative ideal theory can get convinced, that establishing analogues and/or generalizations of many of the deeper results in the classical theory is possible and worth the labor.

---

[4] A new term by us; Griffin and most other authors speak instead of "Prüfer domains of finite character." Gräter uses the term "R-Prüfer rings of finite type." Our term aims at catching more general situations.

The interested reader should first have a look at the Summary of Chap. 3 below (pp. 123–125) to get a more detailed impression, what can be found in this long chapter and what cannot.

Acknowledgments

Among the many mathematicians of present time, whose work is related to this volume, we feel special indebtness to Professors Franz Halter-Koch and Joachim Gräter. Halter-Koch had the insight to define Kronecker function rings axiomatically without using star operations. This has been instrumental for Chap. 3. Gräter obtained already more than 30 years ago, starting with his dissertation TU Braunschweig 1980, a full-fledged deep and general approximation theory for Manis valuations in Prüfer extensions, the essential content of present Chap. 2 (see above). We are sorry to say that there are other important results by Gräter on Prüfer extensions [Gr3], which we could not incorporate in the volume, to keep it in reasonable size.

We are grateful to our friend Digen Zhang, coauthor of Volume I, who has accompanied early versions of Chaps. 1–3 until about 2005, but then has been called to other duties.

Regensburg, Germany Manfred Knebusch
Passau, Germany Tobias Kaiser
December 2012

Contents

Contents of Volume I

Chapter 1
Overrings and PM-Spectra

Summary. If $A \subset R$ is an extension of rings (always commutative with 1), then the set $S(R/A)$, consisting of (the equivalence classes of) all nontrivial PM-valuations v on R (cf. [Vol. I, Chap. I §5]) with $A_v \supset A$, is a poset (= partially ordered set) under the relation

$$v \leq w \Leftrightarrow A_v \subset A_w \Leftrightarrow w \text{ is coarser than } v.$$

We call this poset $S(R/A)$ the *restricted PM-spectrum* of R over A. The poset $S(R/A)$ turns out to be a forest, i.e. has no cycles. {A small point: One has to add to each tree (= connected component of $S(R/A)$) a trivial PM-valuation to obtain a decent forest, where the trees have roots.}

If the extension $A \subset R$ is Prüfer, then we have an order inverting bijection $Y(R/A) \xrightarrow{\sim} S(R/A)$ from the poset $Y(R/A) \subset \operatorname{Spec} A$, consisting of all R-regular prime ideals of A, to $S(R/A)$, which sends each $\mathfrak{p} \in Y(R/A)$ to the valuation $v = v_{\mathfrak{p}}$ with $A_v = A_{[\mathfrak{p}]}$, as is very well known from Volume I. Moreover every overring B of A in R is the intersection of the overrings A_v of A in R with v running through $S(R/A)$, as is again well known from Volume I. Of course, then B is already the intersection of the rings A_v, with v running through the set $\omega(R/B)$ of minimal elements in $S(R/B)$. It corresponds to the set $\Omega(R/B)$ of maximal R-regular ideals of B.

In Chap. 1 we aim at an understanding of the overrings of a Prüfer extension $A \subset R$ in terms of the poset $S(R/A)$ in a combinatorial way. Combinatorial methods seem to grasp as long as the forest $S(R/A)$ is not "too dense". After generalities on PM-spectra in Sects. 1–3 we focus in Sect. 4 on the classes of *"PF-extensions"*, an acronym for "Prüfer extensions with finite avoidance", and in Sect. 5 on the subclass of *"PM-split"* extensions. Here a Prüfer extension $A \subset R$ is called PF, if for any $x \in R$ the set of $v \in S(R/A)$ with $v(x) < 0$ is finite, and $A \subset R$ is PM-split if in addition the trees in $S(R/A)$ have no ramifications, hence are chains.

In any ring extension $A \subset R$ there exists a unique maximal overring B of A such that $A \subset B$ is PF (Theorem 4.8), and the same holds with "PM-split" instead of "PF"

M. Knebusch and T. Kaiser, *Manis Valuations and Prüfer Extensions II*, Lecture Notes in Mathematics 2103, DOI 10.1007/978-3-319-03212-2_1,

(Proposition 5.9[1]). We call these overrings the *PF-hull* and the *PM-split hull* of A in R, denoted by $\mathrm{PF}(A, R)$ and $\mathrm{PMS}(A, R)$ respectively. Of course, $\mathrm{PMS}(A, R) \subset \mathrm{PF}(A, R)$.

If $A \neq \mathrm{PF}(A, R)$, then $A \neq \mathrm{PMS}(A, R)$ (cf. Proposition 5.18). If follows that every PF-extension can be reached by a (perhaps infinite) tower of PM-split extensions. We do not dwell on this point in the book, leaving much space for future research.

A second theme in Chap. 1 is non-trivial factorizations of a Prüfer extension $A \subset R$. By such a factorization we mean a decomposition $R = U \times_A V$, i.e. $UV = R, U \cap V = A$, with overrings U, V in R different from A. Generalities on factorization and the related theory of polars of overrings are already contained in [Vol. I, Chap. II §7 & §8].

We call an overring B of A in R *irreducible*, if B does not admit a non-trivial factorization over A, and we call B *coirreducible*, if R does not admit a non-trivial factorization over B.

The overrings A_v and their polars A_v° with $v \in S(R/A)$ are coirreducible resp. irreducible, and if $A \subset R$ is PM-split then $R = A_v \times_A A_v^\circ$. These are the easiest examples of such overrings and decompositions. In Sects. 6–8 a theory largely parallel to Sects. 4 and 5 is developed, which is based on families of irreducible and coirreducible overrings instead of the families (A_v) and (A_v°). The key property under which the theory works is finite avoidance of these new families, defined in the appropriate way. Here we rely on assumptions concerning $S(R/A)$ to much lesser extent than in Sects. 4 and 5.

In the last Sect. 9 we study the extensions $\tilde{A} \subset R\tilde{A}$ gained from a Prüfer extension $A \subset R$ by an integral ring extension $A \subset \tilde{A}$ in an overring T of A which contains both R and $\tilde{A}$. It is known from Chap. I that $\tilde{A} \subset R\tilde{A}$ is again Prüfer ([Vol. I, Theorem I.5.9]). We analyze the strong relations between the PM-spectra $S(R/A)$ and $S(R\tilde{A}/\tilde{A})$ and then transfer several results about $A \subset R$ from Sects. 4–8 to the extension $\tilde{A} \subset R\tilde{A}$, although the finite avoidance hypotheses assumed for $A \subset R$ in Sects. 4–8 most often are lost in the extension $\tilde{A} \subset R\tilde{A}$. They usually remain valid only if the ring extension $A \subset \tilde{A}$ is finite.

1 A Look at Overrings and Their Polars

In this section $A \subset R$ is a ring extension. In the beginning we will make no further assumption on $A \subset R$, but later we will assume that this extension is Prüfer.

Definition 1. Let B be an R-overring of A such that A is ws (= weakly surjective) in B. We denote by $Z(B/A)$ the image of the set $Y(R/B)$ of R-regular prime ideals

[1]This means Proposition 5.9 in Chap. 1. If we cite this result in its own section we just write Proposition 9. If we cite this result in another chapter we write Proposition 1.5.9.

of B under the restriction map Spec $B \to$ Spec A, $\mathfrak{P} \mapsto \mathfrak{P} \cap A$. If necessary, we write more precisely $Z(B/A, R)$ instead of $Z(B/A)$.[2]

Recall from [Vol. I, Chap. I §4] that, if $A \subset B$ is ws, then, for a given prime ideal $\mathfrak{p}$ in the image $X(B/A)$ of Spec $B \to$ Spec A, the ideal $\mathfrak{p}B$ is the unique prime ideal of B lying over A. Clearly $\mathfrak{p}$ is R-regular iff $\mathfrak{p}B$ is R-regular. Thus $Z(B/A) = X(B/A) \cap Y(R/A) = Y(R/A) \setminus Y(B/A)$. We have an order preserving bijection $\mathfrak{p} \mapsto \mathfrak{p}B$ from $Z(B/A)$ to $Y(R/B)$.

Proposition 1.1. *Assume that B is an R-overring of A and A is ws both in B and R.*

i) If $\mathfrak{p}$ is a prime ideal of A then $\mathfrak{p} \in Z(B/A)$ iff $B \subset A_{[\mathfrak{p}]} \neq R$. In this case $B_{[\mathfrak{p}B]} = A_{[\mathfrak{p}]}$.
ii) $B = \bigcap_{\mathfrak{p} \in Z(B/A)} A_{[\mathfrak{p}]}$.

Proof. i): Let $\mathfrak{p} \in$ Spec A be given. We know that $\mathfrak{p}$ is R-regular iff $A_{[\mathfrak{p}]} \neq R$. ([Vol. I, Lemma III.1.1]). By [Vol. I, Theorem I.3.13] we have $B \subset A_{[\mathfrak{p}]}$ iff $\mathfrak{p}B \neq B$, and then $A_{[\mathfrak{p}]} = B_{[\mathfrak{p}B]}$. This gives the claim.
ii): We know from [Vol. I, Proposition III.1.7] that B is the intersection of the rings $B_{[\mathfrak{P}]}$ with $\mathfrak{P}$ running through $Y(R/B)$. These are the rings $A_{[\mathfrak{p}]}$ with $\mathfrak{p}$ running through $Z(B/A)$. □

Here is another description of $Z(B/A)$ in an important special case.

Proposition 1.2. *Assume that $B = A[I^{-1}]$ with I an R-invertible ideal of A. Then $Y(B/A)$ is the set of all prime ideals $\mathfrak{p}$ of A containing I, hence $Z(B/A)$ is the set of all $\mathfrak{p} \in Y(R/A)$ with $I \not\subset \mathfrak{p}$.*

Proof. Let $\mathfrak{p} \in$ Spec A be given. If $\mathfrak{p}B = B$ there exists some natural number n with $1 \in \mathfrak{p}I^{-n}$, since B is the union of the modules I^{-n} with n running through $\mathbb{N}$. It follows that $I^n \subset \mathfrak{p}$ for some n, hence $I \subset \mathfrak{p}$, since $\mathfrak{p}$ is prime. Conversely, if $I \subset \mathfrak{p}$, then $A = II^{-1} \subset \mathfrak{p}I^{-1} \subset \mathfrak{p}B$, hence $\mathfrak{p}B = B$. □

Corollary 1.3. *Let $\mathfrak{p} \in$ Spec A be given.*

i) An ideal I of A is $A_{[\mathfrak{p}]}$-invertible iff I is R-invertible and $I \not\subset \mathfrak{p}$.
ii) If A is tight in $A_{[\mathfrak{p}]}$, then $A_{[\mathfrak{p}]}$ is the union of the modules I^{-1} with I running through the R-invertible ideals of A not contained in $\mathfrak{p}$.

Proof. i): We may assume that I is R-invertible. Let $B := A[I^{-1}]$. The ideal I is $A_{[\mathfrak{p}]}$-invertible iff $I^{-1} \subset A_{[\mathfrak{p}]}$ iff $B \subset A_{[\mathfrak{p}]}$ iff $\mathfrak{p}$ is not B-regular (cf. [Vol. I, Theorem I.3.13]). By Proposition 2 this means that $I \not\subset \mathfrak{p}$.
Claim ii) now follows from [Vol. I, Theorem II.4.11.i]. □

---

[2] Formally this definition makes sense for an arbitrary ring extension, but the set $Z(B/A)$ will be useful only in the case that A is ws in B.

From now on **we assume that A is Prüfer in R**. Then A is ws in every R-overring B of A. Thus Proposition 1 gives us a description of such a ring B by the subfamily $(v_{\mathfrak{p}} \mid \mathfrak{p} \in Z(B/A))$ of $(v_{\mathfrak{p}} \mid \mathfrak{p} \in Y(R/A))$, and Corollary 3 gives us a more detailed description of the rings $A_{[\mathfrak{p}]}$, since now A is automatically tight in $A_{[\mathfrak{p}]}$.

We state an important general fact.

Theorem 1.4. *Let $(B_i \mid i \in I)$ be a finite family of R-overrings of A. Let further v be a valuation on R with $A_v \supset \bigcap_{i \in I} B_i$. Then there exists an index $i \in I$ with $A_v \supset B_i$.*

Proof. We may assume that $A_v \neq R$. The set $R \setminus A_v$ is closed under multiplication. Thus A_v is PM in R (cf. [Vol. I, Proposition I.5.1.ii]). We have $A_v = A_v(\bigcap_{i \in I} B_i) = \bigcap_{i \in I}(A_v B_i)$, by [Vol. I, Theorem II.1.4(4)]. Now the R-overrings of A_v form a chain ([Vol. I, Theorem III.3.1]). Thus there exists some $k \in I$ with $A_v B_k \subset A_v B_i$ for all $i \in I$. We have $A_v B_k = A_v$, hence $B_k \subset A_v$. □

Corollary 1.5. *Assume again that $(B_i | i \in I)$ is a finite family of R-overrings of A. Then $X(\bigcap_{i \in I} B_i/A) = \bigcup_{i \in I} X(B_i/A)$ and $Z(\bigcap_{i \in I} B_i/A) = \bigcup_{i \in I} Z(B_i/A)$.*

Proof. Observe that, if C is an R-overring of A and $\mathfrak{p} \in \operatorname{Spec} A$, then $\mathfrak{p} \in X(C/A)$ iff $C \subset A_{[\mathfrak{p}]}$. Thus Theorem 4, applied to $v_{\mathfrak{p}}$, gives us $X(B/A) = \bigcup_{i \in I} X(B_i/A)$. Intersecting with $Y(R/A)$ we obtain the second claim. □

Notice that this corollary has already been proved in [Vol. I, Chap. II §8] in a different way ([Vol. I, Theorem II.8.13.a]).

Remark 1.6. If $(B_i \mid i \in I)$ is any family of R-overrings of A and B is the subring of R generated by the B_i, then it is obvious from Proposition 1 that $X(B/A) = \bigcap_{i \in I} X(B_i/A)$ and $Z(B/A) = \bigcap_{i \in I} Z(B_i/A)$. □

Proposition 1.7. *Let $(\mathfrak{p}_i \mid i \in I)$ be a finite family in $Y(R/A)$ with $\mathfrak{p}_i \not\subset \mathfrak{p}_j$ for $i \neq j$, and let $B := \bigcap_{i \in I} A_{[\mathfrak{p}_i]}$. Then $\Omega(R/B) = \{\mathfrak{p}_i B \mid i \in I\}$.*

Proof. We have $Z(A_{[\mathfrak{p}_i]}/A) = \{\mathfrak{p} \in Y(R/A) \mid \mathfrak{p} \subset \mathfrak{p}_i\}$, essentially by Proposition 1, and $Z(B/A)$ is the union of these sets by Corollary 5. Thus the $\mathfrak{p}_i$ are precisely all maximal elements of $Z(B/A)$. Now recall that we have an order preserving bijection $\mathfrak{p} \mapsto \mathfrak{p}B$ from $Z(B/A)$ to $Y(R/B)$. □

Remarks. Proposition 7 is an old result of Griffin [G_2, Prop. 11]. The proof by Griffin is by no means superseded by the present one and is quite interesting. In the next Sect. 2 we will establish generalizations of Theorem 4, Corollary 5 and Proposition 7.

Recall from [Vol. I, Chap. II §7] that the polar I° of an A-submodule I of R with $A \subset I$ is an R-overring of A.

Theorem 1.8. *Assume that I is an A-submodule of R containing A. Then I° is the intersection of the rings $A_{[\mathfrak{m}]}$ with $\mathfrak{m}$ running through all R-regular maximal ideals of A such that $I \not\subset A_{[\mathfrak{m}]}$. If $I = B$ is an R-overring of A, then*

$$B^\circ = \bigcap_{\mathfrak{m}\in\Omega(B/A)} A_{[\mathfrak{m}]}.$$

Proof. Let $B := A[I]$. Then $I^\circ = B^\circ$ (cf. [Vol. I, Corollary II.7.8]), and, for any $\mathfrak{p} \in \operatorname{Spec} A$, we have $I \subset A_{[\mathfrak{p}]}$ iff $B \subset A_{[\mathfrak{p}]}$. Thus we may replace I by B, and we will assume henceforth that $I = B$ is an R-overring of A.

We have $Y(B/A) = \{\mathfrak{p} \in Y(R/A) \mid B \not\subset A_{[\mathfrak{p}]}\}$ by Proposition 1 above. Let $C := \bigcap_{\mathfrak{p}\in Y(B/A)} A_{[\mathfrak{p}]}$. If $\mathfrak{p} \in Y(R/A)$ is given, then $A = B \cap B^\circ \subset A_{[\mathfrak{p}]}$, hence $B \subset A_{[\mathfrak{p}]}$ or $B^\circ \subset A_{[\mathfrak{p}]}$ by Theorem 4. It follows that $B^\circ \subset C$. On the other hand, $B \cap C = \bigcap_{\mathfrak{p}\in Y(R/A)} A_{[\mathfrak{p}]} = A$, hence $C \subset B^\circ$. Thus $C = B^\circ$. If $\mathfrak{p} \in Y(B/A)$, we may choose some $\mathfrak{m} \in \Omega(B/A)$ with $\mathfrak{p} \subset \mathfrak{m}$, hence $A_{[\mathfrak{p}]} \supset A_{[\mathfrak{m}]}$. It follows that $C = \bigcap_{\mathfrak{m}\in\Omega(B/A)} A_{[\mathfrak{m}]}$. □

Corollary 1.9. *Let $\mathfrak{p}$ be a prime ideal of A. The polar $A^\circ_{[\mathfrak{p}]} := (A_{[\mathfrak{p}]})^\circ$ of $A_{[\mathfrak{p}]}$ is the intersection of the rings $A_{[\mathfrak{m}]}$ with $\mathfrak{m}$ running through all elements of $\Omega(R/A)$ different from $\mathfrak{p}$. In particular, if $\mathfrak{p} \notin \Omega(R/A)$ (i.e. $\mathfrak{p}$ is not maximal, or $\mathfrak{p}$ is maximal but not R-regular), then $A^\circ_{[\mathfrak{p}]} = A$.*

Proof. Let $\mathfrak{m} \in \Omega(R/A)$. Then $\mathfrak{m}$ is not $A_{[\mathfrak{p}]}$-regular iff $A_{[\mathfrak{p}]} \subset A_{[\mathfrak{m}]}$ iff $\mathfrak{m} \subset \mathfrak{p}$ iff $\mathfrak{m} = \mathfrak{p}$. Thus $\Omega(A_{[\mathfrak{p}]}/A) = \Omega(R/A) \setminus \{\mathfrak{p}\}$. The claim now follows from Theorem 8. □

Recall from [Vol. I, Chap. II §7] that an R-overring B of A is called a *factor* of R over A if $BB^\circ = R$.

Proposition 1.10. *Let B be an R-overring of A.*

a) Then $Y(B/A) \cap Y(B^\circ/A) = \emptyset$.
b) The following are equivalent:

(1) B is a factor of R over A.
(2) $Y(B/A) \cup Y(B^\circ/A) = Y(R/A)$.
(3) $Z(B^\circ/A) = Y(B/A)$.
(4) $Z(B/A) = Y(B^\circ/A)$.
(5) $BA_{[\mathfrak{p}]} = R$ for every $\mathfrak{p} \in Z(B^\circ/A)$.

Proof. a): By Corollary 5 above we have $X(B/A) \cup X(B^\circ/A) = X(B \cap B^\circ/A) = X(A/A) = \operatorname{Spec} A$. Taking complements in $\operatorname{Spec} A$ gives the claim.
b): By Remark 6 above we have $X(B/A) \cap X(B^\circ/A) = X(BB^\circ/A)$. Taking complements we obtain $Y(B/A) \cup Y(B^\circ/A) = Y(BB^\circ/A)$. Thus (2) means that $Y(BB^\circ/A) = Y(R/A)$. Since an R-overring D of A is uniquely determined by the set $Y(D/A)$, (2) is equivalent to $R = BB^\circ$, i.e. (1). The equivalences (2) $\Leftrightarrow$ (3) and (2) $\Leftrightarrow$ (4) now follow from the fact that, for any R-overring D of A, the set $Z(D/A)$ is the complement of $Y(D/A)$ in $Y(R/A)$.
(1) $\Rightarrow$ (5): We have $R = BB^\circ$. If $\mathfrak{p} \in X(B^\circ/A)$, then $B^\circ \subset A_{[\mathfrak{p}]}$. This gives us $R = BA_{[\mathfrak{p}]}$.

(5) $\Rightarrow$ (1): Suppose that $BB^\circ \neq R$. Then there exists some $\mathfrak{p} \in Y(R/A)$ with $BB^\circ \subset A_{[\mathfrak{p}]}$. It follows from this and the assumption (5) that $R = BA_{[\mathfrak{p}]} = A_{[\mathfrak{p}]}$. But $R \neq A_{[\mathfrak{p}]}$, since $\mathfrak{p}$ is R-regular. This contradiction proves that $BB^\circ = R$. □

Lemma 1.11. *Let $\mathfrak{p}$ be a prime ideal of A and C an R-overring of A. The pair $(A,\mathfrak{p})$ is PM in C iff $A_{[\mathfrak{p}]} \cap C = A$.*

Proof. We have $A_{[\mathfrak{p}]} \cap C = A^C_{[\mathfrak{p}]}$, and the pair $(A^C_{[\mathfrak{p}]}, \mathfrak{p}^C_{[\mathfrak{p}]})$ is PM in C, since A is Prüfer in C. Thus $A_{[\mathfrak{p}]} \cap C = A$ implies, that $(A,\mathfrak{p})$ is PM in C. Conversely if this holds, then certainly $(A,\mathfrak{p})$ is saturated (cf. [Vol. I, Definition 4 in II §5]) in C, i.e. $A^C_{[\mathfrak{p}]} = A$. □

Remark. In this proof we did not need that A is Prüfer in R, but only that A is Prüfer in C.

Theorem 1.12. *Let $\mathfrak{p}$ be a prime ideal of A. The polar of $A_{[\mathfrak{p}]}$ is the PM-hull of the pair $(A,\mathfrak{p})$ (cf. [Vol. I, Chap. III §5]) in R, $A^\circ_{[\mathfrak{p}]} = \mathrm{PM}(A,\mathfrak{p},R)$.*

Proof. This is obvious from the preceding lemma. □

Remark. Our arguments here give a new proof of the existence of the PM-hull of a pair $(A,\mathfrak{p})$ in a Prüfer extension $A \subset R$. Taking the existence of the Prüfer hull $P(A,R)$ for granted ([Vol. I, Chap. I §5), this gives us the existence of the PM-hull $\mathrm{PM}(A,\mathfrak{p},R)$ in general, since the PM-hull of $(A,\mathfrak{p})$ in R is—by definition—the same as the PM-hull of $(A,\mathfrak{p})$ in $P(A,R)$. Taking into account Corollary 9 we see that we also have proved anew [Vol. I, Theorem III.5.5]. We avoided these proofs in [Vol. I, Chap. III §5] since we tried to keep [Vol. I, Chap. III] independent from [Vol. I, Chap. II] as much as possible.

Theorem 12 tells us in particular that $A^\circ_{[\mathfrak{p}]} = A$, if $\mathfrak{p} \in Y(R/A)$ but $\mathfrak{p}$ is not maximal. It will be of interest later on to know more about R-overrings B of A having a trivial polar, i.e. $B^\circ = A$. In particular the following question arises. If $(B_i \mid i \in I)$ is a family of R-overrings of A with $B_i^\circ = A$ for every $i \in I$, under which additional assumptions can we conclude that $(\bigcap_{i\in I} B_i)^\circ = A$? Slightly more generally we may pose this question for A-modules in R containing A instead of overrings. We can give a partial answer.

Definition 2. A family $(X_i \mid i \in I)$ of subsets of a set X *has finite avoidance in* X if for every $x \in X$ the set of indices $i \in I$ such that $x \notin X_i$ is finite. We then also say that $(X_i \mid i \in I)$ is a *family with finite avoidance* in X.

Examples 1.13. a) If A is noetherian (and $A \subset R$ is Prüfer), the family of R-overrings $(A_{[\mathfrak{p}]} \mid \mathfrak{p} \in Y(R/A))$ has finite avoidance in R. This follows from [Vol. I, Scholium II.11.8], applied there to the modules $I = A + Ax$ with x running through R.

b) Of course, every finite family of subsets of a set X has finite avoidance in X.

Theorem 1.14. *Let $(U_i \mid i \in I)$ be a family of A-modules in R having finite avoidance in R. Assume that $U_i \supset A$ and $U_i^\circ = A$ for every $i \in I$. Then $(\bigcap_{i\in I} U_i)^\circ = A$.*

Proof. a) We first deal with the case that I is finite, say $I = \{1, 2, \ldots, r\}$. We may proceed by induction on r, and we see immediately that it suffices to consider the case $r = 2$. Let C be an R-overring of A with $C \cap U_1 \cap U_2 = A$. Then $C \cap U_1 \subset U_2^\circ = A$, hence $C \cap U_1 = A$, hence $C \subset U_1^\circ = A$, i.e. $C = A$. This proves $(U_1 \cap U_2)^\circ = A$.

b) We now prove the proposition for I not finite. Let $x \in R \setminus A$ be given. There exists a finite subset J of I such that $x \in U_i$ for every $i \in I \setminus J$. This implies $A[x] \subset \bigcap_{i \in I \setminus J} U_i$. We know from above that $(\bigcap_{i \in J} U_i)^\circ = A$. We conclude that

$$A[x] \cap \bigcap_{i \in I} U_i = A[x] \cap \bigcap_{i \in J} U_i \neq A.$$

Since this holds for every $x \in R \setminus A$, it follows that $(\bigcap_{i \in I} U_i)^\circ = A$. □

2 Families of Modules and Overrings with Finite Avoidance

In Sect. 1 we introduced the notion of "finite avoidance" for families of subsets of a given set (Definition 2 of Sect. 1), and we applied this notion in the theory of polars of overrings in a given Prüfer extension $A \subset R$ (Theorem 1.14). There are more applications to the theory of R-overrings and, more generally, A-modules in R, as we will see.

In this section we embark on a systematic study of families of A-modules with finite avoidance. To a large extend this study can be regarded as a prolongation of parts of the "multiplicative ideal theory" developed in [Vol. I, Chap. II].

Lemma 2.1. *Let A be a ring and M an A-module. For any subset Z of M let AZ denote the A-submodule of M generated by Z. Now let X be a subset of M and $(X_\lambda \mid \lambda \in \Lambda)$ a family of subsets of X with finite avoidance in X. Then $(AX_\lambda \mid \lambda \in \Lambda)$ has finite avoidance in AX.*

Proof. Let $z \in AX$ be given. We can write $z = \sum_{i=1}^{n} a_i x_i$ with elements a_i of A and x_i of X. For every $i \in \{1, \ldots, n\}$ there exists a finite subset S_i of Λ such that $x_i \in X_\lambda$ for every $\lambda \in \Lambda \setminus S_i$. Let $S := \bigcup_{i=1}^{n} S_i$, which is again finite. For every $\lambda \in \Lambda \setminus S$ we have $z \in AX_\lambda$. □

Sometimes we will need the following slightly more general fact. It can be proved in the same way as Lemma 1.

Lemma 2.2. *Let $A \subset R$ be a ring extension, further M an R-module and J an A-submodule of R. For any subset Z of M let JZ denote the A-submodule of M generated by the elements az with $a \in J$, $z \in Z$ (i.e. the set of finite sums of such elements). Assume that X is a subset of M and $(X_\lambda \mid \lambda \in \Lambda)$ a family of subsets of X with finite avoidance in X. Then $(JX_\lambda \mid \lambda \in \Lambda)$ has finite avoidance in JX.* □

Lemma 2.3. *Let $A \subset R$ be a ring extension and $(I_\lambda \mid \lambda \in \Lambda)$ a family of A-submodules of R. Further let J be an A-submodule of R which is invertible in R. Then*

$$\bigcap_{\lambda\in\Lambda}(I_\lambda J) = (\bigcap_{\lambda} I_\lambda)J.$$

Proof. This is evident, since the map $K \mapsto KJ$ from the set of A-submodules K of R to itself is bijective and order preserving (with respect to inclusion), the inverse mapping being $K \mapsto KJ^{-1}$. □

Proposition 2.4. *Assume that $A \subset R$ is a Prüfer extension. Let M be an A-submodule of R and let $(I_\lambda \mid \lambda \in \Lambda)$ be a family of A-submodules of M with finite avoidance in M. Further let J be an R-regular A-submodule of R. Then the family $(I_\lambda J \mid \lambda \in \Lambda)$ has finite avoidance in MJ and*

$$\bigcap_{\lambda\in\Lambda}(I_\lambda J) = (\bigcap_{\lambda\in\Lambda} I_\lambda)J.$$

Proof. Lemma 2 tells us that $(I_\lambda J \mid \lambda \in \Lambda)$ has finite avoidance in MJ. Let $I := \bigcap_{\lambda\in\Lambda} I_\lambda$, and let $x \in \bigcap_{\lambda\in\Lambda}(I_\lambda J)$ be given. We have to verify that $x \in IJ$.

We choose some index $\omega \in \Lambda$. Then we choose an R-regular finitely generated A-module $J_0 \subset J$ such that $x \in I_\omega J_0$. The A-module J_0 is invertible in R, since $A \subset R$ is Prüfer. We have $J_0^{-1}x \subset I_\omega$. The set of indices $S := \{\lambda \in \Lambda \mid J_0^{-1}x \not\subset I_\lambda\}$ is finite, since J_0^{-1} is finitely generated, $J_0^{-1}x \subset M$, and $(I_\lambda \mid \lambda \in \Lambda)$ has finite avoidance in M. We have $x \in I_\lambda J_0$ for every $\lambda \in \Lambda \setminus S$.

We now can choose a finitely generated A-submodule J_1 of J with $J_0 \subset J_1$ and $x \in I_\lambda J_1$ for every $\lambda \in S$. Then $x \in \bigcap_{\lambda\in\Lambda}(I_\lambda J_1)$. Since $A \subset R$ is Prüfer, the module J_1 is again R-invertible. Lemma 3 tells us that $x \in IJ_1 \subset IJ$. □

Corollary 2.5. *Let $A \subset R$ be a Prüfer extension and $(I_\lambda \mid \lambda \in \Lambda)$ a family of R-regular A-submodules of R. Assume that $(I_\lambda \mid \lambda \in \Lambda)$ has finite avoidance in $\sum_{\lambda\in\Lambda} I_\lambda$. Then $\bigcap_{\lambda\in\Lambda} I_\lambda$ is again R-regular.*

Proof. Apply Proposition 4 with $J := R$ and $M := \sum_{\lambda\in\Lambda} I_\lambda$. □

We now can prove a far reaching generalization of Theorem 1.4.

Theorem 2.6. *Let $A \subset R$ be a Prüfer extension and $(I_\lambda \mid \lambda \in \Lambda)$ a family of R-regular A-submodules of R with finite avoidance in $\sum_{\lambda\in\Lambda} I_\lambda$. Let v be a valuation on R with $A_v \supset A$ and $A_v \supset \bigcap_{\lambda\in\Lambda} I_\lambda$. Then there exists some $\mu \in \Lambda$ with $A_v \supset I_\mu$.*

Proof. We may replace v by its special restriction $v|_R$. Then v is PM. Let $M := \sum_{\lambda\in\Lambda} I_\lambda$. Lemma 1 tells us that the family $(I_\lambda A_v \mid \lambda \in \Lambda)$ has finite avoidance in MA_v. Now, for every $\lambda \in \Lambda$, the A_v-submodule MA_v of R is R-regular, hence is v-convex in R (cf. [Vol. I, Theorem III.2.2]). Thus the set $\{I_\lambda A_v \mid \lambda \in \Lambda\}$ is a chain (i.e. totally ordered by inclusion).

We claim that this chain has a smallest element. Indeed, suppose that this is not true. Then there exists a sequence $(\lambda_n \mid n \in \mathbb{N})$ in Λ such that the sequence $(I_{\lambda_n} A_v \mid n \in \mathbb{N})$ of A_v-modules is strictly decreasing. Choosing an element $x \in I_{\lambda_1} A_v$ not contained in $I_{\lambda_2} A_v$, we have $x \notin I_{\lambda_n} A_v$ for $n \geq 2$. This contradicts the fact that $(I_\lambda A_v \mid \lambda \in \Lambda)$ has finite avoidance in MA_v.

We have proved that there exists some $\mu \in \Lambda$ with $I_\mu A_v \subset I_\lambda A_v$ for every $\lambda \in \Lambda$. Using Proposition 4, there with $J = A_v$, we obtain

$$I_\mu A_v = \bigcap_{\lambda \in \Lambda} (I_\lambda A_v) = (\bigcap_{\lambda \in \Lambda} I_\lambda) A_v = A_v,$$

since $\bigcap_{\lambda \in \Lambda} I_\lambda \subset A_v$. Thus $I_\mu \subset A_v$. □

This theorem, in the special case of R-overrings instead of R-regular modules, implies the following corollary by exactly the same argument as used in Sect. 1 to obtain Corollary 1.5 as a consequence of Theorem 1.4.

Corollary 2.7. *Assume that $A \subset R$ is Prüfer and $(B_i \mid i \in I)$ is a family of R-overrings of A with finite avoidance in R. Let $B := \bigcap_{i \in I} B_i$. Then $X(B/A) = \bigcup_{i \in I} X(B_i/A)$ and $Z(B/A) = \bigcup_{i \in I} Z(B_i/A)$.* □

Here is another application of Theorem 6. If $A \subset R$ is a ring extension and U a subset of R we denote by $A[U]$ the subring of R generated by $A \cup U$.

Proposition 2.8. *Assume that $A \subset R$ is Prüfer. Let $(I_\lambda \mid \lambda \in \Lambda)$ be a family of R-regular A-submodules of R with finite avoidance in $\sum_{\lambda \in \Lambda} I_\lambda$. Then*

$$\bigcap_{\lambda \in \Lambda} A[I_\lambda] = A[\bigcap_{\lambda \in \Lambda} I_\lambda].$$

Proof. Every R-overring B of A is the intersection of the rings A_v with v running through all PM-valuations on R such that $A_v \supset B$ (cf. Proposition 1.1). Thus it suffices to prove the following: Let v be a PM-valuation on R with $A_v \supset A[\bigcap_{\lambda \in \Lambda} I_\lambda]$. Then $A_v \supset \bigcap_{\lambda \in \Lambda} A[I_\lambda]$.

A_v contains the A-module $\bigcap_{\lambda \in \Lambda} I_\lambda$. Theorem 6 tells us that there exists some $\mu \in \Lambda$ with $A_v \supset I_\mu$. It follows that $A_v \supset A[I_\mu]$, hence $A_v \supset \bigcap_{\lambda \in \Lambda} A[I_\lambda]$. □

Remark 2.9. If $A \subset R$ is Prüfer and $(I_\lambda \mid \lambda \in \Lambda)$ is any family of subsets of R then

$$A[\bigcup_{\lambda \in \Lambda} I_\lambda] = \sum_{\lambda \in \Lambda} A[I_\lambda].$$

Indeed, it follows from [Vol. I, Proposition II.1.6] that the set on the right hand side is a subring of A. □

Remark 2.10. Let $A \subset R$ be a ring extension and $(I_\lambda \mid \lambda \in \Lambda)$ a family of subsets of R with finite avoidance in the set $\bigcup_{\lambda \in \Lambda} I_\lambda =: J$. Then the family $(A[I_\lambda] \mid \lambda \in \Lambda)$ has finite avoidance in $A[J]$.

Proof. Let $x \in A[J]$ be given. We choose a finite subset U of J such that $x \in A[U]$. There exists a finite subset S of Λ such that $U \subset I_\lambda$ for every $\lambda \in \Lambda \setminus S$. It follows that $x \in A[I_\lambda]$ for every $\lambda \in \Lambda \setminus S$. □

We look at the behavior of families of modules with finite avoidance under localisation.

Proposition 2.11. *Let $A \subset R$ be any ring extension and S a multiplicative subset of A. Assume that $(I_\lambda \mid \lambda \in \Lambda)$ is a family of A-submodules of R having finite avoidance in $M := \sum_{\lambda\in\Lambda} I_\lambda$. Then the family $(S^{-1}I_\lambda \mid \lambda \in \Lambda)$ of $S^{-1}A$-modules has finite avoidance in $S^{-1}M$, and*

$$\bigcap_{\lambda\in\Lambda}(S^{-1}I_\lambda) = S^{-1}(\bigcap_{\lambda\in\Lambda} I_\lambda).$$

Proof. Let $\xi \in S^{-1}M$ be given. We write $\xi = x/s$ with $x \in M$ and $s \in S$. There exists a finite subset Φ of Λ such that $x \in I_\lambda$ for every $\lambda \in \Lambda \setminus \Phi$. This implies that $\xi \in S^{-1}I_\lambda$ for every $\lambda \in \Lambda \setminus \Phi$. Thus $(S^{-1}I_\lambda \mid \lambda \in \Lambda)$ has finite avoidance in $S^{-1}M$.

Assume now that $\xi \in \bigcap_{\lambda\in\Lambda}(S^{-1}I_\lambda)$. We have $\xi = x/s$ as above with $x \in M$, $s \in S$ and $x \in I_\lambda$ for $\lambda \in \Lambda \setminus \Phi$, Φ finite. Let $U := \bigcap_{\lambda\in\Phi} I_\lambda$, $V := \bigcap_{\lambda\in\Lambda\setminus\Phi} I_\lambda$.

Since Φ is finite, we have $S^{-1}U = \bigcap_{\lambda\in\Phi} S^{-1}I_\lambda$. Also

$$S^{-1}U \cap S^{-1}V = S^{-1}(U \cap V) = S^{-1}(\bigcap_{\lambda\in\Phi} I_\lambda).$$

Now $x \in V$, hence $\xi \in S^{-1}V$, and also

$$\xi \in \bigcap_{\lambda\in\Phi} S^{-1}I_\lambda = S^{-1}U.$$

Thus $\xi \in S^{-1}(\bigcap_{\lambda\in\Lambda} I_\lambda)$. □

We now can extend a distributivity result from [Vol. I, Chap. II §1] from finite families of modules to families with finite avoidance.

Proposition 2.12. *Assume that $A \subset R$ is Prüfer, J is an A-submodule of R, and $(I_\lambda \mid \lambda \in \Lambda)$ is a family of A-submodules of R with finite avoidance in $\sum_{\lambda\in\Lambda} I_\lambda$. Moreover assume either that J is R-regular or all the modules I_λ are R-regular. Then*

$$(*) \qquad J + \bigcap_{\lambda\in\Lambda} I_\lambda = \bigcap_{\lambda\in\Lambda}(J + I_\lambda).$$

Proof. Given a prime ideal $\mathfrak{p}$ of A, it suffices to verify that $(*)$ holds after localizing both sides with respect to $\mathfrak{p}$. Proposition 11 tells us that $(I_{\lambda\mathfrak{p}} \mid \lambda \in \Lambda)$ has finite avoidance in $\sum_\lambda I_{\lambda\mathfrak{p}}$, and also that

$$(J + \bigcap_{\lambda\in\Lambda} I_\lambda)_{\mathfrak{p}} = J_{\mathfrak{p}} + \bigcap_{\lambda\in\Lambda} I_{\lambda\mathfrak{p}}$$

and

$$(\bigcap_{\lambda\in\Lambda}(J + I_\lambda))_{\mathfrak{p}} = \bigcap_{\lambda\in\Lambda}(J_{\mathfrak{p}} + I_{\lambda\mathfrak{p}}).$$

Thus, replacing A, R, J, I_λ by their localizations $A_{\mathfrak{p}}, R_{\mathfrak{p}}, J_{\mathfrak{p}}, I_{\lambda\mathfrak{p}}$, we may assume that $A = A_v$ with v a local Manis valuation on R.

Now the proof runs essentially as in the special case where Λ is finite, done in [Vol. I, Chap. II] (cf. [Vol. I, Theorem II.1.4(2)]). The R-regular A-modules are v-convex. Thus they form a chain and they all contain the support $\mathfrak{q}$ of A, cf. [Vol. I, Lemma II.1.2]. On the other hand, if an A-submodule M of R is not R-regular, then $M \subset \mathfrak{q}$ by the same lemma.

We first consider the case that there exists some $\lambda \in \Lambda$, such that I_λ is not R-regular, hence $I_\lambda \subset \mathfrak{q}$. By hypothesis, J is R-regular, hence $J \supset \mathfrak{q}$. Now both sides of the equation $(*)$ are equal to J.

It remains to consider the case that all I_λ are R-regular, hence v-convex. Since the set $\{I_\lambda \mid \lambda \in \Lambda\}$ is a chain and has finite avoidance in ΣI_λ, it follows that this set has a smallest element I_μ (cf. the proof of Theorem 6). Now both sides of $(*)$ are equal to $J + I_\mu$. □

Remark. Another distributivity law coming to mind is

$$J \cap (\sum_{\lambda\in\Lambda} I_\lambda) = \sum_{\lambda\in\Lambda}(J \cap I_\lambda).$$

This holds in much greater generality than Proposition 12, cf. [Vol. I, Lemma II.7.1]. □

3 The PM-Spectrum as a Partially Ordered Set

We insert some more terminology, relevant for everything to follow. Let R be any ring (as always, commutative with 1).

Definition 1. The *PM-spectrum* of R is the set of equivalence classes of PM-valuations on R. We denote this set by $\mathrm{pm}(R)$, and we denote the subset of equivalence classes of non-trivial PM-valuations on R by $S(R)$. We call $S(R)$ the *restricted PM-spectrum* of the ring R.

Usually we are sloppy and think of the elements of $\mathrm{pm}(R)$ as valuations instead of classes of valuations, replacing an equivalence class by one of its members. We introduce on $\mathrm{pm}(R)$ a partial ordering as follows.

Definition 2. Let $v: R \to \Gamma \cup \infty$ and w be PM-valuations of R. We decree that $v \leq w$ if either both v and w are nontrivial and $A_v \subset A_w$, which means that w is a coarsening of v (i.e. $w \sim v/H$ with H a convex subgroup of Γ, cf. [Vol. I, Chap. I §1]), or w is trivial and $\text{supp}\, v \subset \text{supp}\, w$. □

Remarks 3.1. a) We have a map supp: $\text{pm}(R) \to \text{Spec}\, R$ from $\text{pm}(R)$ to the Zariski spectrum Spec R, sending a PM-valuation on R to its support. This map is compatible with the partial orderings on $\text{pm}(R)$ and Spec R: If $v \leq w$ then $\text{supp}\, v \subset \text{supp}\, w$.

b) The restriction of the support map supp: $\text{pm}(R) \to \text{Spec}\, R$ to the subset $\text{pm}(R) \setminus S(R)$ of trivial valuations on R is an isomorphism of this poset with Spec R. {"poset" is an abbreviation of "partially ordered set."}

c) Notice that $S(R)$ is something like a "forest". For every $v \in S(R)$ the set of all $w \in S(R)$ with $v \leq w$ is a chain (i.e. totally ordered). Indeed, these valuations w correspond uniquely with the R-overrings B of A_v such that $B \neq R$. Perhaps this chain does not have a maximal element. We should add on top of the chain the trivial valuation v^* on R with $\text{supp}\, v^* = \text{supp}\, v$. The valuations v^* should be regarded as the roots of the trees of our forest. {The trees are growing in downward direction. To get a decent forest, one should add all the valuations v^* and reverse the ordering.} □

This last remark indicates that it is not completely silly to include the trivial valuations in the PM-spectrum, although we are interested in nontrivial valuations. Other reasons will be indicated later.

Usually we will not use the full PM-spectrum $\text{pm}(R)$ but only the part consisting of those valuations $v \in \text{pm}(R)$ such that $A_v \supset A$ for a given subring A.

Definition 3. Let $A \subset R$ be a ring extension.

a) A *valuation on R over A* is a valuation v on R with $A_v \supset A$. In this case the *center of v on A* is the prime ideal $\mathfrak{p}_v \cap A$. We denote it by $\text{cent}_A(v)$.

b) The PM-*spectrum of R over A* (or: of $A \subset R$) is the partially ordered subset consisting of the PM-valuations v on R over A. We denote this poset by $\text{pm}(R/A)$. The *restricted* PM-*spectrum of R over A* is the subposet $S(R) \cap \text{pm}(R/A)$ of $\text{pm}(R/A)$. We denote it by $S(R/A)$.

Remarks 3.2. a) Notice that, if v and w are elements of $\text{pm}(R/A)$ and $v \leq w$, then $\text{cent}_A(v) \supset \text{cent}_A(w)$. Also, if $v \in \text{pm}(R/A)$ and $\mathfrak{p} := \text{cent}_A(v)$, then $A_{[\mathfrak{p}]} \subset A_v$ and $\mathfrak{p}_v \cap A_{[\mathfrak{p}]} = \mathfrak{p}_{[\mathfrak{p}]}$. In the special case that $A \subset R$ is Prüfer the pair $(A_{[\mathfrak{p}]}, \mathfrak{p}_{[\mathfrak{p}]})$ is Manis in R. Since this pair is dominated by $(A_v, \mathfrak{p}_v)$, we have $(A_{[\mathfrak{p}]}, \mathfrak{p}_{[\mathfrak{p}]}) = (A_v, \mathfrak{p}_v)$ (cf. [Vol. I, Theorem I.2.4]). It follows that, for $A \subset R$ Prüfer, the center map cent_A: $\text{pm}(R/A) \to \text{Spec}\, A$ is an anti-isomorphism (i.e., an order reversing bijection) from the poset $\text{pm}(R/A)$ to the poset Spec A. {Of course, we know this for long.} It maps $S(R/A)$ onto the set $Y(R/A)$ of R-regular prime ideals of A.

b) In [HK] the set of (equivalence classes of) valuations on R has been equipped with a topology, which makes it a spectral space (cf. [Ho], [HK]), called the *valuation spectrum* $\mathrm{Spv}(R)$ of R.[3] It is not difficult to verify that $\mathrm{pm}(R/A)$ as a subspace of $\mathrm{Spv}(R)$ is again spectral, and that the center map cent_A *is a homeomorphism from* pm(R/A) *to the space* Spec A with its usual Zariski topology. If $v, w \in \mathrm{Spv}(R)$ are given, then $v \leq w$ iff $v \in \overline{\{w\}}$ in the space $\mathrm{pm}(R/A)$ (i.e., in $\mathrm{Spv}(R)$). In the following we will not exploit these facts, albeit they are important. *We will be content to work with pm(R/A) as a partially ordered set.* □

Definition 4. If $A \subset R$ is Prüfer and $\mathfrak{p} \in \mathrm{Spec}\, A$, we denote the PM-valuation v of R over A with $\mathrm{cent}_A(v) = \mathfrak{p}$ by $v_{\mathfrak{p}}$. {We used this notation already before if $\mathfrak{p}$ is R-regular.} If necessary, we more precisely write $v_{\mathfrak{p}}^R$ instead of $v_{\mathfrak{p}}$.

For a Prüfer extension $A \subset R$ the posets $\mathrm{pm}(R/A)$ and $S(R/A)$ are nothing new for us. Here it is only a question of taste and comfort, whether we use the posets Spec A and $Y(R/A)$ (as we did in Sect. 1), or work directly with $\mathrm{pm}(R/A)$ and $S(R/A)$.

Theorem 3.3. *Let $A \subset R$ be a Prüfer extension and B an R-overring.*

i) For every PM-valuation w of R over A the special restriction $w|_B$ of w to B is a PM-valuation of B over A.
ii) The map $w \mapsto w|_B$ from $\mathrm{pm}(R/A)$ *to* $\mathrm{pm}(B/A)$ *is an isomorphism of posets.*

Proof. a) Let w be a PM-valuation on R over A. Then $v := w|_B$ is a special valuation on B with $A_v = A_w \cap B$ and $\mathfrak{p}_v = \mathfrak{p}_w \cap B$. In particular v is a valuation over A. The set $B \setminus A_v$ is closed under multiplication. Thus A_v is PM in B (cf. [Vol. I, Proposition I.5.1.iii]). [Vol. I, Proposition III.6.6] tells us that v is Manis, hence PM. We have $\mathrm{cent}_A(w) = \mathrm{cent}_A(v)$.

b) Since the center maps from $\mathrm{pm}(R/A)$ to Spec A and $\mathrm{pm}(B/A)$ to Spec A both are anti-isomorphisms of posets, we have a unique isomorphism of posets

$$\alpha\colon \mathrm{pm}(R/A) \xrightarrow{\sim} \mathrm{pm}(B/A)$$

such that $\mathrm{cent}_A(w) = \mathrm{cent}_A(\alpha(w))$ for every $w \in \mathrm{pm}(R/A)$. From $\mathrm{cent}_A(w) = \mathrm{cent}_A(w|_B)$ we conclude that $\alpha(w) = w|_B$. □

The theorem shows well, that we sometimes should work with the full PM-spectrum $\mathrm{pm}(R/A)$ instead of $S(R/A)$: In the situation of the proposition, whenever $R \neq B$, there exist nontrivial PM-valuations w on R over A such that $w|_B$ is trivial. (All PM-valuations w of R over B have this property.) Thus we do not have a decent map from $S(R/A)$ to $S(B/A)$.

Proposition 3.4a. *Assume that $A \subset B$ is an arbitrary ring extension and $B \subset R$ is a Prüfer extension. For every PM-valuation v on B over A there exists (up to equivalence) a unique PM-valuation w on R (over A) with $w|_B = v$.*

[3]There are various topologies on this set which give spectral spaces relevant for applications. Here $\mathrm{Spv}(R)$ means the same as in [HK].

Proof. If w is a PM-valuation on R and $v := w|_B$ then $A_w \supset A_v$. Thus, given a PM-valuation v on B over A it suffices to consider valuations w on R such that $A_w \supset A_v$. Then, replacing A by A_v, we may assume in advance that $A \subset B$ is PM. Now $A \subset R$ is Prüfer, and the claim follows from Theorem 3. □

Definition 5. In the situation of Proposition 4 we denote the PM-valuation w on R with $w|_B = v$ by v^R, and we call v^R the valuation *induced on R by v*.

Proposition 3.4b. *If v_1 is a second PM-valuation on B and $v \leq v_1$ then $v^R \leq v_1^R$. Thus $v \mapsto v^R$ is an isomorphism from* $\text{pm}(B/A)$ *onto a sub-poset of* $\text{pm}(R/A)$. *It consists of all $w \in \text{pm}(R/A)$ such that $A_w \cap B$ is PM in B.*

Proof. Given v and $v_1 \geq v$ we obtain the first claim by applying Theorem 3 to the extensions $A_v \subset B \subset R$. The second claim is obvious. □

If M is a subset of $\text{pm}(B/A)$ we denote the set $\{v^R \mid v \in M\}$ by M^R.

Theorem 3.5. *Assume that $A \subset B$ is a convenient extension (cf. [Vol. I, Definition 2 in I §6]) and $B \subset R$ a Prüfer extension. Then the map $S(B/A) \to S(B/A)^R$, $v \mapsto v^R$ is an isomorphism of posets, the inverse map being $w \mapsto w|_B$. The set $S(R/A)$ is the disjoint union of $S(B/A)^R$ and $S(R/B)$. The extension $A \subset R$ is again convenient.*

Proof. a) Let $w \in S(R/A)$ be given. If $A_w \supset B$ then $w \in S(R/B)$, and $w|_B$ is trivial. Otherwise $A_w \cap B \neq B$, and the extension $A_w \cap B \subset B$ is PM, since $A \subset B$ is convenient. Now Proposition 4b tells us that $w = v^R$ for some $v \in S(B/(A_w \cap B)) \subset S(B/A)$. Of course, $v = w|_B$. Conversely, if $v \in S(B/A)$ is given, then by Proposition 4a we have a unique $w = v^R \in S(R/A)$ with $w|_B = v$. The first two claims are now obvious.

b) Let C be an R-overring of A such that $R \setminus C$ is closed under multiplication. We have to verify that C is PM in R.

The set $B \setminus (C \cap B)$ is closed under multiplication. Thus $C \cap B$ is PM in B. It follows that $C \cap B$ is Prüfer in R, hence convenient in R. Since $C \cap B \subset C \subset R$, and $R \setminus C$ is closed under multiplication, we conclude that C is PM in R (cf. [Vol. I, Proposition I.5.1.iii]). □

Recall that various examples of convenient extensions have been given in [Vol. I, Chap. I §6]. In particular, Prüfer extensions are convenient. In the case that $A \subset B$ is Prüfer, Theorem 5 boils down to Theorem 3.

We now look at the minimal elements in the poset $S(R/A)$ for $A \subset R$ a ring extension. If Λ is any poset, let $\Lambda_{\min}$ denote the set of minimal elements of Λ.

Definition 6. We call the set $S(R/A)_{\min}$ the *minimal restricted* PM-spectrum of R over A (or: of the extension $A \subset R$). We denote this set by $\omega(R/A)$. □

We write down a consequence of Proposition 5 for minimal restricted PM-spectra.

Corollary 3.6. *Let $A \subset B$ be a convenient extension and $B \subset R$ a Prüfer extension. Then*

$$\omega(B/A)^R \subset \omega(R/A) \subset \omega(B/A)^R \cup \omega(R/B).$$

Proof. a) Let $v \in \omega(B/A)^R$ be given. If $w \in S(R/A)$ and $w \leq v^R$ then

$$B \cap A_w \subset B \cap A_{v^R} = A_v \subsetneq B.$$

We conclude, say by Theorem 5, that $w = u^R$ for some $u \in S(B/A)$. Then $u = w|_B \leq v^R|_B = v$. Since v is minimal, we have $u = v$, and $w = v^R$. Thus v^R is minimal in $S(R/A)$.

b) Let $w \in \omega(R/A)$ be given. Then either $w \in S(R/B)$ or $w = v^R$ for some $v \in S(B/A)$. In the first case certainly $w \in \omega(R/B)$ and in the second case $v \in \omega(B/A)$. {N.B. It may happen that a given $w \in \omega(R/B)$ is not minimal in $S(R/A)$.} □

We give a reformulation of part of [Vol. I, Theorem III.11.9] in the language of PM-spectra, together with some immediate consequences.

Scholium 3.7. Let $A \subset R$ and $A \subset C$ be subextensions of a ring extension $A \subset T$. Assume that $A \subset R$ is convenient and $A \subset C$ is ws. Then $C \subset RC$ is convenient. For every PM-valuation w on RC over C the restriction $w|R$ to R is again PM. The restriction map $w \mapsto w|R$ from $\mathrm{pm}(RC/C)$ to $\mathrm{pm}(R/A)$ is an isomorphism of the poset $\mathrm{pm}(RC/C)$ to the subposet U of $\mathrm{pm}(R/A)$ consisting of all $v \in \mathrm{pm}(R/A)$, whose center on A is not R-regular, $(\mathfrak{p}_v \cap A)C \neq C$. It maps $S(RC/C)$ onto $U \cap S(R/A)$. If $v \in U$ then $w \in U$ for every $w \in S(R/A)$ with $v \leq w$.

Assume in addition that $R \cap C$ is Prüfer in R. Then it follows from [Vol. I, Theorem III.11.4] (with $R \cap C$ playing the role of the ring A there), that $U = \mathrm{pm}(R/R \cap C)$. □

Let $A \subset R$ be a ring extension. We ask whether for every $v \in S(R/A)$ there exists some $u \in \omega(R/A)$ with $u \leq v$. This certainly holds if $A \subset R$ is Prüfer, since every R-regular prime ideal of A is contained in a maximal ideal of A, which then is again R-regular. The center map $\mathrm{cent}_A \colon S(R/A) \xrightarrow{\sim} Y(R/A)$ restricts to a bijection $\omega(R/A) \xrightarrow{\sim} \Omega(R/A)$. How about more general ring extensions?

We will not get very far. We start with a technical definition which will be also used in later sections.

Definition 7. Let Λ be a poset. We say that Λ *has enough minimal elements* (resp. *has enough maximal elements*), if for every chain Δ in Λ there exists some $\mu \in \Lambda$ with $\mu \leq \lambda$ (resp. $\mu \geq \lambda$) for every $\lambda \in \Delta$. □

Notice that, if Λ has enough minimal elements, then, by Zorn's lemma, for every $\lambda \in \Lambda$ there exists a minimal element μ of Λ with $\mu \leq \lambda$. More generally, for every chain Δ in Λ there exists some $\mu \in \Lambda_{\min}$ with $\mu \leq \lambda$ for every $\lambda \in \Delta$.

Proposition 3.8. *If $A \subset R$ is convenient, the restricted PM-spectrum $S(R/A)$ has enough minimal elements.*

Proof. If $A \subset R$ is convenient we can identify $S(R/A)$ with the set of all subrings B of R such that $A \subset B \subsetneq R$ and $R \setminus B$ is closed under multiplication, via $B = A_v$ for $v \in S(R/A)$.

Now let $\mathfrak{C}$ be a chain of such subrings of R. Then the intersection C of the family of rings $\mathfrak{C}$ is again a subring of R with $A \subset C \subsetneq R$ and $R \setminus C$ multiplicatively closed. □

There exist other ring extensions $A \subset R$ than the convenient ones such that $S(R/A)$ has enough minimal elements. For example let R be any finitely generated ring and A the prime ring $\mathbb{Z}\cdot 1_R$. Then all chains in $S(R/A)$ are finite. Thus certainly $S(R/A)$ has enough minimal elements.

We also mention that, for any ring extension $A \subset R$, the poset $S(R/A)$ has enough minimal elements iff $\text{pm}(R/A)$ has enough minimal elements, as is easily seen.

Proposition 3.9. *Let $A \subset B \subset R$ be ring extensions. Assume that $S(R/A)$ has enough minimal elements. Then $S(R/B)$ has enough minimal elements.*

Proof. Let Δ be a chain in $S(R/B)$. This is also a chain in $S(R/A)$. Thus there exists some $u \in S(R/A)$ with $u \leq w$ for every $w \in \Delta$. The ring $A_u B$ is PM in R, since A_u is PM in R. We have an element v in $S(R/B)$ with $A_u B = A_v$. Now $A_v \subset A_w$ for every $w \in \Delta$, hence $v \leq w$ for every $w \in \Delta$. □

We return to Prüfer extensions.

Definition 8. If Λ is a poset and $(\Lambda_i \mid i \in I)$ a family of subposets, then we say, that Λ is the *direct sum of the family* $(\Lambda_i \mid i \in I)$, if $\Lambda = \bigcup_{i \in I} \Lambda_i$, and elements of different sets Λ_i are incomparable, i.e. $v \not\leq w$ if $v \in \Lambda_i$, $w \in \Lambda_j$ and $i \neq j$. {In particular, $\Lambda_i \cap \Lambda_j = \emptyset$ if $i \neq j$.} We then write $\Lambda = \bigsqcup_{i \in I} \Lambda_i$. If the set I is finite, say $I = \{1, \dots, n\}$, we also write $\Lambda = \Lambda_1 \sqcup \Lambda_2 \sqcup \cdots \sqcup \Lambda_n$. □

If $\Lambda = \bigsqcup_{i \in I} \Lambda_i$ then, of course, $\Lambda_{\min}$ is the disjoint union of the sets $(\Lambda_i)_{\min}$.

Theorem 3.10. *Let $A \subset R$ be a Prüfer extension and $(B_i \mid i \in I)$ a family of R-overrings of A with finite avoidance in R. Assume that R is generated by this family of subrings, and $B_i \cap B_j = A$ for any two different indices $i, j \in I$.*

i) Then $S(R/A) = \bigsqcup_{i \in I} S(B_i/A)^R$.
ii) For every $i \in I$ let B_i' denote the subring of A generated by the B_j with $j \neq i$, i.e. $B_i' = \sum_{j \neq i} B_j$. Then $S(B_i/A)^R = S(R/B_i')$.

N.B. By Theorem 5 we also have an isomorphism $S(B_i/A) \xrightarrow{\sim} S(B_i/A)^R$, sending an element v of $S(B_i/A)$ to v^R. Thus $S(R/A)$ is isomorphic to the direct sum of the posets $S(B_i/A)$.

Proof. a) We first deal with the case that I consists of two elements, $I = \{1, 2\}$. In this case the claim can be easily deduced from [Vol. I, Chap. III §11]. We prefer to give a direct proof in the present framework.

If $v \in S(R/A)$ then $A_v \supset B_1 \cap B_2$. This implies that $A_v \supset B_1$ or $A_v \supset B_2$ (cf. Theorem 1.4). But A_v cannot contain both rings, since this would imply $A_v = R$. Thus $S(R/A)$ is the disjoint union of the subsets $S(R/B_1)$ and $S(R/B_2)$. If $v \in S(R/B_i)$, $w \in S(R/A)$ and $v \leq w$ then also $w \in S(R/B_i)$ $(i = 1, 2)$. Thus no element of $S(R/B_1)$ is comparable with an element of $S(R/B_2)$. This proves $S(R/A) = S(R/B_1) \sqcup S(R/B_2)$. Theorem 5 tells us that $S(R/A)$ is the disjoint union of $S(B_1/A)^R$ and $S(R/B_1)$. It follows that $S(R/B_2) = S(B_1/A)^R$. For the same reason $S(R/B_1) = S(B_2/A)^R$.

b) Let $i \in I$ be given. Since the family $(B_j \mid j \in I \setminus \{i\})$ has finite avoidance in R, we conclude by Proposition 2.12 that $B_i \cap B_i' = A$. We also have $B_i + B_i' = R$. Thus the second claim ii) in the theorem is already covered by the proof above. Moreover $S(R/A) = S(R/B_i) \sqcup S(R/B_i')$.

c) We now can prove the first claim (i) in general. If $w \in S(R/A)$ is given then certainly there exists some $i \in I$ with $B_i \not\subset A_w$. This implies that $w \in S(B_i/A)^R$ by Theorem 5. Thus $S(R/A)$ is the union of the sets $S(B_i/A)^R = S(R/B_i')$, with i running through I.

Let now indices $i \neq j$ in I and elements $v \in S(B_i/A)^R$, $w \in S(B_j/A)^R$ be given. We want to prove that $v \not\leq w$, and then will be done. We have $v = u_1^R$, $w = u_2^R$ with uniquely determined elements $u_1 \in S(B_i/A)$, $u_2 \in S(B_j/A)$. Let $C := B_i + B_j$. Applying what has been done in step a) to A, C, B_1, B_2, we see that $u_1^C \not\leq u_2^C$. Since we have an isomorphism $S(C/A) \xrightarrow{\sim} S(C/A)^R$, $w \mapsto w^R$, we conclude that $u_1^R \not\leq u_2^R$. {Notice that $(u_i^C)^R = u_i^R$.} □

Corollary 3.11. *In the situation of Theorem 10 the minimal restricted PM-spectrum $\omega(R/A)$ is the disjoint union of the sets $\omega(B_i/A)^R$, with i running through I.* □

4 Prüfer Extensions with Finite Avoidance

The general idea behind the following definition and much study later on is to think of a Prüfer extension $A \subset R$ as a family of PM-valuations, namely the minimal restricted PM-spectrum $\omega(R/A)$, which coincides with the family $(v_\mathfrak{p} \mid \mathfrak{p} \in \Omega(R/A))$. {We often also take the non-trivial coarsenings of these valuations in account, arriving at the restricted PM-spectrum $S(R/A)$.}

Definition 1. We say that a Prüfer extension $A \subset R$ has *finite avoidance*, if the family $(A_v \mid v \in \omega(R/A))$ has finite avoidance in R. We call such an extension a *Prüfer extension with finite avoidance* or a *PF-extension* for short. (The letters P and F stand for "Prüfer" and "finite avoidance".) We then also say that *A is PF in R* or that *R is PF over A*.

In order to prove that a given Prüfer extension $A \subset R$ has finite avoidance, it sometimes is not necessary to know the set $\omega(R/A)$ in advance. We state a generalisation of Proposition 1.7.

Theorem 4.1. *Let $A \subset R$ be a Prüfer extension. Assume that Λ is a set of nontrivial PM-valuations on R over A such that $v \nleq w$ for any two different elements v, w of Λ. Assume also that the family $(A_v \mid v \in \Lambda)$ has finite avoidance in R. Let $B := \bigcap_{v \in \Lambda} A_v$. Then $\Lambda = \omega(R/B)$, and thus the extension $B \subset R$ is PF.*

Proof. Let $w \in \omega(R/B)$ be given. Then $A_w \supset \bigcap_{v \in \Lambda} A_v$. Theorem 2.6 tells us that there exists some $v \in \Lambda$ with $A_w \supset A_v$, i.e. $v \leq w$. Since w is a minimal element of $S(R/B)$ it follows that $w = v$. Thus $w \in \Lambda$.

It remains to be shown that a given element v of Λ is minimal in $S(R/B)$. Suppose there exists some $w \in S(R/B)$ with $w < v$. Again we conclude from $A_w \supset B$ by Theorem 2.6 that there exists some $u \in \Lambda$ with $w \geq u$. We have $v \geq u$, hence $v = u$ by our assumption on Λ. It follows that $v = w$, a contradiction. □

Theorem 4.2. *Let $A \subset R$ be a Prüfer extension with finite avoidance. Then the R-overrings B of A correspond bijectively with the subsets X of $S(R/A)$ consisting of pairwise incomparable elements, via $B = \bigcap_{v \in X} A_v$, $X = \omega(R/B)$. {For $X = \emptyset$ we have to read $B = R$.}*

Proof. The only problem is to prove that, if X is a set as above and $B = \bigcap_{v \in X} A_v$, then $X = \omega(R/B)$. This follows from Theorem 1, once we have verified that $(A_v \mid v \in X)$ has finite avoidance.

For every $v \in X$ we choose an element $\alpha(v) \in \omega(R/A)$ with $\alpha(v) \leq v$, which is possible since $S(R/A)$ has enough minimal elements (cf. Sect. 3). We claim that the map $\alpha: X \to \omega(R/A)$, $v \mapsto \alpha(v)$, is injective. Indeed, let $v_1, v_2 \in X$ be given with $\alpha(v_1) = \alpha(v_2) = u$. From $u \leq v_1$, $u \leq v_2$ we conclude that $v_1 \leq v_2$ or $v_2 \leq v_1$. Since X consists of incomparable elements, we have $v_1 = v_2$.

The family $(A_v \mid v \in \omega(R/A))$ has finite avoidance in R. Since α is injective, also $(A_{\alpha(v)} \mid v \in X)$ has finite avoidance in R. Since $A_{\alpha(v)} \subset A_v$ for every $v \in X$, we conclude that $(A_v \mid v \in X)$ has finite avoidance in R. □

This theorem gives us a good hold on the overrings in a PF extension. It makes it plausible that such Prüfer extensions deserve special interest.

Remark. Translating back from PM-valuations to regular prime ideals we can state Theorem 2 as follows: If $A \subset R$ is a PF extension, the R-overrings B of A correspond bijectively with the subsets X' of $Y(R/A)$ consisting of pairwise non-comparable elements, via

$$B = \bigcap_{\mathfrak{p} \in X'} A_{[\mathfrak{p}]}, \qquad X' = \{\mathfrak{p} \in Y(R/A) \mid \mathfrak{p}B \in \Omega(R/B)\}.$$

□

The class of PF extensions has a close relation to the smaller class of Prüfer extensions with finite minimal restricted PM-spectrum, as we will see. We give a name to this smaller class.

Definition 2. We say that a ring extension $A \subset R$ is *PM-finite* if $A \subset R$ is Prüfer and $\omega(R/A)$ is finite. We then also say that R is *PM-finite over* A.

We prove some permanence properties of the notions "finite avoidance" and "PM-finite".

Proposition 4.3. *Assume that $A \subset R$ is Prüfer with finite avoidance. Let B be an R-overring of A. Then the extensions $A \subset B$ and $B \subset R$ both have finite avoidance. If $A \subset R$ is PM-finite, then $A \subset B$ and $B \subset R$ both are PM-finite.*

Proof. a) The set $\omega(R/B)$ consists of pairwise incomparable elements of $S(R/A)$. As shown in the proof of Theorem 2, we have an injection $\alpha: \omega(R/B) \hookrightarrow \omega(R/A)$ such that $A_{\alpha(v)} \subset A_v$ for every $v \in \omega(R/B)$. Thus, if $\omega(R/A)$ is finite, then $\omega(R/B)$ is finite. If $A \subset R$ has finite avoidance, then $(A_v \mid v \in \omega(R/B))$ has finite avoidance, as has been also shown in the proof of Theorem 2.

b) If $v \in \omega(B/A)$, the induced valuation v^R is an element of $\omega(R/A)$ (cf. Corollary 3.6), and $A_v = B \cap A_{v^R}$. The family $(A_{v^R} \mid v \in \omega(B/A))$ is a subfamily of $(A_w \mid w \in \omega(R/A))$, and thus has finite avoidance in R. It follows that $(A_v \mid v \in \omega(B/A))$ has finite avoidance in B. If $\omega(R/A)$ is finite, then $\omega(B/A)$ is finite. □

As a partial converse to Proposition 3 we have

Proposition 4.4. *Assume that $A \subset B$ and $B \subset R$ are Prüfer extensions.*

i) If $A \subset B$ and $B \subset R$ are PM-finite, $A \subset R$ is PM-finite.
ii) If $A \subset B$ is PM-finite and $B \subset R$ has finite avoidance, $A \subset R$ has finite avoidance.

Proof. Corollary 3.6 tells us that $\omega(R/A) \subset \omega(B/A)^R \cup \omega(R/B)$. Thus, if both $\omega(B/A)$ and $\omega(R/B)$ are finite, $\omega(R/A)$ is finite. Assume now that $\omega(B/A)$ is finite and $(A_v \mid v \in \omega(R/B))$ has finite avoidance in R. We have $\omega(R/A) = \omega(B/A)^R \cup X$ with $X := \omega(R/A) \cap S(R/B) \subset \omega(R/B)$. The family $(A_v \mid v \in X)$ has finite avoidance in R, since $(A_v \mid v \in \omega(R/B))$ has this property. The family $(A_v \mid v \in \omega(R/A))$ consists only of finitely many more R-overrings than $(A_v \mid v \in X)$. Thus this family has again finite avoidance in R. □

Proposition 4.5. *Assume that $A \subset R$ is a PF extension and the ring R is finitely generated over A. Then $A \subset R$ is PM-finite.*

Proof. We choose elements $x_1, \dots, x_n \in R$ such that $R = A[x_1, \dots, x_n]$. For every $i \in \{1, \dots, n\}$ the set U_i consisting of all $v \in \omega(R/A)$ with $x_i \notin A_v$ is finite. If $v \in \omega(R/A)$, there exists at least one $i \in \{1, \dots, n\}$ with $x_i \notin A_v$, since $A_v \neq R$. We conclude that $\omega(R/A) = U_1 \cup \dots \cup U_n$. Thus $\omega(R/A)$ is finite. □

We now are ready to prove

Theorem 4.6. *Let $A \subset R$ be a Prüfer extension. The following are equivalent.*

(1) $A \subset R$ has finite avoidance.
(2) Every R-overring B of A, which is finitely generated over A, is PM-finite over A.

(2′) *For every* $x \in R \setminus A$ *the ring* $A[x]$ *is PM-finite over* A.
(3) *For every* R*-invertible ideal* I *of* A *the set of maximal ideals* $\mathfrak{p} \supset I$ *of* A *is finite.*
(3′) *For every* $x \in R \setminus A$ *the set of maximal ideals* $\mathfrak{p} \supset (A{:}\,x)$ *of* A *is finite.*

Proof. (1) ⇒ (2): Let B be an R-overring of A which is finitely generated over A. Proposition 3 tells us that $A \subset B$ has finite avoidance. Then Proposition 5 tells us that $\omega(B/A)$ is finite.

(2) ⇒ (2′): trivial

(2′) ⇒ (1): Let $x \in R$ be given and $U := \{v \in \omega(R/A) \mid x \notin A_v\}$. Let $B := A[x]$. Then $U = \{v \in \omega(R/A) \mid B \not\subset A_v\}$. It follows from Corollary 3.6 that $U = \omega(B/A)^R$. By assumption $\omega(B/A)$ is finite. Thus U is finite.

(2) ⇔ (3): This follows from Proposition 1.2, since the R-overrings of A which are finitely generated over A are precisely the rings $A[I^{-1}]$ with I running through the R-invertible ideals of A.

(2′) ⇔ (3′): This is clear by the same proposition, since for every $x \in R \setminus A$ we have $(A + Ax)^{-1} = (A{:}\,x)$, hence $A[x] = A[I^{-1}]$ with $I = (A{:}\,x)$. □

Lemma 4.7. *Let* $A \subset R$ *be a Prüfer extension, and let* $(B_i \mid 1 \le i \le n)$ *be a finite family of* R*-overrings of* A*. Assume that each* B_i *is PM-finite over* A*. Then the ring* $B_1 \dots B_n = B_1 + \cdots + B_n$ *is PM-finite over* A.

Proof. It suffices to consider the case $n = 2$, since then we can proceed by induction on n. We further may replace R by $B_1 B_2$, hence assume $R = B_1 B_2$. Let $A' := B_1 \cap B_2$. Since $A \subset B_1$ is PM-finite, we conclude by Proposition 3 that both $A \subset A'$ and $A' \subset B_1$ are PM-finite. Also $A' \subset B_2$ is PM-finite. Corollary 3.11 tells us that

$$\omega(R/A') = \omega(B_1/A')^R \,\dot{\cup}\, \omega(B_2/A')^R.$$

Since $\omega(B_1/A')$ and $\omega(B_2/A')$ are finite, the set $\omega(R/A')$ is finite. Also $\omega(A'/A)$ is finite. We conclude by Proposition 4 that $\omega(R/A)$ is finite. □

Definition 3. Let $A \subset R$ be any ring extension. The *PF-hull of* A *in* R is the set of all $x \in R$ such that the extension $A \subset A[x]$ is PM-finite. {In particular, $A \subset A[x]$ has to be Prüfer, cf. Definition 2.} We denote this set by PF(A, R). □

The reason for this terminology is apparent from the following theorem.

Theorem 4.8. *PF*(A, R) *is a subring of* R *containing* A*. It is PF over* A*. If* B *is any* R*-overring, the extension* $A \subset B$ *is PF iff* B *is contained in PF*(A, R).

Proof. We may replace R by the Prüfer hull $P(A, R)$, and thus we assume without loss of generality that $A \subset R$ is Prüfer. If $A \subset B$ is a subextension of $A \subset R$ which has finite avoidance, it follows from Theorem 6 that $B \subset \mathrm{PF}(A, R)$. Let $(B_\lambda \mid \lambda \in \Lambda)$ denote the set of all R-overrings B of A, which are PF over A, indexed in some way. Let further $B := \sum_{\lambda \in \Lambda} B_\lambda$ denote the subring of R generated by this family. We will verify that $A \subset B$ has finite avoidance. Then we will know that $B = \mathrm{PF}(A, R)$, and all claims of the theorem are evident.

Let $x \in B$ be given. We choose a finite subset U of Λ such that $x \in \sum_{\lambda \in U} B_\lambda$. We write $x = \sum_{\lambda \in U} x_\lambda$ with $x_\lambda \in B_\lambda$. Since $A \subset B_\lambda$ has finite avoidance, the subextension $A \subset A[x_\lambda]$ is PM-finite ($\lambda \in U$). Now Lemma 7 tells us that the ring C generated over A by the finite family $(x_\lambda \mid \lambda \in U)$ is PM-finite over A. Since $A[x] \subset C$, also $A[x]$ is PM-finite over A. Since this holds for every $x \in B$, it follows by Theorem 6, that $A \subset B$ has finite avoidance. □

We write down some properties of PF-hulls.

Remarks 4.9. i) If $A \subset B \subset R$ are ring extensions, then $\mathrm{PF}(A, B) = B \cap \mathrm{PF}(A, R)$.

ii) If $A \subset B$ is a PM-finite extension and $B \subset R$ is any ring extension, then $\mathrm{PF}(A, R) = \mathrm{PF}(B, R)$.

All this is evident from Theorem 8 and Proposition 3. □

Proposition 4.10. *Let $\varphi\colon R \twoheadrightarrow \overline{R}$ be a surjective ring homomorphism. Let A be a subring of R and $\overline{A} := \varphi(A)$.*

i) If $A \subset R$ is PF (resp. PM-finite), the same holds for $\overline{A} \subset \overline{R}$.
ii) If $\overline{A} \subset \overline{R}$ is PF (resp. PM-finite), the same holds for $\varphi^{-1}(\overline{A}) \subset R$.

Proof. We first assume that $\varphi^{-1}(\overline{A}) = A$. Now [Vol. I, Proposition I.5.8] and [Vol. I, Chap. III §11] tell us that $A \subset R$ is Prüfer iff $\overline{A} \subset \overline{R}$ is Prüfer, further, that in this case we have an isomorphism of posets $S(\overline{R}/\overline{A}) \xrightarrow{\sim} S(R/A)$, $w \mapsto w \circ \varphi$. This gives us a bijection $\omega(\overline{R}/\overline{A}) \xrightarrow{\sim} \omega(R/A)$. Thus $A \subset R$ is PM-finite iff $\overline{A} \subset \overline{R}$ is PM-finite. If $\varphi^{-1}(\overline{A}) \neq A$, but $A \subset R$ is PM-finite, then also $\varphi^{-1}(\overline{A}) \subset R$ is PM-finite (cf. Proposition 3), and we conclude again that $\overline{A} \subset \overline{R}$ is PM-finite. Now the assertions about the property PF follow easily by use of Theorem 8. □

This proposition immediately implies

Corollary 4.11. *Let $\varphi\colon R \to \overline{R}$ be a surjective ring homomorphism, A a subring of R and $\overline{A} := \varphi(A)$. Then $\varphi(\mathrm{PF}(A, R)) \subset \mathrm{PF}(\overline{A}, \overline{R})$, and $\varphi^{-1}(\mathrm{PF}(\overline{A}, \overline{R})) = \mathrm{PF}(\varphi^{-1}(\overline{A}), R)$.* □

Proposition 4.12. *Let $A \subset R$ and $A \subset C$ be subextensions of a ring extension $A \subset T$. Assume that $A \subset C$ is ws and $A \subset R$ is Prüfer. {N.B. We then know that $RC = R \otimes_A C$, cf. [Vol. I, Proposition I.4.2].}*

i) If $A \subset R$ is PF (resp. PM-finite), the same holds for the extension $C \subset RC$.
ii) If $C \subset RC$ is PF (resp. PM-finite), the same holds for the extension $R \cap C \subset R$.

Proof. We first assume that $A = R \cap C$. [Vol. I, Theorem III.11.4] (or Scholium 3.7) tells us that we have an isomorphism of posets $S(RC/C) \xrightarrow{\sim} S(R/A)$, sending a valuation $w \in S(RC/C)$ to its restriction $w|R$. This isomorphism restricts to a bijection $\omega(RC/C) \xrightarrow{\sim} \omega(R/A)$. In particular $\omega(RC/C)$ is finite iff $\omega(R/A)$ is finite. Applying this to finite subextensions $A \subset B$ of $A \subset R$ and to $A \subset C$, we obtain by Theorem 8 that C is PF in RC iff A is PF in R.

Now all claims are proved if $A = R \cap C$. If $A \neq R \cap C$ and $A \subset R$ is PF (resp. PM-finite), we know by Proposition 3 that $R \cap C \subset R$ has the same property. As proved, this implies that $C \subset RC$ has this property. □

Corollary 4.13. *Assume again that $A \subset R$ and $A \subset C$ are subextensions of a ring extension $A \subset T$ with $A \subset R$ Prüfer and $A \subset C$ ws. Then* $\mathrm{PF}(A, R)C \subset \mathrm{PF}(C, RC)$ *and* $R \cap \mathrm{PF}(C, RC) = \mathrm{PF}(R \cap C, R)$. □

Proposition 4.14. *Let $A \subset R$ be a ring extension and S a multiplicative subset of A. Assume that $A \subset R$ is PF (resp. PM-finite). Then the same holds for the extension $S^{-1}A \subset S^{-1}R$.*

Proof. Let $\overline{A}$ and $\overline{R}$ denote the images of the localisation maps $A \to S^{-1}A$, $R \to S^{-1}R$, and let $\overline{S}$ denote the image of S in $\overline{A}$. We have $S^{-1}A = \overline{S}^{-1}\overline{A}$, $S^{-1}R = \overline{S}^{-1}\overline{R}$. Proposition 10 tells us that the extension $\overline{A} \subset \overline{R}$ is PF (resp. PM-finite). Then Proposition 12 tells us that the same holds for the extension $\overline{S}^{-1}\overline{A} \to \overline{S}^{-1}\overline{R}$, since $\overline{A}$ is ws in $\overline{S}^{-1}\overline{A}$. □

Corollary 4.15. *If $A \subset R$ is a ring extension and S is a multiplicative subset of R, then*

$$S^{-1}\mathrm{PF}(A, R) \subset \mathrm{PF}(S^{-1}A, S^{-1}R).$$

□

Remark 4.16. Notice that it may well happen that $S^{-1}\mathrm{PF}(A,R) \neq \mathrm{PF}(S^{-1}A, S^{-1}R)$. For example, if $A \subset R$ is Prüfer, and $\mathfrak{p}$ is any prime ideal of A, then $A_\mathfrak{p} \subset R_\mathfrak{p}$ is PM, hence certainly PF. □

We hasten to give an example of an extension which is PM-finite, and also of an extension which is PF but not PM-finite.

Example 4.17. Let R be a ring with large Jacobson radical ([Vol. I, Definition 3 in I §6]) and let $(v_i \mid 1 \leq i \leq r)$ be a family of finitely many special nontrivial valuations on R. Let $A := \bigcap_{i=1}^r A_{v_i}$ and $\mathfrak{p}_i := \mathfrak{p}_{v_i} \cap A \quad (1 \leq i \leq r)$. [Vol. I, Theorem I.6.10] tells us that the v_i are PM and A is Prüfer in R. Then Proposition 1.7 tells us that $\Omega(R/A) = \{\mathfrak{p}_1, \ldots, \mathfrak{p}_r\}$, hence $\omega(R/A) = \{v_1, \ldots, v_n\}$. Thus R is PM-finite over A.

Example 4.18. Every Dedekind domain A is PF in its quotient field. We refer the reader to Bourbaki [Bo, Chap.VII, §2] for a proof of this fact in the broad context of Krull domains. □

There exist very many Dedekind domains which have infinitely many maximal ideals, for example the ring of integers in a number field or the ring of holomorphic functions on $X \setminus S$ for X a compact Riemann surface and S a nonempty finite set of points of X. All these rings are PF but not PM-finite in their quotient fields.

We now give an example of a Prüfer extension which is not PF.

Example 4.19. We equip the set $\mathbb{N}$ of natural numbers with the discrete topology. Let $\beta\mathbb{N}$ denote the Stone-Čech compactification of this space (cf. [GJ]). The ring $A := C(\beta\mathbb{N})$ of continuous $\mathbb{R}$-valued functions on $\beta\mathbb{N}$ can be identified by restriction with a subring of the ring $R := C(\mathbb{N})$ of (continuous) $\mathbb{R}$-valued functions on $\mathbb{N}$, namely the subring of bounded $\mathbb{R}$-valued functions on $\mathbb{N}$, cf. [GJ]. For every $f \in R$ we have $1 + f^2 \in R^*$ and $1/(1 + f^2) \in A$. Thus A is Prüfer in R (cf. [Vol. I, Theorem I.6.16]; we could also invoke [Vol. I, Theorem I.6.14], since also $f/(1 + f^2) \in A$ for every $f \in R$).

We now pick the function $h \in \mathbb{R}$ with $h(n) = n$ for every $n \in \mathbb{N}$. Let $x \in (\beta\mathbb{N}) \setminus \mathbb{N}$. Every neighbourhood of x contains an infinite subset of $\mathbb{N}$, since $\mathbb{N}$ is dense in $\beta\mathbb{N}$. Thus $1/(1+h^2)(x) = 0$. We conclude that the ideal $I := 1/(1+h^2)A$ of A is contained in all maximal ideals $\mathfrak{m}_x := \{g \in A \mid g(x) = 0\}$ of A, with x running through the infinite set $(\beta\mathbb{N}) \setminus \mathbb{N}$. But I is R-invertible. We conclude by Theorem 6, that A is not PF in R. We can say slightly more: Since I is invertible in $A[h]$, already the extension $A \subset A[h]$ is not PF. □

Any Prüfer extension $A \subset R$, which is *not* PF, contains very many overrings B which are PF, but not PM-finite in R. This will follow from a theorem which may be viewed as a generalization of Theorem 1.

Theorem 4.20. *Let $A \subset R$ be a Prüfer extension and $(B_\lambda \mid \lambda \in \Lambda)$ a family of overrings of A in R, which has finite avoidance in R. Assume that every B_λ is PF in R. Then the intersection $B := \bigcap_{\lambda\in\Lambda} B_\lambda$ is PF in R and*

$$S(R/B) = \bigcup_{\lambda\in\Lambda} S(R/B_\lambda).$$

Proof. For every $\lambda \in \Lambda$ the family $(A_v \mid v \in \omega(R/B_\lambda))$ has finite avoidance in R. The intersection of this family is the ring B_λ. The family $(B_\lambda \mid \lambda \in \Lambda)$ has again finite avoidance in R. It now follows easily that the family $(A_v \mid v \in \omega(R/B_\lambda), \lambda \in \Lambda)$ has finite avoidance in R. This family has the intersection $\bigcap_{\lambda\in\Lambda} B_\lambda = B$. Let

$$T := \bigcup_{\lambda\in\Lambda} \omega(R/B_\lambda) \subset S(R/B).$$

Every chain in T has a minimal element, since otherwise we could find an infinite sequence $v_1 > v_2 > v_3 > \dots$ in T. But this would contradict the fact that $(A_v \mid v \in T)$ has finite avoidance in R (cf. our argument in the proof of Theorem 2.6). Thus T has enough minimal elements. It follows that the intersection of the family $(A_v \mid v \in T_{\min})$ is again B. This family has finite avoidance in R, since it is a subfamily of $(A_v \mid v \in T)$. Now Theorem 1 tells us that B is PF in R and $T_{\min} = \omega(R/B)$. For every $v \in S(R/B)$ there exists some $u \in T_{\min}$ with $u \le v$. We have $u \in \omega(R/B_\lambda)$ for some $\lambda \in \Lambda$, hence $v \in S(R/B_\lambda)$. This proves that $S(R/B) = \bigcup_{\lambda\in\Lambda} S(R/B_\lambda)$. □

Example 4.21. We start with a Prüfer extension $A \subset R$ which is not PF, e.g. the extension given in Example 19. We choose a sequence $(v_n \mid n \in \mathbb{N})$ of pairwise different elements in $\omega(R/A)$. Let $B := \bigcap_{n\in\mathbb{N}} A_{v_n}$. We claim that the extension $B \subset R$ is PF and $\omega(R/B) = \{v_n \mid n \in \mathbb{N}\}$.

In order to prove this we introduce for every $n \in \mathbb{N}$ the subring $B_n = A_{v_1} \cap \cdots \cap A_{v_n}$ of R. We know by [Vol. I, Theorem I.6.10] and Proposition 1.7 that B_n is PM-finite in R and $\omega(R/B_n) = \{v_1, \ldots, v_n\}$ (cf. Example 17). The family $(B_n \mid n \in \mathbb{N})$ obviously has finite avoidance in R. Theorem 20 gives the claim. □

5 PM-Split Extensions

In the category of Prüfer extensions of a given ring A the PM-extensions of A should be regarded as the local objects. We now introduce a class of Prüfer extensions of A, called "PM-split extensions" of A, which can be regarded as the direct sums of these local objects. (We refuse to make this a formal statement.)

Let $A \subset R$ be a Prüfer extension and $\Omega := \Omega(R/A)$. For $\mathfrak{p} \in \Omega$ we denote now the PM-hull $\mathrm{PM}(A, \mathfrak{p}, R)$ of $(A, \mathfrak{p})$ in R more briefly by $A^{\mathfrak{p}}$. We recall from Sect. 1 that

$$A^{\mathfrak{p}} = A^{\circ}_{[\mathfrak{p}]} = \bigcap_{\mathfrak{m}\in\Omega\setminus\{\mathfrak{p}\}} A_{[\mathfrak{m}]}.$$

Thus $A^{\mathfrak{p}} \cap A_{[\mathfrak{p}]} = A$ and $A^{\mathfrak{p}} \subset A_{[\mathfrak{m}]}$ for every $\mathfrak{m} \in \Omega \setminus \{\mathfrak{p}\}$.

We denote the subring of R generated by the family $(A^{\mathfrak{p}} \mid \mathfrak{p} \in \Omega)$ of rings in R by R_0. Notice that

$$R_0 = \sum_{\mathfrak{p}\in\Omega} A^{\mathfrak{p}}$$

(cf. [Vol. I, Proposition II.1.6]). Notice also that $A^{\mathfrak{p}} \cap A^{\mathfrak{m}} = A$ for $\mathfrak{p}, \mathfrak{m}$ different elements of Ω.

Definition 1. We call a ring extension $A \subset R$ *PM-split*, if $A \subset R$ is Prüfer and $R = R_0$. We then also say that *A is PM-split in R*.

Remark 5.1. Assume that $A \subset R$ is Prüfer and there is given a family $(B_i \mid i \in I)$ of R-overrings of A, such that A is PM in every B_i and the ring R is generated by the family of subrings $(B_i \mid i \in I)$. Then A is PM-split in R.

Indeed, we may assume without loss of generality that $A \neq B_i$ for every $i \in I$. For each $i \in I$ there exists a unique R-regular maximal ideal $\mathfrak{p}_i$ of A such that $(A, \mathfrak{p}_i)$ is PM in B_i, hence $B_i \subset A^{\mathfrak{p}_i}$. We have $R = \sum_{i\in I} A^{\mathfrak{p}_i}$, hence $R = R_0$. □

Proposition 5.2. *Let $A \subset R$ be any Prüfer extension, and $A \subset B$ a subextension of $A \subset R$. Then $B_0 = B \cap R_0$. The extension $A \subset B$ is PM-split iff $B \subset R_0$.*

Proof. We may assume $A \neq B$. If $\mathfrak{p} \in \Omega(R/A) = \Omega$ then clearly $\mathrm{PM}(A, \mathfrak{p}, B) = B \cap \mathrm{PM}(A, \mathfrak{p}, R)$. It follows, say by [Vol. I, Lemma II.7.1], that

$$B \cap R_0 = \sum_{\mathfrak{p} \in \Omega} \mathrm{PM}(A, \mathfrak{p}, B).$$

Now $\Omega \supset \Omega(B/A)$. If $\mathfrak{p} \in \Omega \setminus \Omega(B/A)$, then $\mathfrak{p}$ is a maximal ideal of A which is not B-regular, hence $\mathrm{PM}(A, \mathfrak{p}, B) = A$. We conclude that

$$B \cap R_0 = \sum_{\mathfrak{p} \in \Omega(B/A)} \mathrm{PM}(A, \mathfrak{p}, B) = B_0.$$

In particular $B = B_0$ iff $B \subset R_0$. □

Theorem 5.3. *Assume that $A \subset R$ is a PM-split extension. Let Λ be a non empty subset of $\Omega := \Omega(R/A)$. Then the R-overring $B := \sum_{\mathfrak{p} \in \Lambda} A^{\mathfrak{p}}$ of A is a factor*[4] *of R/A with the complement $C := \sum_{\mathfrak{m} \in \Omega \setminus \Lambda} A^{\mathfrak{m}}$, i.e. $R = B \times_A C$. We also have $C = \bigcap_{\mathfrak{p} \in \Lambda} A_{[\mathfrak{p}]}$.*

Proof. Let $B := \sum_{\mathfrak{p} \in \Lambda} A^{\mathfrak{p}}$ and $C := \sum_{\mathfrak{m} \in \Omega \setminus \Lambda} A^{\mathfrak{m}}$. We have $R = B + C = BC$. Moreover

$$B \cap C = \sum_{\mathfrak{p} \in \Lambda, \mathfrak{m} \in \Omega \setminus \Lambda} (A^{\mathfrak{p}} \cap A^{\mathfrak{m}}) = A,$$

by the distributivity law [Vol. I, Lemma II.7.1]. This proves $R = B \times_A C$.

Let $C' := \bigcap_{\mathfrak{p} \in \Lambda} A_{[\mathfrak{p}]}$. For every $\mathfrak{p} \in \Lambda$ we have $A^{\mathfrak{p}} \cap A_{[\mathfrak{p}]} = A$. Thus $B \cap C' = A$, again by [Vol. I, Lemma II.7.1]. Further

$$A^{\mathfrak{m}} = \bigcap_{\mathfrak{p} \in \Omega \setminus \{\mathfrak{m}\}} A_{[\mathfrak{p}]} \subset C'$$

for every $\mathfrak{m} \in \Omega \setminus \Lambda$. Thus $C \subset C'$. It follows that $B + C' = R$. We have proved that C' is a complement of B in R over A. This forces $C' = C$ (cf. [Vol. I, Chap. II §7]). □

Proposition 5.4. *If $A \subset R$ is PM-split and $\mathfrak{p} \in \Omega(R/A)$, then $R = A^{\mathfrak{p}} \times_A A_{[\mathfrak{p}]}$, and $A_{[\mathfrak{p}]}^{\circ\circ} = A_{[\mathfrak{p}]}$. The restriction $v_{\mathfrak{p}}|A^{\mathfrak{p}}$ of the PM-valuation $v_{\mathfrak{p}}$ corresponding to the PM-extension $A_{[\mathfrak{p}]} \subset R$ is the PM-valuation corresponding to the PM-extension $A \subset A^{\mathfrak{p}}$.*

Proof. The first claim is covered by Theorem 3, applied to $\Lambda := \{\mathfrak{p}\}$. The second claim is clear by the following general fact: If $A \subset R$ is Prüfer and has a

[4]Recall [Vol. I, Definition 3 in II §7].

decomposition $R = B \times_A C$ into factors, then $B^\circ = C$ and $C^\circ = B$, hence $B^{\circ\circ} = B$. Finally, the last claim follows from [Vol. I, Theorem III.11.4]. $\square$

We now can state a characterization theorem for PM-split extensions.

Theorem 5.5. *Let $A \subset R$ be a Prüfer extension and $\Omega := \Omega(R/A)$. The following are equivalent.*

(1) $A \subset R$ is PM-split.
(2) For every $\mathfrak{p} \in \Omega$ the ring $A_{[\mathfrak{p}]}$ is a factor of R over A.
(3) For every $\mathfrak{p} \in \Omega$ the ring $A^{\mathfrak{p}}$ is a factor of R over A and $A_{[\mathfrak{p}]}^{\circ\circ} = A$.
(4) $A \subset R$ is PF (cf. Sect. 4, Definition 1), and $A_{[\mathfrak{p}]} + A_{[\mathfrak{m}]} = R$ for any two different elements $\mathfrak{p}, \mathfrak{m}$ of Ω.

Proof. The implications (1) ⇒ (2) and (1) ⇒ (3) are covered by Proposition 4, and (2) ⇒ (3) is evident from the theory of factors ([Vol. I, Chap. II §7]), since $A_{[\mathfrak{p}]}^{\circ} = A^{\mathfrak{p}}$.

(2), (3) ⇒ (1): Suppose that R is not PM-split, i.e. $R_0 \neq R$. Then there exists an R-regular prime ideal $\mathfrak{p}_1$ of A with $R_0 \subset A_{[\mathfrak{p}_1]}$. We choose a maximal ideal $\mathfrak{p}$ of A containing $\mathfrak{p}_1$. Then $\mathfrak{p} \in \Omega$. We have $A_{[\mathfrak{p}]} \subset A_{[\mathfrak{p}_1]}$ and $A^{\mathfrak{p}} \subset R_0 \subset A_{[\mathfrak{p}_1]}$. Since $A_{[\mathfrak{p}]} + A^{\mathfrak{p}} = R$ by assumption (2), we conclude that $A_{[\mathfrak{p}_1]} = R$. This means that $\mathfrak{p}_1$ is not R-regular, a contradiction. Thus $R_0 = R$.

(1) ⇒ (4): It is trivial that PM-extensions are PF. We conclude by Theorem 4.8 that $A \subset R$ is PF. Let $\mathfrak{p} \in \Omega$ be fixed. For every $\mathfrak{m} \in \Omega \setminus \{\mathfrak{p}\}$ we have $A^{\mathfrak{p}} \subset A_{[\mathfrak{m}]}$. It follows from $A_{[\mathfrak{p}]} + A^{\mathfrak{p}} = R$ that $A_{[\mathfrak{p}]} + A_{[\mathfrak{m}]} = R$.

(4) ⇒ (1): Since the family $(A_{[\mathfrak{p}]} \mid \mathfrak{p} \in \Omega)$ has finite avoidance in R, we have, for every $\mathfrak{p} \in \Omega$,

$$A_{[\mathfrak{p}]} + A^{\mathfrak{p}} = A_{[\mathfrak{p}]} + \bigcap_{\mathfrak{m} \in \Omega \setminus \{\mathfrak{p}\}} A_{[\mathfrak{m}]} = \bigcap_{\mathfrak{m} \in \Omega \setminus \{\mathfrak{p}\}} (A_{[\mathfrak{p}]} + A_{[\mathfrak{m}]}) = R,$$

due to the distributivity law stated in Proposition 2.12. $\square$

Remark 5.6. If $A \subset R$ is Prüfer and $\mathfrak{p} \in Y(R/A)$, then the R-overrings B of $A_{[\mathfrak{p}]}$ correspond uniquely with the valuations w of R coarsening $v_{\mathfrak{p}}$, via $B = A_w$. Thus the condition $A_{[\mathfrak{p}]} + A_{[\mathfrak{m}]} = R$ in Theorem 5, (4) means, that in $S(R/A)$ there is no element w with $v_{\mathfrak{p}} \leq w$ and $v_{\mathfrak{m}} \leq w$. It follows that the poset $S(R/A)$ is the direct sum of the chains $\{w \in S(R/A) \mid w \geq u\}$ with u running through $\omega(R/A)$. In more imaginative terms, our "forest" $S(R/A)$ has no branchings. $\square$

Proposition 5.7. *Let $A \subset B \subset R$ be ring extensions, and assume that $A \subset R$ is PM-split. Then both $A \subset B$ and $B \subset R$ are PM-split.*

Proof. We read off from Proposition 2 that $A \subset B$ is PM-split. In order to prove that $B \subset R$ is PM-split we use criterion (4) in Theorem 5. If v, w are different elements of $\omega(R/A)$ then $A_v + A_w = R$. Thus the same holds for different elements v, w of $S(R/A)$, hence of $\omega(R/B)$. Proposition 4.3 tells us that the extension $B \subset R$ is PF. Criterion (4) is fulfilled for $B \subset F$. $\square$

Notice, that conversely there is *no reason*, that $A \subset R$ is PM-split if $A \subset B$ and $B \subset R$ both have this property, since the forest $S(R/A)$ may well have branchings if neither $S(B/A)$ nor $S(R/B)$ has branchings.

Proposition 5.8. *Assume that $A \subset R$ is a Prüfer extension and $(B_\lambda \mid \lambda \in \Lambda)$ is a family of R-overrings of A, which has finite avoidance in R. Assume also that for every $\lambda \in \Lambda$ the ring B_λ is PM-split in R. Assume finally that $B_\lambda + B_\mu = R$ for any two different $\lambda, \mu \in \Lambda$. Then the intersection $B := \bigcap_{\lambda \in \Lambda} B_\lambda$ is PM-split in R, and the restricted PM-spectrum $S(R/B)$ is the direct sum (cf. Sect. 3, Definition 8) of the posets $S(R/B_\lambda)$, $S(R/B) = \bigsqcup_{\lambda \in \Lambda} S(R/B_\lambda)$.*

Proof. We know by Theorem 5 that all the extension $B_\lambda \subset R$ are PF. Theorem 4.20 tells us that $B \subset R$ is PF and that $S(R/B)$ is the union of the subsets $S(R/B_\lambda)$ with λ running through Λ. If $v \in S(R/B_\lambda)$ and $w \in S(R/B_\mu)$ with $\lambda \neq \mu$, then $A_v + A_w = R$, since $B_\lambda + B_\mu = R$. In particular, v and w are not comparable. Thus $S(R/B)$ is the direct sum of the posets $S(R/B_\lambda)$, and $\omega(R/B)$ is the disjoint union of the sets $\omega(R/B_\lambda)$. If v and w are different elements of $\omega(R/B_\lambda)$ for some $\lambda \in \Lambda$, then again $A_v + A_w = R$, since B_λ is PM-split in R. Thus $A_v + A_w = R$ for any two different elements v, w of $\omega(R/B)$. Condition (4) in Theorem 5 holds. We conclude that B is PM-split in R. □

Definition 2. a) Let $A \subset R$ be any ring extension. The *PM-split hull*, or *PMS-hull* for short, *of A in R* is the ring $P(A, R)_0$ derived from the Prüfer extension $A \subset P(A, R)$ in the way indicated at the beginning of this section. {Recall that $P(A, R)$ denotes the Prüfer hull of A in R, defined in [Vol. I, Chap. I §5].} We denote this ring by $\mathrm{PMS}(A, R)$.

b) If A is any ring, we define the *PMS-hull* $\mathrm{PMS}(A)$ *of* A by $\mathrm{PMS}(A) := \mathrm{PMS}(A, Q(A))$[5] □

Of course, $\mathrm{PMS}(A, R) = \mathrm{PMS}(A, P(A, R)) = \mathrm{PMS}(A, \mathrm{PF}(A, R))$, and $\mathrm{PMS}(A) = \mathrm{PMS}(A, M(A)) = \mathrm{PMS}(A, P(A)) = \mathrm{PMS}(A, \mathrm{PF}(A))$. If $A \subset R$ is Prüfer, we could retain the notation R_0 for $\mathrm{PMS}(A, R)$, but the new notation is more precise, since R_0 depends on the ring extension $A \subset R$, not just on R alone.

The terminology established in Definition 2 is justified by the following fact.

Proposition 5.9. *Let $A \subset R$ be any ring extension. A subextension $A \subset B$ of $A \subset R$ is PM-split iff B is contained in* $\mathrm{PMS}(A, R)$.

Proof. We may replace R by the Prüfer hull $P(A, R)$ and thus assume that $A \subset R$ is Prüfer. Now the claim is covered by Proposition 2. □

Remark 5.10. If $A \subset B$ is a PM-split extension and $A \subset R$ is any ring extension, then there exists at most one homomorphism $\varphi: B \to R$ over A, and φ is injective, due to the fact that the inclusion mapping $A \hookrightarrow B$ is a flat epimorphism (cf. [Vol. I, Chap. I §3 & §4]). We conclude by Proposition 9 that

[5] Recall the notations $Q(A), M(A), P(A)$ from [Vol. I, Chap. I §4 & §5].

$\varphi(B) \subset \mathrm{PMS}(A, R)$, if φ exists. By the way, the same holds for a PF-extension $A \subset B$ and the PF-hull $\mathrm{PF}(A, R)$. □

Proposition 5.11. *If $A \subset B \subset R$ are ring extensions then* $\mathrm{PMS}(A, B) = B \cap \mathrm{PMS}(A, R)$.

Proof. Proposition 9 (or Proposition 2) tells us that A is PM-split in $B \cap \mathrm{PMS}(A, R)$. Thus $B \cap \mathrm{PMS}(A, R) \subset \mathrm{PMS}(A, B)$. On the other hand, since A is PM-split in the R-overring $\mathrm{PMS}(A, B)$, we have $\mathrm{PMS}(A, B) \subset \mathrm{PMS}(A, R)$, hence $\mathrm{PMS}(A, B) \subset B \cap \mathrm{PMS}(A, B)$. □

We derive some results on PMS-hulls similar to those obtained for PF-hulls in Sect. 4 (Proposition 4.10—Corollary 4.15).

Proposition 5.12. *Let $\varphi\colon R \twoheadrightarrow \overline{R}$ be a surjective ring homomorphism. Let A be a subring of R and $\overline{A} := \varphi(A)$.*

a) If $A \subset R$ is PM-split, the same holds for $\overline{A} \subset \overline{R}$.
b) If $\overline{A} \subset \overline{R}$ is PM-split, the same holds for $\varphi^{-1}(\overline{A}) \subset R$.

Proof. a): We know that $\overline{A} \subset \overline{R}$ is Prüfer. R is generated by the subrings $\mathrm{PM}(A, \mathfrak{p}, R)$ with $\mathfrak{p}$ running through $\Omega(R/A)$. Thus $\overline{R}$ is generated by the subrings $\varphi(\mathrm{PM}(A, \mathfrak{p}, R))$. Now A is PM in each of these rings by [Vol. I, Proposition III.9.8]. It follows by Remark 1 above that A is PM-split in $\overline{R}$.

b): Replacing A by $\varphi^{-1}(\overline{A})$, we may assume that $A = \varphi^{-1}(\overline{A})$. We now use criterion (2) in Theorem 5. We read off from [Vol. I, Proposition III.9.1] that we have a bijection $\Omega(\overline{R}/\overline{A}) \xrightarrow{\sim} \Omega(R/A)$ mapping a prime ideal $\overline{\mathfrak{p}} \in \Omega(\overline{R}/\overline{A})$ to $\varphi^{-1}(\overline{\mathfrak{p}}) \in \Omega(R/A)$ (cf. proof of Proposition 4.10). Further $\varphi^{-1}(\overline{A}^{\overline{\mathfrak{p}}}) = A^{\mathfrak{p}}$ with $\mathfrak{p} := \varphi^{-1}(\overline{\mathfrak{p}})$ by [Vol. I, Corollary III.9.2]. Also $\varphi^{-1}(\overline{A}_{[\overline{\mathfrak{p}}]}) = A_{[\mathfrak{p}]}$, as is easily checked. Suppose that $A^{\mathfrak{p}} + A_{[\mathfrak{p}]} \neq R$ for one of these prime ideals $\mathfrak{p} = \varphi^{-1}(\overline{\mathfrak{p}})$. Then there exists a valuation $v \in S(R/A)$ with $A^{\mathfrak{p}} + A_{[\mathfrak{p}]} \subset A_v$. But $v = w \circ \varphi$ for some $w \in S(\overline{R}/\overline{A})$, again by [Vol. I, Proposition III.9.1]. Since $\varphi(A^{\mathfrak{p}}) = \overline{A}^{\overline{\mathfrak{p}}}$ and $\varphi(A_{[\mathfrak{p}]}) = \overline{A}_{[\overline{\mathfrak{p}}]}$, we arrive at the contradiction $\overline{A}^{\overline{\mathfrak{p}}} + \overline{A}_{[\overline{\mathfrak{p}}]} \subset A_w$. Thus $A^{\mathfrak{p}} + A_{[\mathfrak{p}]} = R$ for every $\mathfrak{p} \in \Omega(R/A)$. Criterion (2) in Theorem 5 is fulfilled. □

Corollary 5.13. *Assume again that $\varphi\colon R \twoheadrightarrow \overline{R}$ is a surjective ring homomorphism and that A is a subring of R. Let $\overline{A} := \varphi(A)$.*

i) Then $\varphi(\mathrm{PMS}(A, R)) \subset \mathrm{PMS}(\overline{A}, \overline{R})$.
ii) If $A = \varphi^{-1}(\overline{A})$, then $\mathrm{PMS}(A, R) = \varphi^{-1}(\mathrm{PMS}(\overline{A}, \overline{R}))$.

□

Proposition 5.14. *Let $A \subset R$ and $A \subset C$ be subextensions of a ring extension $A \subset T$. Assume that $A \subset C$ is ws and $A \subset R$ is Prüfer.*

i) If $A \subset R$ is PM-split, the same holds for the extension $C \subset RC$.
ii) If $C \subset RC$ is PM-split, the same holds for the extension $R \cap C \subset R$.

Proof. i): Let $A' := R \cap C$. Then $A' \subset R$ is PM-split by Proposition 7. Moreover $A' \subset C$ is ws and $AC = A'C$. Thus we may replace A by A'. We assume henceforth that $R \cap C = A$.

Let $\Omega := \Omega(R/A)$. Then $R = \sum_{\mathfrak{p}\in\Omega} A^{\mathfrak{p}}$ with $A^{\mathfrak{p}} = \mathrm{PM}(A, \mathfrak{p}, R)$. This implies $RC = \sum_{\mathfrak{p}\in\Omega} A^{\mathfrak{p}}C$. We know from [Vol. I, Chap. III §11], say [Vol. I, Corollary III.11.5], that C is PM in $A^{\mathfrak{p}}C$ for every $\mathfrak{p} \in \Omega$. It follows (cf. Remark 1) that C is PM-split in RC.

ii): We assume again that $A = R \cap C$. [Vol. I, Theorem III.11.4] tells us that the R-overrings B of A, which are PM over A, correspond bijectively with the RC-overrings D of C, which are PM over C, via $B = D \cap R$, $D = BC$. If RC is the sum of these rings D then R is the sum of the corresponding rings B, as follows from the distributivity law [Vol. I, Lemma II.7.1]. □

Corollary 5.15. *Assume again that $A \subset R$ and $A \subset C$ are subextensions of a common ring extension. Then*

$$\mathrm{PMS}(A, R)C \subset \mathrm{PMS}(C, RC) \text{ and } R \cap \mathrm{PMS}(C, RC) = \mathrm{PMS}(R \cap C, R).$$

□

Proposition 5.16. *Assume that $A \subset R$ is a PM-split extension and S is a multiplicative subset of A. Then $S^{-1}A \subset S^{-1}R$ is PM-split.*

Proof. This follows from Propositions 11 and 13 by the same reasoning as in the proof of Proposition 4.14. □

Corollary 5.17. *If $A \subset R$ is a ring extension and S a multiplicative subset of A, then*

$$S^{-1}\mathrm{PMS}(A, R) \subset \mathrm{PMS}(S^{-1}A, S^{-1}R).$$

□

How about examples of PM-split extensions?

If A is a Dedekind domain, then A is PM-split in its quotient field Quot(A). Indeed, we noticed in Example 4.18, that A is PF in $R :=$ Quot(A). For every $v \in \omega(R/A) = S(R/A)$, the valuation group in $\mathbb{Z}$. Thus $A_v + A_w = R$ for $v \neq w$. Condition (4) in Theorem 5 is fulfilled.

If we choose in Example 4.17 the valuations v_i, $1 \leq i \leq r$, independent (i.e. $A_{v_i} + A_{v_j} = R$ for $i \neq j$), then we know that the extension $A \subset R$ there is PM-split.

Later we will meet more complicated and more interesting examples of PM-split extensions. But already now we can say that for every nontrivial PF-extension $A \subset R$ the PM-split subextension $A \subset \mathrm{PMS}(A, R)$ is again nontrivial. More precisely we have the following

Proposition 5.18. *Let $A \subset R$ be a PF-extension. Let R_0 denote the PM-split hull* PMS(A, R). *Then for every valuation $v \in \omega(R/A)$ the special restriction $v|_{R_0}$ is not trivial.*

Proof. We have $v = v_{\mathfrak{p}}$ for some $\mathfrak{p} \in \Omega(R/A)$, and $A_v = A^R_{[\mathfrak{p}]}$. A_v has the polar $A^{\mathfrak{p}}$ in R over A (in the notation from the beginning of the section). The overring $A^{\mathfrak{p}}$ is the intersection of the rings A_w with w running through $\omega(R/A) \setminus \{v\}$. This intersection is not contained in A_v, since the family $(A_w \mid w \in \omega(R/A) \setminus \{v\})$ has finite avoidance in R (cf. Theorem 2.6). Thus certainly $A^{\mathfrak{p}} \neq A$. Now $A^{\mathfrak{p}} \subset R_0$, and $A^{\mathfrak{p}}$ is also the polar of $A_v \cap R_0 = A^{R_0}_{[\mathfrak{p}]}$ in R_0 over A. It follows that $A_v \cap R_0 = A^{R_0}_{[\mathfrak{p}]} \neq R_0$. This is the claim. □

The result can be sharpened as follows.

Theorem 5.19. *Assume again that A is PF in R and R_0 =* PMS(A, R). *Then $\omega(R/A) = \omega(R_0/A)^R$. Thus the restriction map $v \mapsto v|_{R_0}$ is a bijection from $\omega(R/A)$ to $\omega(R_0/A)$.*

Proof. By Corollary 3.6 we have

$$\omega(R_0/A)^R \subset \omega(R/A) \subset \omega(R_0/A)^R \cup \omega(R/R_0).$$

Proposition 18 tells us that in $S(R/A)$ the subset $\omega(R/A)$ does not meet $\omega(R/R_0)$. Thus $\omega(R_0/A)^R = \omega(R/A)$. □

6 Irreducible and Coirreducible Overrings

In this section *we are given a Prüfer extension $A \subset R$. Usually the word "overring" means "R-overring of A"*, if nothing else is said.

Definition 1. An overring B is called *irreducible* (more precisely: *irreducible over* A), if the extension $A \subset B$ is irreducible, as defined in [Vol. I, Chap. II §7], i.e. B has no nontrivial factor over A and $B \neq A$. The overring B is called *coirreducible*, (more precisely: *coirreducible in* R) if the extension $B \subset R$ is irreducible.

Example. If $\mathfrak{p}$ is an R-regular prime ideal of A, then $A_{[\mathfrak{p}]}$ is coirreducible, and $A^{\mathfrak{p}}$ = PM$(A, \mathfrak{p}, R) = A^{\circ}_{[\mathfrak{p}]}$ is irreducible, provided $A^{\mathfrak{p}} \neq A$. □

Our first goal is to prove that every irreducible overring is contained in a unique maximal irreducible overring and that every coirreducible overring contains a unique minimal coirreducible overring. Here we will only exploit the fact that the lattice of overrings is distributive. We refrain to develop the arguments in arbitrary distributive lattices, since this does not really pay for our study here. But it will be no surprise that to every result for irreducible overrings there corresponds a "dual" result for coirreducible overrings, since inverting the ordering in a distributive lattice gives us another distributive lattice.

Later the symmetry between results for irreducible and for coirreducible overrings will be broken, due to the fact that the overrings $A_{[\mathfrak{p}]}$ are more basic and important than the overrings $A^{\mathfrak{p}}$.

Lemma 6.1. *Assume we are given a factorization $R = B \times_A C$ (cf. [Vol. I, Definition 3 in II §7]).*

a) If D is an irreducible overring then $D \subset B$ or $D \subset C$. In the first case $D \cap C = A$, while in the second case $D \cap B = A$.
b) If D is a coirreducible overring then $B \subset D$ or $C \subset D$. In the first case $D + C = R$, while in the second case $D + B = R$.

Proof. a): [Vol. I, Proposition II.7.15] tells us that $D = (B \cap D) \times_A (C \cap D)$. This implies $D = B \cap D$, i.e. $D \subset B$, or $D = C \cap D$, i.e. $D \subset C$. In the first case it follows from $B \cap C = A$ that $D \cap C = A$, in the second that $D \cap B = A$.
b): [Vol. I, Proposition II.7.15] tells us that $R = BD \times_D CD$. Thus $BD = D$, i.e. $B \subset D$, or $CD = D$, i.e. $C \subset D$. In the first case it follows from $B + C = R$, that $D + C = R$, in the second, that $D + B = R$. □

Lemma 6.2. *a) If D_1 and D_2 are irreducible overrings and $D_1 \cap D_2 \neq A$, then $D_1 + D_2$ is again irreducible.*
b) If D_1 and D_2 are coirreducible overrings and $D_1 + D_2 \neq R$, then $D_1 \cap D_2$ is again coirreducible.

Proof. We prove part (a), leaving the "dual" proof of (b) to the reader. We may replace R by the subring $D_1 + D_2$. Thus we assume that $R = D_1 + D_2$, and now we have to prove that R is irreducible over A. Suppose there exists a nontrivial factorization $R = B \times_A C$. By Lemma 1 the ring D_1 is contained in either B or in C, and also D_2 is contained in B or C. Neither B nor C can contain both D_1 and D_2, since $D_1 + D_2 = R$. From $B \cap C = A$ we conclude that $D_1 \cap D_2 = A$. This contradicts our hypothesis that $D_1 \cap D_2 \neq A$. Thus R is irreducible. □

Lemma 6.3. *Let $\mathscr{D}$ be a chain of overrings.*

a) If every $D \in \mathscr{D}$ is irreducible then the union E of all members of $\mathscr{D}$ is irreducible.
b) If every $D \in \mathscr{D}$ is coirreducible then the intersection F of all members of $\mathscr{D}$ is coirreducible.

Proof. We prove claim (b), leaving the dual proof of (a) to the reader. We may replace A by the overring F. Thus we assume that $A = F$, and now we have to prove that R is irreducible over A. Let $R = B \times_A C$ be a factorization of R over A. By Lemma 1(b) every $D \in \mathscr{D}$ contains either B or C. We distinguish two cases.

1. *Case.* Every $D \in \mathscr{D}$ contains B. It follows that $B \subset F$, i.e. $B = A$.
2. *Case.* There exists some $D \in \mathscr{D}$ not containing B. For every $D' \in \mathscr{D}$ with $D' \subset D$ we have $B \not\subset D'$, hence $C \subset D'$. It follows that $C \subset F$, i.e. $C = A$.

Our factorization of R is trivial. □

Theorem 6.4. *a) For every irreducible overring D there exists a unique maximal irreducible overring E containing D. If E and E' are two different maximal irreducible overrings, then $E \cap E' = A$.*
b) For every coirreducible overrings D there exists a unique minimal coirreducible overring F contained in D. If F and F' are two different minimal coirreducible overrings, then $F + F' = R$.

Proof. We again prove part (b) leaving the dual proof of (a) to the reader.

i) Let D be a coirreducible overring. By Lemma 3(b) and Zorn's lemma we conclude that D contains a minimal coirreducible overring F.
ii) Let F and F' be two minimal coirreducible overrings, and assume that $F + F' \neq R$. Lemma 2(b) tells us that $F \cap F'$ is again coirreducible. This implies $F = F'$, due to the minimality of F and F'. In particular, if F and F' are both contained in some coirreducible overring D, then $F = F'$, since every such overring is different from R. □

Definition 2. a) If D is an irreducible overring, we denote the maximal irreducible overring containing D by $H(D)$. We write more precisely $H_{R/A}(D)$ if necessary. We call $H(D)$ the *irreducible hull of D (in R over A)*.
b) If D is a coirreducible overring, we denote the minimal coirreducible overring contained in D by $C(D)$. We write more precisely $C_{R/A}(D)$ if necessary. We call $C(D)$ the *coirreducible core of D (in R over A)*.
c) We denote the set of irreducible overrings of A in R by $\mathrm{Ir}(R/A)$ and the set of coirreducible overrings of A in R by $\mathrm{Coir}(R/A)$. We regard both $\mathrm{Ir}(R/A)$ and $\mathrm{Coir}(R/A)$ as posets (= partially ordered sets), the ordering being the inclusion relation. □

Lemma 3 tells us that $\mathrm{Ir}(R/A)$ has enough maximal elements and $\mathrm{Coir}(R/A)$ has enough minimal elements in the sense previously defined (Sect. 3, Definition 7).

Definition 3. Let Λ be a poset and λ, μ elements of Λ.

a) A *path in Λ from λ to μ* is a finite sequence $(\lambda_0, \lambda_1, \dots, \lambda_n)$ of elements of Λ with λ_{i-1} and λ_i comparable $(1 \leq i \leq n)$ and $\lambda_0 = \lambda$, $\lambda_n = \mu$. If such a path exists, we say that λ and μ are *connectable in Λ*.
b) A poset is called *connected*, if any two elements of it are connectable.
c) The maximal connected subposets of Λ are called the *connected components* of Λ. □

Clearly Λ is the direct sum (cf. Sect. 3, Definition 3) of its connected components. Also, if $\lambda \in \Lambda$ is given, the *connected component of λ*, i.e. the unique connected component of Λ containing λ, is the set of all $\mu \in \Lambda$ connectable to λ.

Theorem 4 gives us a good hold on the connected components of the posets $\mathrm{Ir}(R/A)$ and $\mathrm{Coir}(R/A)$.

Corollary 6.5. *a) Two irreducible overrings D, D' are connectable in* $\mathrm{Ir}(R/A)$ *iff $H(D) = H(D')$. Thus every connected component of* $\mathrm{Ir}(R/A)$ *contains a unique maximal element of* $\mathrm{Ir}(R/A)$.

b) *Any two coirreducible overrings* D, D' *are connectable in* $\mathrm{Coir}(R/A)$ *iff* $C(D) = C(D')$. *Thus every connected component of* $\mathrm{Coir}(R/A)$ *contains a unique minimal element of* $\mathrm{Coir}(R/A)$.

Proof of claim (b).. If $C(D) = C(D') = F$ then (D, F, D') is a path in $\mathrm{Coir}(R/A)$ from D to D'. Conversely, if there is given a path $(D_0, D_1, \dots, D_n)$ in $\mathrm{Coir}(R/A)$ from D to D', then D_{i-1} and D_i are comparable, hence $C(D_{i-1}) = C(D_i)$ for $1 \le i \le n$. It follows that $C(D) = C(D')$. □

One way to get insight into a minimal coirreducible overring F is a description of the set $\omega(R/F)$, since F is the intersection of the rings A_v with v running through $\omega(R/F)$. We will pursue this theme in Sect. 7. Already now we state the following fact.

Proposition 6.6. *Let* F *be a minimal coirreducible overring. Then*

$$\omega(R/F) = S(R/F) \cap \omega(R/A).$$

Proof. Of course, $\omega(R/F)$ contains $S(R/F) \cap \omega(R/A)$. Let $v \in \omega(R/F)$ be given. We choose some $u \in \omega(R/A)$ with $u \le v$. Then $A_u \subset A_v$ and $F \subset A_v$. Invoking Theorem 4, we see that $C(A_u) = C(A_v) = F$. Thus $u \in S(R/F)$. It follows that $u = v$. We have proved that $v \in \omega(R/A)$. Thus $\omega(R/F) \subset S(R/F) \cap \omega(R/A)$. □

In a—rather general—favorable situation the minimal coirreducible overrings are responsible for all factorizations (cf. [Vol. I, Definition 3 in II §7]) of R over A, as we explain now.

Definition 4. We say that the extension $A \subset R$ has *coarse finite avoidance*, if the set $\mathrm{Coir}(R/A)_{\min}$ of minimal coirreducible overrings has finite avoidance in R.

We choose an indexing $(F_i \mid i \in I)$ of the set $\mathrm{Coir}(R/A)_{\min}$, with $F_i \ne F_j$ for $i \ne j$ of course. Our terminology makes sense due to the following proposition.

Proposition 6.7. *If* $A \subset R$ *has finite avoidance then* $A \subset R$ *has coarse finite avoidance.*

Proof. Let $x \in R$ be given and $J := \{i \in I \mid \text{with } x \notin F_i\}$. For every $i \in J$ we choose some $v_i \in \omega(R/F_i)$ with $x \notin A_{v_i}$. This is possible, since F_i is the intersection of all rings A_v with v running through $\omega(R/F_i)$.

The valuations v_i are elements of $\omega(R/A)$ by Proposition 6. From $F_i \subset A_{v_i}$ we conclude that $F_i = C(A_{v_i})$. Thus certainly $v_i \ne v_j$ for any two elements $i \ne j$ of J. The family $(A_v \mid v \in \omega(R/A))$ has finite avoidance in R. Since $x \notin A_{v_i}$ for every $i \in J$, the set J is finite. □

For every subset J of I we define $F_J := \bigcap_{i \in J} F_i$. {Read $F_\emptyset = R$.}

Theorem 6.8. *a)* $\bigcap_{i \in I} F_i = A$.

b) *Assume that* A *has coarse finite avoidance in* R. *Then the factors* B *of* R *over* A *(cf. [Vol. I, Definition 3 in II §7]) correspond bijectively with the subsets* J *of* I *via* $B = F_J$. *For each* $J \subset I$ *we have* $R = F_J \times_A F_{I \setminus J}$.

Proof. a): For every $v \in S(R/A)$ the overring A_v is coirreducible, hence contains some $F \in \mathrm{Coir}(R/A)$. Since the intersection of all rings A_v is A, the same holds for the rings $F \in \mathrm{Coir}(R/A)$.

b): Assume that $(F_i \mid i \in I)$ has finite avoidance in R. Let J be a non-empty proper subset of I. If $j \in J$ and $k \in I \setminus J$ then $F_j + F_k = R$. Applying the distributivity law Proposition 2.12 twice, we obtain $F_j + F_{I\setminus J} = R$ for every $j \in J$ and then $F_J + F_{I\setminus J} = R$. By our claim (a), already proved, we have $F_J \cap F_{I\setminus J} = A$. Thus $R = F_J \times_A F_{I\setminus J}$. Trivially this also holds if $J = I$ or $J = \emptyset$. Conversely, if a factorization $R = B \times_A C$ is given, then, for every $i \in I$, either $B \subset F_i$ or $C \subset F_i$ by Lemma 1. Let J be the set of all $i \in I$ with $B \subset F_i$. Then $B \subset F_J$ and $C \subset F_{I\setminus J}$. it follows that $B = C^\circ \supset F^\circ_{I\setminus J} = F_J$ and $C = B^\circ \supset F^\circ_J = F_{I\setminus J}$. Thus $B = F_J$ and $C = F_{I\setminus J}$. □

For every $i \in I$ we introduce the overring $F^i := \bigcap_{j \in I\setminus\{i\}} F_j$.

Theorem 6.9. *a) For every $i \in I$ the overring F^i is contained in the polar F_i° of F_i.*

b) Assume that A has coarse finite avoidance in R. Then $F^i = F_i^\circ$, and the F^i are precisely all maximal irreducible overrings of A. Also $F_J = \sum_{i \in I\setminus J} F^i$ for every subset J of I.

Proof. Part (a) of Theorem 8 tells us that $F_i \cap F^i = A$, which means that $F^i \subset F_i^\circ$. Assume now that $(F_i \mid i \in I)$ has finite avoidance in R. Then $R = F_i \times_A F^i$ by part (b) of Theorem 8. This implies that $F^i = F_i^\circ$ {and $F_i = (F^i)^\circ$}. By the transfer theorem for overrings ([Vol. I, Corollary II.6.6]) the lattice of overrings of A in F^i is isomorphic to the lattice of overrings of F_i in R. Since R is irreducible over F_i it follows that F^i is irreducible over A. (In particular $F^i \neq A$. This is also clear by Theorem 8.) Let E_i denote the irreducible hull $H(F^i)$. Applying Lemma 1(a) to E_i and the factorization $R = F^i \times_A F_i$, we learn that $E_i \subset F^i$, since certainly $E_i \not\subset F_i$. Thus $E_i = F^i$.

Let now E be any maximal irreducible overring. For every $i \in I$ we deduce from the factorization $R = F^i \times_A F_i$, again by Lemma 1(a), that either $E \subset F^i$ or $E \subset F_i$. We cannot have $E \subset F_i$ for every $i \in I$, since the intersection of the F_i is A. Thus there exists some $i \in I$ with $E \subset F^i$. Since E is maximal irreducible, it follows that $E = F^i$.

It is clear by [Vol. I, Chap. II §7] that the set of factors of R over A is a sublattice of the lattice of overrings of A in R (cf. [Vol. I, Proposition II.7.14]), and that the map $B \mapsto B^\circ$ is an antiautomorphism of this sublattice. Given a subset J of I, we have defined $F_J = \bigcap_{i\in J} F_i$. It follows that $F_J^\circ = \sum_{i\in J} F_i^\circ = \sum_{i\in J} F^i$. On the other hand, Theorem 8 tells us that $F_J^\circ = F_{I\setminus J}$. This gives us the last claim. □

Large parts of our study of PM-split extensions in Sect. 5 fit well into the present framework. Recall that a PM-split extension has finite avoidance (cf. Theorem 5.5), hence certainly has coarse finite avoidance. We now characterize PM-split extensions within the class of Prüfer extensions with coarse finite avoidance.

Proposition 6.10. *Assume that our Prüfer extension $A \subset R$ has coarse finite avoidance. The following conditions are equivalent.*

(1) R is PM-split over A.
(2) $\mathrm{Coir}(R/A) = \{A_{[\mathfrak{p}]} \mid \mathfrak{p} \in Y(R/A)\}$.
(2′) $\mathrm{Coir}(R/A)_{\min} = \{A_{[\mathfrak{p}]} \mid \mathfrak{p} \in \omega(R/A)\}$.

Proof. The equivalence (2) ⇔ (2′) is evident, since every overring $B \neq R$ of some ring $A_{[\mathfrak{p}]}$ in R, $\mathfrak{p} \in Y(R/A)$, is again of this type: $B = A_{[\mathfrak{p}']}$ with $\mathfrak{p}' \in Y(R/A)$ and $\mathfrak{p}' \subset \mathfrak{p}$.

(1) ⇒ (2′): Let $F \in \mathrm{Coir}(R/A)_{\min}$ be given. We choose some $v \in \omega(R/F)$. By Proposition 6 we know that $v \in \omega(R/A)$, i.e. $A_v = A_{[\mathfrak{p}]}$ with $\mathfrak{p} \in \Omega(R/A)$. Now Proposition 5.4 tells us, that $R = A_{[\mathfrak{p}]} \times_A A^{\mathfrak{p}}$. Since $F \subset A_{[\mathfrak{p}]}$, we certainly have $F + A_{[\mathfrak{p}]} \neq R$. Lemma 1(b) tells us that $F \supset A_{[\mathfrak{p}]}$, i.e. $F = A_{[\mathfrak{p}]}$. On the other hand, every $A_{[\mathfrak{p}]}$, $\mathfrak{p} \in \Omega(R/A)$, is coirreducible, and thus contains some $F \in \mathrm{Coir}(R/A)_{\min}$. As just proved, this implies $A_{[\mathfrak{p}]} = F$. Thus $A_{[\mathfrak{p}]}$ is minimally coirreducible.

(2′) ⇒ (1): Let $\mathfrak{p} \in \Omega(R/A)$ be given. Then $A^{\circ}_{[\mathfrak{p}]} = A^{\mathfrak{p}}$, and Theorem 8 tells us that $R = A_{[\mathfrak{p}]} \times_A A^{\mathfrak{p}}$. Condition (4) in Theorem 5.5 is fulfilled. Thus R is PM-split over A. □

Corollary 6.11. *If $A \subset R$ is PM-split, then the overrings $A^{\mathfrak{p}}$ with $\mathfrak{p}$ running through $\Omega(R/A)$ are precisely all maximal irreducible overrings of A in R.*

Proof. This follows from Theorem 8 and Proposition 10. □

We return to the study of the poset $\mathrm{Coir}(R/A)$ for any Prüfer extension $A \subset R$.

Proposition 6.12. *Assume that $(B_\lambda \mid \lambda \in \Lambda)$ is a family of overrings with finite avoidance in R and $B_\lambda + B_\mu = R$ for $\lambda \neq \mu$. Let $B := \bigcap_{\lambda \in \Lambda} B_\lambda$.*

i) The poset $\mathrm{Coir}(R/B)$ is the direct sum (cf. Sect. 3, Definition 8) of the posets $\mathrm{Coir}(R/B_\lambda)$,

$$\mathrm{Coir}(R/B) = \bigsqcup_{\lambda \in \Lambda} \mathrm{Coir}(R/B_\lambda).$$

ii) If every B_λ has coarse finite avoidance in R, then B has coarse finite avoidance in R.

Proof. Replacing A by the overring B we assume that $A = \bigcap_{\lambda \in \Lambda} B_\lambda$. For every $\lambda \in \Lambda$ we introduce the overring $B^\lambda = \bigcap_{\mu \neq \lambda} B_\mu$. Then $B_\lambda \cap B^\lambda = A$, and

$$B_\lambda + B^\lambda = \bigcap_{\mu \in \Lambda \setminus \{\lambda\}} (B_\lambda + B_\mu) = R,$$

due to our assumption and the distributivity law Proposition 2.12. Thus $R = B_\lambda \times_A B^\lambda$ for every $\lambda \in \Lambda$. Let now $D \in \mathrm{Coir}(R/A)$ be given. Lemma

1(b) tells us that, for every $\lambda \in \Lambda$ either $D \supset B_\lambda$ or $D \supset B^\lambda$, and $D + B_\lambda = R$ in the second case. Suppose that $D \supset B^\lambda$ for every $\lambda \in \Lambda$. Then, again by Proposition 2.12,

$$R = \bigcap_{\lambda \in \Lambda} (D + B_\lambda) = D + \bigcap_{\lambda \in \Lambda} B_\lambda = D + A = D.$$

But $D \neq R$, since D is coirreducible. Thus there exists some $\lambda \in \Lambda$ with $D \supset B_\lambda$. We conclude that $\mathrm{Coir}(R/A)$ is the union of the subsets $\mathrm{Coir}(R/B_\lambda)$ with λ running through Λ.

Suppose there exist indices $\lambda \neq \mu$ in Λ and overrings $D \in \mathrm{Coir}(R/B_\lambda)$, $D' \in \mathrm{Coir}(R/B_\mu)$, which are comparable, say $D' \subset D$. Then $D \supset B_\lambda$ and $D \supset B_\mu$, hence $D \supset B_\lambda + B_\mu = R$, i.e. $D = R$. This is a contradiction. Thus $\mathrm{Coir}(R/A)$ is the direct sum of the posets $\mathrm{Coir}(R/B_\lambda)$.

It follows that $\mathrm{Coir}(R/A)_{\min}$ is the disjoint union of the sets $\mathrm{Coir}(R/B_\lambda)_{\min}$. Assume now that for every $\lambda \in \Lambda$ the set of overrings $\mathrm{Coir}(R/B_\lambda)_{\min}$ has finite avoidance in R. We claim that $\mathrm{Coir}(R/B)_{\min}$ has finite avoidance in R. Let $x \in R$ be given. We have a finite subset U of Λ such that $x \in B_\lambda$ for every $\lambda \in \Lambda \setminus U$. Thus $x \in F$ for every $F \in \mathrm{Coir}(R/B_\lambda)_{\min}$ with $\lambda \in \Lambda \setminus U$. For every $\lambda \in U$ we have a finite subset Φ_λ of $\mathrm{Coir}(R/B_\lambda)_{\min}$ such that $x \in F$ for every $F \in \mathrm{Coir}(R/B_\lambda)_{\min} \setminus \Phi_\lambda$. Let $\Phi := \bigcup_{\lambda \in U} \Phi_\lambda$. This is a finite subset of $\mathrm{Coir}(R/A)_{\min}$, and $x \in F$ for every $F \in \mathrm{Coir}(B/A) \setminus \Phi$. □

Corollary 6.13. *Let $(B_\lambda \mid \lambda \in \Lambda)$ be a family of coirreducible overrings with finite avoidance in R and $B_\lambda + B_\mu = R$ for $\lambda \neq \mu$. Let $B := \bigcap_{\lambda \in \Lambda} B_\lambda$. Then the B_i are precisely all minimal coirreducible overrings of B in R. In particular, B has coarse finite avoidance in R.*

Proof. Each poset $\mathrm{Coir}(R/B_\lambda)$ has a unique minimal element B_λ. Proposition 12 applies and tells us, that the B_λ are the minimal elements of $\mathrm{Coir}(R/B)$. □

Procedure 6.14. Let again $(B_\lambda \mid \lambda \in \Lambda)$ be a family of coirreducible overrings with finite avoidance in R, but now we do *not* assume that $B_\lambda + B_\mu = R$ for $\lambda \neq \mu$. {We even allow repetitions in our family $(B_\lambda \mid \lambda \in \Lambda)$.} We prove again that R has coarse finite avoidance over $B := \bigcap_{\lambda \in \Lambda} B_\lambda$, and we develop a somewhat constructive method to obtain the set $\mathrm{Coir}(R/B)_{\min}$ from the given family $(B_\lambda \mid \lambda \in \Lambda)$.

We call two indices $\lambda, \mu \in \Lambda$ *connectable*, if there exists a finite sequence $\lambda_0 = \lambda, \lambda_1, \dots, \lambda_n = \mu$ in Λ such that $B_{\lambda_{i-1}} + B_{\lambda_i} \neq R$ for each $i \in \{1, \dots, n\}$. We then have a partition $(\Lambda_i \mid i \in I)$ of Λ into "connected components", i.e. subsets Λ_i such that any two elements of Λ_i are connectable, but no element of Λ_i is connectable with any element of $\Lambda \setminus \Lambda_i$. For each $i \in I$ we define

$$F_i := \bigcap_{\lambda \in \Lambda_i} B_\lambda.$$

First Claim: For every $i \in I$ the overring F_i is coirreducible.

Proof. Let $i \in I$ be given. For each subset S of Λ_i let $B_S := \bigcap_{\lambda \in S} B_\lambda$. If $\mathfrak{S}$ is a chain of subsets of Λ and B_S is coirreducible for each $S \in \mathfrak{S}$, then B_T is coirreducible for T the union of all $S \in \mathfrak{S}$, as follows from Lemma 3(b). Thus there exists a maximal subset Δ of Λ_i such that B_Δ is coirreducible. Suppose that $\Delta \neq \Lambda_i$. We choose an index $\rho \in \Lambda_i \setminus \Delta$. There exists a sequence $\lambda_0, \lambda_1, \dots, \lambda_n = \rho$ of indices in Λ_i with $\lambda_0 \in \Delta$ and $B_{\lambda_{i-1}} + B_{\lambda_i} \neq R$ for each $i \in \{1, \dots, n\}$. Let r denote the smallest number in $\{1, \dots, n\}$ with $\lambda_r \notin \Delta$, and let $\lambda := \lambda_{r-1}$, $\mu := \lambda_r$. We have $B_\lambda \in \Delta$, $B_\mu \notin \Delta$ and $B_\lambda + B_\mu \neq R$. A fortiori $B_\Delta + B_\mu \neq R$. Now Lemma 2(b) tells us that $B_{\Delta \cup \{\mu\}}$ is coirreducible. This contradicts the maximality of Δ. We conclude that $\Delta = \Lambda_i$, hence F_i is coirreducible.

Second Claim: If i, j are different indices in I then $F_i + F_j = R$.

Proof. If $\lambda \in \Lambda_i$ and $\mu \in \Lambda_j$ then $B_\lambda + B_\mu = R$. Now observe that the families $(B_\lambda \mid \lambda \in \Lambda_i)$ and $(B_\mu \mid \mu \in \Lambda_j)$ both have finite avoidance in R, since $(B_\lambda | \lambda \in \Lambda)$ has finite avoidance in R. Applying the distributive law Proposition 2.12 twice we first obtain $B_\lambda + F_j = R$ for every $\lambda \in \Lambda_i$ and then $F_i + F_j = R$.

Third Claim: The family $(F_i \mid i \in I)$ has finite avoidance in R.

Proof. Let $x \in R$ be given. There exists a finite subset U of Λ such that $x \in B_\lambda$ for every $\lambda \in \Lambda \setminus U$. We further have a finite subset J of I with $U \subset \bigcup_{i \in J} \Lambda_i$. It follows that $x \in F_i$ for every $i \in I \setminus J$.

Having proved the three claims we now conclude by Proposition 12, that $(F_i \mid i \in I)$ is the family of all coirreducible overrings of B in R, and that R has coarse finite avoidance over B. □

Theorem 6.15. *Let $(B_\lambda \mid \lambda \in \Lambda)$ be a family of overrings, which has finite avoidance in R. Assume that every B_λ has coarse finite avoidance in R. Then the intersection $B := \bigcap_{\lambda \in \Lambda} B_\lambda$ has coarse finite avoidance in R.*

Proof. For every $\lambda \in \Lambda$ we choose an indexing $(F_{\lambda i} \mid i \in I_\lambda)$ of the set $\mathrm{Coir}(R/B_\lambda)_{\min}$, without repetitions of course. Each of these families has finite avoidance in R, and $\bigcap_{i \in J_\lambda} F_{\lambda i} = B_\lambda$. Further $(B_\lambda \mid \lambda \in \Lambda)$ has finite avoidance in R. It follows easily that the total family $(F_{\lambda i} \mid \lambda \in \Lambda, i \in I_\lambda)$ has finite avoidance in R. The intersection of all these rings $F_{\lambda i}$ is B, and they all are coirreducible in R. As proved in Procedure 14, this implies that R has coarse finite avoidance over B. □

It is possible to obtain results on the poset $\mathrm{Ir}(R/A)$ somewhat dual to what we just have done (Proposition 12—Theorem 15). We now only prove, among other things, a counterpart to Proposition 12. A counterpart to Theorem 15 will become clear in the next section without extra effort (Remark 7.7).

Proposition 6.16. *Let $(B_\lambda \mid \lambda \in \Lambda)$ be a family of overrings. Assume that $R = \sum_{\lambda \in \Lambda} B_\lambda$ and $B_\lambda \cap B_\mu = A$ for $\lambda \neq \mu$. For every $\lambda \in \Lambda$ let $B^\lambda := \sum_{\mu \in \Lambda \setminus \{\lambda\}} B_\mu$.*

i) $R = B_\lambda \times_A B^\lambda$ *for every* $\lambda \in \Lambda$.

ii) *For every* $\lambda \in \Lambda$ *we have an isomorphism of posets* $\mathrm{Ir}(B_\lambda/A) \xrightarrow{\sim} \mathrm{Ir}(R/B^\lambda)$, *and an isomorphism of posets* $\mathrm{Coir}(B_\lambda/A) \xrightarrow{\sim} \mathrm{Coir}(R/B^\lambda)$, *which sends a ring* $D \in \mathrm{Ir}(B_\lambda/A)$ *(resp.* $D \in \mathrm{Coir}(B_\lambda/A)$*) to* $D + B^\lambda$. *The inverse map is given by* $D' \mapsto D' \cap B_\lambda$.

iii) *The poset* $\mathrm{Ir}(R/A)$ *is the direct sum of the posets* $\mathrm{Ir}(B_\lambda/A)$,

$$\mathrm{Ir}(R/A) = \bigsqcup_{\lambda \in \Lambda} \mathrm{Ir}(B_\lambda/A).$$

iv) *The family* $(B^\lambda \mid \lambda \in \Lambda)$ *has finite avoidance in* R, *and*

$$\mathrm{Coir}(R/A) = \bigsqcup_{\lambda \in \Lambda} \mathrm{Coir}(R/B^\lambda).$$

Proof. i): By the law of distributivity [Vol. I, Lemma II.7.1] we have $B_\lambda \cap B^\lambda = \sum_{\mu \neq \lambda}(B_\lambda \cap B_\mu) = A$.

Clearly $B_\lambda + B^\lambda = R$. Thus $R = B_\lambda \times_A B^\lambda$.

ii): Claim (ii) follows from (i) immediately by the transfer theory for overrings in [Vol. I, Chap. II], cf. [Vol. I, Corollary II.6.6] and [Vol. I, Theorem II.7.16].

iii): Let $D \in \mathrm{Ir}(R/A)$ be given. Lemma 1 tells us that $D \subset B_\lambda$ or $D \subset B^\lambda$ for every $\lambda \in \Lambda$. Suppose that $D \subset B^\lambda$ for every $\lambda \in \Lambda$. Then $D \cap B_\lambda = A$ for every $\lambda \in \Lambda$. By the law of distributivity [Vol. I, Lemma II.7.1] we conclude that

$$D = D \cap (\sum_{\lambda \in \Lambda} B_\lambda) = \sum_{\lambda \in \Lambda}(D \cap B_\lambda) = A,$$

a contradiction. Thus there exists some $\lambda \in \Lambda$ with $D \subset B_\lambda$. This proves that $\mathrm{Ir}(R/A)$ is the union of the sets $\mathrm{Ir}(B_\lambda/A)$.

Let $D \in \mathrm{Ir}(B_\lambda/A)$ and $D' \in \mathrm{Ir}(B_\mu/A)$ be given with $\lambda \neq \mu$. Suppose $D \subset D'$. Then $D \subset B_\lambda \cap B_\mu = A$, a contradiction. Thus the poset $\mathrm{Ir}(R/A)$ is the direct sum of the posets $\mathrm{Ir}(R/B_\lambda)$.

iv): Let $x \in R$ be given. There exists a finite subset U of Λ with $x \in \sum_{\lambda \in U} B_\lambda$. If $\lambda \in \Lambda \setminus U$ then $x \in B^\lambda$. Thus the family $(B^\lambda \mid \lambda \in \Lambda)$ has finite avoidance in R. Due to the distributivity law Proposition 2.12 we have $B^\lambda + B^\mu = R$ for $\lambda \neq \mu$. Proposition 12 tells us that $\mathrm{Coir}(R/B)$ is the direct sum of the posets $\mathrm{Coir}(R/B^\lambda)$. □

7 Completely Reducible Prüfer Extensions

Let $A \subset R$ be a Prüfer extension. As before, "overring" means "overring of A in R", if nothing else is said. We choose an indexing $(E_i \mid i \in I)$ of the set $\mathrm{Ir}(R/A)_{\max}$ of maximal irreducible overrings of A in R, with $E_i \neq E_j$ for $i \neq j$, of course. Recall from Sect. 6 that $E_i \cap E_j = A$ if $i \neq j$.

Definition 1. We call the Prüfer extension $A \subset R$ *completely reducible*, if the ring R is generated by the set $\mathrm{Ir}(R/A)_{\max}$ of maximal irreducible overrings, i.e. $R = \sum_{i \in I} E_i$. {If $\mathrm{Ir}(R/A)$ is empty, this shall mean $R = A$.} We then also say that R is *completely reducible over* A, or that the extension $A \subset R$ is PCR for short, or that R is PCR over A[6] □

The PM-split extensions, studied in Sect. 5, are just the PCR-extensions in which the (maximal) irreducible overrings are PM. We now try to prove similar results for PCR-extensions as obtained in Sect. 5 for PM-split extensions.

Remark 7.1. If $(B_\lambda \mid \lambda \in \Lambda)$ is a family of irreducible overrings and $R = \sum_{\lambda \in \Lambda} B_\lambda$, then R is completely irreducible over A. Indeed, for every $\lambda \in \Lambda$ the irreducible hull $H(B_\lambda)$ is one of the overrings E_i, and thus $R = \sum_{i \in I} E_i$. □

Proposition 7.2. *Let $(E'_i \mid i \in I')$ be a family of irreducible overrings such that $E'_i \cap E'_j = A$ for $i \neq j$. Assume that $\sum_{i \in I'} E'_i = R$. Then R is completely reducible over A, and $(E'_i \mid i \in I')$ runs through the set $\mathrm{Ir}(R/A)_{\max}$ of all maximal irreducible overrings.*

Proof. This follows from Proposition 6.16(iii). □

Notice that Proposition 2 is a statement dual to Corollary 6.13.

Starting from our family $(E_i \mid i \in I)$ of all maximal irreducible overrings, we define for every $i \in I$

$$F_i := \sum_{j \in I \setminus \{i\}} E_j,$$

for every $i \in I$, and

$$E_J := \sum_{i \in J} E_i \quad , \quad F_J := \bigcap_{i \in J} F_i$$

for every subset J of I. {Read $E_\emptyset = A$ and $F_\emptyset = R$.}

Theorem 7.3. *Assume that R is completely reducible over A.*

a) $R = E_J \times_A E_{I \setminus J}$ for every subset J of I. These are all factorizations of R over A.
b) $(F_i \mid i \in I)$ is the family of all minimal coirreducible overrings without repetitions (i.e. $F_i \neq F_j$ for $i \neq j$). R has coarse finite avoidance (cf. Sect. 6, Definition 4) over A.
c) $F_J = E_{I \setminus J}$ for every subset J of I.

Proof. Various arguments will be similar to those in the proof of Proposition 6.16.

[6]The letter P stands for "Prüfer", CR stands for "completely reducible".

1) Let J be a subset of R different from $\emptyset$ and I. We have $E_j \cap E_k = A$ for $j \in J$ and $k \in I \setminus J$. Applying the distributive law [Vol. I, Lemma II.7.1] twice we first obtain $E_j \cap E_{I\setminus J} = A$ for every $j \in J$ and then $E_J \cap E_{I\setminus J} = A$. Since $E_J + E_{I\setminus J} = R$, this proves that $R = E_J \times_A E_{I\setminus J}$. Trivially this also holds if $J = \emptyset$ or $J = I$.
2) By definition we have $F_i = E_{I\setminus\{i\}}$. Thus $R = E_i \times_A F_i$ for every $i \in I$. The lattice of overrings of A in E_i is isomorphic to the lattice of overrings of F_i in R, due to the transfer theorem for overrings ([Vol. I, Corollary II.6.6]). Now E_i is irreducible over A. Thus R is irreducible over F_i, i.e. F_i is a coirreducible overring. Let F_i' denote the coirreducible core $C(F_i)$ (cf. Sect. 6, Definition 2). We apply Lemma 6.1(b) to F_i' and the factorization $R = E_i \times_A F_i$. The ring F_i' does not contain E_i, since $E_i \cap F_i = A$ and $F_i' \subset F_i$. Thus $F_i' \supset F_i$. We conclude that $F_i' = F_i$. This proves that all the F_i are minimal coirreducible overrings. If $i \neq j$, then $E_j \subset F_i$ but $E_j \cap F_j = A$. Thus $F_i \neq F_j$ for $i \neq j$.
3) Let F be a minimal coirreducible overring. We apply Lemma 6.1(b) to F and all factorizations $R = E_i \times_A F_i$. We see that $E_i \subset F$ or $F_i \subset F$ for every $i \in I$. We cannot have $E_i \subset F$ for every $i \in I$, since the ring generated by all E_i is R. Thus $F_i \subset F$ for some $i \in I$. Since F is minimal coirreducible, it follows $F_i = F$. We have proved that $(F_i \mid i \in I)$ is the family of all minimal coirreducible overrings without repetitions.
4) We verify that the family $(F_i \mid i \in I)$ has finite avoidance in R. Let $x \in R$ be given. There exists a finite subset U of I with $x \in \sum_{i \in U} E_i = E_U$. For every $j \in I \setminus U$ we have $E_U \subset F_j$, hence $x \in F_j$.
5) We now know that R has coarse finite avoidance over A. Thus we can invoke Theorem 6.8. The theorem tells us, that $R = F_{I\setminus J} \times_A F_J$ for every subset J of R, and that these are all the factorizations of R over A. Clearly $E_J \subset F_{I\setminus J}$ and $E_{I\setminus J} \subset F_J$. Since also $R = E_J \times_A E_{I\setminus J}$, we conclude easily that $E_J = F_{I\setminus J}$ and $E_{I\setminus J} = F_J$. Indeed, $E_{I\setminus J} = E_J^\circ \supset F_{I\setminus J}^\circ = F_J$, hence $E_{I\setminus J} = F_J$. □

The following Theorem 5 states in particular that the properties "coarse finite avoidance" and "complete reducibility" are the same thing.

This might come as a surprise at first glance. But observe that Theorem 5 is the off hand analogue of Theorem 5.5 above on PM-split extensions, if we replace PM-extensions by irreducible extensions and hence PM-split extensions by completely irreducible Prüfer extensions. The good news is that the analogy really works. {Notice that $F + F' = R$ for any two different minimal coirreducible overrings F, F' (cf. Theorem 6.4(b)), while not necessarily $A_v + A_{v'} = R$ for different $v, v' \in \omega(R/A)$.}

Theorem 7.5. *The following conditions are equivalent (for any Prüfer extension $A \subset R$).*

(1) R is completely reducible over A.
(2) Every minimal coirreducible overring is a factor of R over A.
(3) The extension $A \subset R$ has coarse finite avoidance.

Proof. The implications (1) ⇒ (2) and (1) ⇒ (3) are covered by Theorem 3.

(2) ⇒ (1): Let $(F_\lambda \mid \lambda \in \Lambda)$ denote the family of all minimal coirreducible overrings (without repetitions, of course). For every $\lambda \in \Lambda$ we have $R = F_\lambda^\circ \times_A F_\lambda$. Since the extension $F_\lambda \subset R$ is irreducible, also $A \subset F_\lambda^\circ$ is irreducible (cf. [Vol. I, Corollary II.6.6]). Let $R' := \sum_{\lambda\in\Lambda} F_\lambda^\circ$. We verify that $R' = R$, and then will be done.

Suppose that $R' \neq R$. We choose some $v \in S(R/R')$. There exists an index $\mu \in \Lambda$ such that $C(A_v) = F_\mu$. We have $F_\mu^\circ \subset R' \subset A_v$ and $F_\mu \subset A_v$, hence $F_\mu^\circ + F_\mu \subset A_v$. This is a contradiction, since $F_\mu^\circ + F_\mu = R$. Thus indeed $R' = R$.

(3) ⇒ (1): Let again $(F_\lambda \mid \lambda \in \Lambda)$ denote the family of all minimal coirreducible overrings. We now assume that $A \subset R$ has coarse finite avoidance. We know by Theorem 6.10 that the rings

$$F^\lambda := \bigcap_{\mu\in\Lambda\setminus\{\lambda\}} F_\mu \qquad (\lambda \in \Lambda)$$

are precisely all maximal irreducible overrings. Let $R_0 := \sum_{\lambda\in\Lambda} F^\lambda$. We verify that $R_0 = R$, and then will be done.

Suppose that $R_0 \neq R$. We choose a valuation $v \in S(R/R_0)$. There exists an index $\mu \in \Lambda$ with $C(A_v) = F_\mu$. Now $F_\mu + F_\lambda = R$ for every $\lambda \neq \mu$. Since the family $(F_\lambda \mid \lambda \in \Lambda)$ has finite avoidance in R, it follows that $F_\mu + F^\mu = R$ (cf. Proposition 2.12). But $F_\mu \subset A_v$ and $F^\mu \subset R_0 \subset A_v$, hence $F_\mu + F^\mu \subset A_v$. This is a contradiction, and we conclude that $R_0 = R$. □

Corollary 7.6. *Every PF-extension is completely reducible.*

Proof. We know by Proposition 6.7 that every PF-extension has coarse finite avoidance. □

We can state a counterpart of Theorem 6.15, which in the light of Theorem 5 becomes a triviality.

Remark 7.7. Let $(B_\lambda \mid \lambda \in \Lambda)$ be a family of overrings of A in R. Assume that every B_λ has coarse finite avoidance over A. Then the ring $B := \sum_{\lambda\in\Lambda} B_\lambda$ has coarse finite avoidance over A. □

Definition 2. a) We call the Prüfer extension $A \subset R$ *finitely reducible* if the set $\mathrm{Coir}(R/A)_{\min}$ of minimal coirreducible overrings is finite. We then also say that R is *finitely reducible over* A, or that the extension $A \subset R$ is PFR (= finitely reducible Prüfer), or that R is PFR over A.

b) We denote the cardinality of the set $\mathrm{Coir}(R/A)_{\min}$ by $\rho(R/A)$. This cardinal number is finite iff R is PFR over A. □

Remark 7.8. a) If $A \subset R$ is finitely reducible, then certainly $A \subset R$ has coarse finite avoidance, and thus $A \subset R$ is completely reducible by Theorem 4. Moreover, as for any PCR extension, we have a bijection $\mathrm{Coir}(R/A)_{\min} \xrightarrow{\sim} \mathrm{Ir}(R/A)_{\max}$ by mapping a minimal coirreducible overring F to its polar F°, as is clear from Theorem 3. Thus the PFR-extensions are just the PCR-extensions with finitely many maximal irreducible overrings. This justifies the terminology in Definition 2.

b) If there exist a finite family of irreducible R-overrings $(B_i \mid 1 \leq i \leq n)$ of A generating R over A, then R is finitely reducible over A and $\rho(R/A) \leq n$. This follows from the argument given in Remark 1.
c) If R_1 and R_2 are overrings of A in R, which both are completely reducible over A, then also $R_1 R_2$ is completely reducible over A, and

$$\rho(R_1 R_2/A) \leq \rho(R_1/A) + \rho(R_2/A).$$

This is again clear by the argument in Remark 1. In particular, if R_1 and R_2 are finitely reducible over A, the same holds for $R_1 R_2$. □

We now strive for an analogue of Proposition 4.4, with coarse finite avoidance (= complete irreducibility) instead of finite avoidance and finite reducibility instead of PM-finiteness.

Lemma 7.9. *Assume that D is an R-overring of A such that both extensions $A \subset D$ and $D \subset R$ are irreducible. Then either $A \subset R$ is irreducible or $R = D \times_A C$ with $A \subset C$ irreducible.*

Proof. Assume that $A \subset R$ is reducible, $R = B \times_A C$ with $B \neq A$ and $C \neq A$. Lemma 6.1(b) tells us that $B \subset D$ or $C \subset D$. Without loss of generality we assume that $B \subset D$. Then $D = D \cap (B + C) = B + (D \cap C)$, and $B \cap (D \cap C) = A$. Thus $D = B \times_A (D \cap C)$. Since D is irreducible over A and $B \neq A$, this forces $B = D$, and $R = D \times_A C$. The extension $A \subset C$ is irreducible, since $D \subset R$ is irreducible. □

Theorem 7.10. *Assume that $A \subset B$ is a finitely reducible subextension of $A \subset R$, and that $B \subset R$ is completely reducible. Then $A \subset R$ is completely reducible and $\rho(R/A) \leq [\rho(B/A) + 1]\rho(R/B)$.*

Proof. a) We first deal with the case that $\rho(R/B) = 1$, i.e. B is coirreducible in R. We proceed by induction on $n := \rho(B/A)$. The case $n = 1$ is covered by Lemma 9. Assume that $n > 1$. We choose a factorization $B = D \times_A F$ with D irreducible over A. The extensions $F \subset B$ is irreducible, and $A \subset F$ is finitely reducible with $\rho(F/A) = n - 1$.

1. *Case:* R is irreducible over F. By induction hypothesis we conclude that R is finitely reducible over A and $\rho(R/A) \leq n$.
2. *Case:* R is reducible over F. Since the extensions $F \subset B$ and $B \subset R$ both are irreducible, we conclude by Lemma 9 that $R = B \times_F C$ with C irreducible over F. By induction hypothesis, C is finitely reducible over A with $\rho(C/A) \leq n$. We have $R = B + C = D + F + C = D + C$, and D is irreducible over A. Thus R is finitely reducible over A and $\rho(R/A) \leq 1 + \rho(C/A) \leq n + 1$. This proves our claim in the case that $B \subset R$ is irreducible.

b) Assume now that $B \subset R$ is completely reducible, and let $(E_i \mid i \in I)$ an indexing (without repetitions) of the set of maximal irreducible overrings of B in R. We have $R = \sum_{i \in I} E_i$. As just proved, every E_i is finitely reducible over A with $\rho(E_i/A) \le \rho(B/A) + 1$. Write $E_i = \sum_{\lambda \in \Lambda_i} D_{i\lambda}$, with $(D_{i\lambda} \mid \lambda \in \Lambda_i)$ the family of all irreducible factors of E_i over A. We have

$$R = \sum_{i \in I} \sum_{\lambda \in \Lambda_i} D_{i\lambda} = \sum_{i \in I} \sum_{\lambda \in \Lambda_i} H(D_{i\lambda}),$$

where $H(D_{i\lambda})$ denotes the irreducible hull of $D_{i\lambda}$ in R over A. Omitting repetitions in the right hand sum, we see that R is completely reducible over A with

$$\rho(R/A) \le \sum_{i \in I} \rho(E_i/A) \le [\rho(B/A) + 1]\rho(E/B),$$

since I has the cardinality $\rho(E/B)$. {N.B. The argument is also valid if the cardinal number $\rho(E/B)$ is infinite.} □

Remark 7.11. The bound in Theorem 10 seems to be rough. Let us consider the special case that $\rho(B/A) = 1$ more closely. Theorem 10 gives the bound $\rho(R/A) \le 2\rho(B/A)$. But now a much better bound exists.

Let again $(E_i \mid i \in I)$ be the family of irreducible factors of R over B. Let I_1 denote the set of all indices of I such that E_i is irreducible over A, and let $I_2 = I \setminus I_1$. Lemma 9 tells us that for $i \in I_2$ we have $E_i = B \times_A C_i$ with C_i irreducible over A. If I_2 is empty, we have $R = \sum_{i \in I_1} E_i$, hence $\rho(R/A) \le \rho(R/B)$. If $I_2 \ne \emptyset$, we have

$$R = \sum_{i \in I_1} E_i + \sum_{i \in I_2} (B + C_i) = B + \sum_{i \in I_1} E_i + \sum_{i \in I_2} C_i,$$

hence $\rho(R/A) \le \rho(R/B) + 1$. □

Proposition 7.12. *Let $A \subset T$ be any ring extension. There exists a unique overring U of A in T with the following properties:*

(1) $A \subset U$ is PCR.
(2) If $A \subset C$ is a PCR subextension of $A \subset T$ then $C \subset U$.

Proof. Let $R := P(A,T)$, the Prüfer hull of A in T. We may replace $A \subset T$ by the Prüfer extension $A \subset R$. We choose an indexing $(E_i \mid i \in I)$ of the set $\mathrm{Ir}(R/A)_{\max}$. Clearly the ring $U := \sum_{i \in I} E_i$ has the properties (1), (2) above. □

Definition 3. a) We call the ring U in Proposition 12 the *completely reducible Prüfer hull*, or *PCR-hull* for short, *of A in T*, and write $U = \mathrm{PCR}(A,T)$.

b) If A is any ring, we define $\mathrm{PCR}(A) := \mathrm{PCR}(A, P(A))$,[7] and call this ring the *completely reducible Prüfer hull*, or *PCR-hull*, of the ring A. □

Remark 7.13. Given a ring extension $A \subset T$, we have the following inclusions for the various "hulls" introduced in the present chapter up to now:

$$A \subset \mathrm{PMS}(A,T) \subset \mathrm{PF}(A,T) \subset \mathrm{PCR}(A,T) \subset P(A,T).$$

Here the second inclusion from left reflects the fact that every PM-split extension is PF, while the third one reflects the fact that every PF-extension is completely reducible (Corollary 6 above). In particular

$$A \subset \mathrm{PMS}(A) \subset \mathrm{PF}(A) \subset \mathrm{PCR}(A) \subset P(A).$$

□

Remark 7.14. If A is any ring and $A \subset B$ is a completely reducible Prüfer extension, there exists a unique ring homomorphism $\varphi\colon B \to \mathrm{PCR}(A)$ over A, and φ is injective. Indeed, we have such a map φ from B to $P(A)$, as has been proved in [Vol. I, Chap. I §5], and $\varphi(B) \subset \mathrm{PCR}(A)$, since $\varphi(B)$ is completely reducible over A.

Concerning the hulls $\mathrm{PMS}(A)$ and $\mathrm{PF}(A)$ the analogue of all this is also true, with "completely reducible" replaced by "PM-split" and "PF" respectively. □

Proposition 7.15 (Transfer principle for irreducible and coirreducible factors). *Let $A \subset R$ be Prüfer (as before) and let $A \subset C$ be a ws extension. We regard R and C as subrings of $R \otimes_A C = RC$. Assume that $R \cap C = A$.*

Then R is completely reducible over A iff RC is completely reducible over C. In this case the irreducible (resp. coirreducible) factors D of R over A correspond bijectively with the irreducible (resp. coirreducible) factors D' of RC over C via $D' = DC$, $D = D' \cap R$.

Proof. All this follows from the fact that we have an isomorphism $D \mapsto DC$ from the lattice of overrings of A in R to the lattice of overrings of C in RC, the inverse isomorphism being given by $D' \mapsto D' \cap C$ (cf. [Vol. I, Chap. II §6 & §7]). □

Corollary 7.16. *Let again $A \subset R$ be Prüfer, $A \subset C$ be ws, and $R \cap C = A$. Then*

$$\mathrm{PCR}(C, RC) = \mathrm{PCR}(A, R) \cdot C.$$

□

Looking again at analogues between PMS-extensions and PCR-extensions we are sorry to say that there seems to exist no analogue of Proposition 5.14(i): If we drop the condition $R \cap C = A$ in Proposition 15, we cannot prove that $C \subset RC$

---

[7] Recall that $P(A)$ denotes the Prüfer hull of A, cf. [Vol. I, Definition 3 in I §5].

is completely reducible, if $A \subset R$ is completely reducible. Also the analogue to Proposition 5.16 on localizations seems to be wrong for PCR-extensions.

8 Connectedness in the Restricted PM-Spectrum

As before we are given a Prüfer extension $A \subset R$, and "overring" means "overring of A in R", if nothing else is said. We have an isomorphism from the restricted PM-spectrum $S(R/A)$ to a subposet of the set of coirreducible overrings $\mathrm{Coir}(R/A)$, mapping a valuation $v \in S(R/A)$ to the associated overring A_v. We will often regard $S(R/A)$ as a subposet of $\mathrm{Coir}(R/A)$ in this way. Notice that then $S(R/A)$ is an upper set in $\mathrm{Coir}(R/A)$, hence for any overring B the set

$$S(R/B) = \mathrm{Coir}(R/B) \cap S(R/A)$$

is again an upper set in $\mathrm{Coir}(R/A)$.

Every element D of $\mathrm{Coir}(R/A)$ is uniquely determined by the subset $S(R/D)$ of $S(R/A)$. Indeed, D is the minimum of $S(R/D)$ in $\mathrm{Coir}(R/A)$, and

$$S(R/D) = \{\xi \in S(R/A) \mid D \leq \xi\}.$$

How to characterize these subsets $S(R/D)$ of $S(R/A)$ within the poset $S(R/A)$?

Concerning the minimal coirreducible overrings F the best answer seems to be, that the sets $S(R/F)$ are just the connected components (cf. Sect. 6, Definition 4) of the poset $S(R/A)$. We will prove this now in the case that $A \subset R$ has finite avoidance.

Lemma 8.1. *Let Δ be a connected component of the poset $S(R/A)$. Then there exists a unique minimal $F \in \mathrm{Coir}(R/A)$ with $C(A_v) = F$ for every $v \in \Delta$.*

Proof. Regarding $S(R/A)$ as a subposet of $\mathrm{Coir}(R/A)$, it is evident that Δ is contained in a connected component of $\mathrm{Coir}(R/A)$. Corollary 6.5(b) gives the claim. □

Proposition 8.2. *If $R \neq A$ and $S(R/A)$ is connected, R is irreducible over A.*

Proof. By Lemma 1 there exists a coirreducible overring F with $C(A_v) = F$ for every $v \in S(R/A)$. In particular $A_v \supset F$ for every v. Since the intersection of the rings A_v is A, we have $F = A$. Thus the extension $A \subset R$ is irreducible. □

Theorem 8.3. *Assume that the extension $A \subset R$ has finite avoidance. Let D be a coirreducible overring. Then the poset $S(R/D)$ is connected.*

Proof. Suppose that $S(R/D)$ is not connected. We write $S(R/D)$ as a direct sum $S_1 \sqcup S_2$ with both S_1 and S_2 not empty. Let $\omega := \omega(R/D)$, $\omega_1 := (S_1)_{\min}$, $\omega_2 := (S_2)_{\min}$. Then $\omega = \omega_1 \dot\cup \omega_2$. We introduce the rings

$$B_i := \bigcap_{v \in S_i} A_v = \bigcap_{v \in \omega_i} A_v \quad (i = 1, 2).$$

We have $B_1 \cap B_2 = D$. If $v \in \omega_1$ and $v' \in \omega_2$ then $A_v + A_{v'} = R$, since otherwise there would exist some $w \in S(R/D)$ with $A_v + A_{v'} \subset A_w$, i.e. $v \leq w$, $v' \leq w$. But v and v' are not connectable in $S(R/D)$. Now both the families $(A_v | v \in \omega_1)$, $(A_v | v \in \omega_2)$ have finite avoidance in R, since the extensions $B_1 \subset R$, $B_2 \subset R$ have finite avoidance (cf. Proposition 4.3). Using twice the distributivity law Proposition 2.12, we deduce that

$$A_v + B_2 = \bigcap_{v' \in \omega_2} (A_v + A_{v'}) = R$$

for every $v \in \omega_1$, and then, that

$$B_1 + B_2 = \bigcap_{v \in \omega_1} (A_v + B_2) = R.$$

Since also $B_1 \cap B_2 = D$, we have a decomposition $R = B_1 \times_D B_2$. This contradicts our assumption that the extension $D \subset R$ is irreducible. It follows that $S(R/D)$ is connected. □

Later we will meet Prüfer extensions which do not have finite avoidance, but where nevertheless the conclusion in Theorem 3 is true. Since this conclusion has very agreeable consequences, as we will see, it pays to isolate this property.

Definition 1. We call a Prüfer extension $A \subset R$ *connective*, if, for every coirreducible overring D, the restricted PM-spectrum $S(R/D)$ is connected. □

In this terminology Theorem 3 states that every PF-extension is connective.

Proposition 8.4. *Assume that the extension $A \subset R$ is connective. Then the sets $S(R/F)$ with F a minimal coirreducible overring are precisely all connected components of $S(R/A)$.*

Proof. Lemma 1 tells us that every connected component of $S(R/A)$ is contained in such a set $S(R/F)$, and by assumption these sets $S(R/F)$ are connected. □

We add some observations on the combinatorics in the poset $S(R/A)$. These will imply an improvement of our knowledge of coirreducible overrings in the case that $A \subset R$ is PF.

It is intuitively clear that the connected components of $S(R/A)$ are the "trees" of the "forest" $S(R/A)$ (cf. Remark 3.1(c)). Thus the following lemma does not come as a surprise.

Lemma 8.5. *Let Δ be a connected component of $S(R/A)$ and let V be a finite subset of Δ. Then there exists some $w \in \Delta$ with $v \leq w$ for every $v \in V$.*

Proof. a) We prove that, given a path $(v_0, v_1, \dots, v_n)$ in $S(R/A)$, there exists some $w \in S(R/A)$ with $v_i \le w$ for every $i \in \{0, \dots, n\}$. We proceed by induction on n, the case $n = 0$ being trivial.

$(n-1) \to n$: By induction hypothesis there exists some $u \in S(R/A)$ with $v_i \le u$ for $0 \le i \le n-1$. If $v_n \le v_{n-1}$ then also $v_n \le u$, and we are done. There remains the case that $v_n > v_{n-1}$. Since $u \ge v_{n-1}$, the elements v_n and u are comparable. If $v_n \le u$, we take $w := u$. If $u \le v_n$, we take $w := v_n$.

b) We now prove the claim of the lemma. Write $\Delta = \{v_1, v_2, \dots, v_n\}$. We may assume that $n \ge 2$. Since v_1 is connectable to every v_i, $i \ge 2$, we have elements w_i, $2 \le i \le n$, such that $w_i \ge v_1$ and $w_i \ge v_i$. The elements w_i form a chain, since they all dominate v_1. Let w denote the maximum of $w_2, \dots, w_n$. Then $w \ge v$ for every $v \in \Delta$. □

Proposition 8.6. *Assume that $S(R/A)$ is connected. Then for every overring B the restricted PM-spectrum $S(R/B)$ is again connected.*

Proof. Let two elements v_1, v_2 of $S(R/B)$ be given. They are connectable in $S(R/A)$. Lemma 5 tells us that there exists some $w \in S(R/A)$ with $v_1 \le w$ and $v_2 \le w$. We have $B \subset A_{v_1} \subset A_w$, hence $w \in S(R/B)$. Thus v_1 and v_2 are connectable in $S(R/B)$. □

Remark 8.7. It now is evident, that a Prüfer extension $A \subset R$ is connective iff the poset $S(R/F)$ is connected for every *minimal* coirreducible overring F. In particular, $A \subset R$ is connective if $S(A/R)$ is connected. □

Corollary 8.8. *If $S(R/A)$ is connected, every overring different from R is coirreducible.*

Proof. This is clear by Propositions 6 and 2. □

Proposition 8.9. *Assume that $A \subset R$ is connective. Let D and D' be overrings with $D \subset D' \ne R$. If D is coirreducible, then also D' is coirreducible.*

Proof. By assumption, $S(R/D)$ is connected. Corollary 8 gives the claim. □

Proposition 8.10. *Assume that $A \subset R$ is connective. Let D_1 and D_2 be coirreducible R-overrings of A. Then $C(D_1) = C(D_2)$ iff $D_1 \cap D_2$ is coirreducible.*

Proof. If $D_1 \cap D_2$ is coirreducible, then $C(D_1) = C(D_1 \cap D_2) = C(D_2)$. Conversely, assume that $C(D_1) = C(D_2) = F$. We have $F \subset D_1 \cap D_2 \subsetneq R$. Proposition 9 tells us that $D_1 \cap D_2$ is coirreducible. □

Problem. Is it possible, to prove under a reasonable hypothesis, that for irreducible overrings D_1, D_2 with $H(D_1) = H(D_2)$ the overring $D_1 + D_2$ is again irreducible? We do not know the answer.

Lemma 8.11. *Assume that w is a maximal element of the poset $S(R/A)$. Then the set $\Delta_w := \{v \in S(R/A) \mid v \le w\}$ is a connected component of $S(R/A)$.*

Proof. Of course, Δ_w is connected. Let Δ' denote the connected component of $S(R/A)$ containing Δ_w. We have to verify that $\Delta_w = \Delta'$. Let $u \in \Delta'$ be given.

Lemma 5 tells us that there exists some $w' \in S(R/A)$ with $u \leq w'$ and $w \leq w'$. Since w is maximal in $S(R/A)$, we have $w = w'$. Thus $u \in \Delta_w$. □

If Λ is any poset, let $\Lambda_{\max}$ denote the set of maximal elements of Λ. We now impose on our Prüfer extension $A \subset R$ the condition, that *for every $v \in S(R/A)$ there exists an element $w \in S(R/A)_{\max}$ with $v \leq w$*. This element w then is unique, and we denote it by v^*.

In the terminology of Sect. 3 this condition means that the poset $S(R/A)$ *has enough maximal elements* (cf. Sect. 3, Definition 7), since clearly the maximal chains in $S(R/A)$ are the sets $\{u \in S(R/A) \mid u \geq v\}$ with v running through $S(R/A)_{\min} = \omega(R/A)$. Notice also that our condition means that for every $v \in S(R/A)$ (or $v \in \omega(R/A)$) the group Γ_v contains a maximal proper convex subgroup.

Example 8.12. Let k be a real closed field and M a semialgebraic subset of k^n. Let R be the ring of continuous semialgebraic functions on M and A the subring consisting of all $f \in R$ which are bounded on M. Then A is Prüfer in R (cf. [Vol. I, Chap. I §6 p. 71f]). Assume that $A \neq R$. Every valuation $v \in S(R/A)$ has height $\leq n$, i.e. the group Γ_v contains at most $n + 1$ convex subgroups, as is well known. Thus all chains in $S(R/A)$ contain at most $n + 1$ elements. $S(R/A)$ has certainly enough maximal elements. We remark that most often this Prüfer extension $A \subset R$ does not have finite avoidance, cf. [KZ_1, Ex. 1.3, Ex. 2.1]. □

Proposition 8.13. *Assume that $S(R/A)$ has enough maximal elements. Then every connected component Δ of $S(R/A)$ contains a unique element $w \in S(R/A)_{\max}$, and Δ coincides with the set Δ_w introduced in Lemma 11.*

Proof. It suffices to find an element $w \in S(R/A)_{\max}$ such that $v \leq w$ for every $v \in \Delta$. We then have $\Delta \subset \Delta_w$ and conclude by Lemma 11 that $\Delta = \Delta_w$.

Lemma 5 tells us that for every finite set $V \subset \Delta$ there exists a unique element $w_V \in S(R/A)_{\max}$, such that $v \leq w_V$ for every $v \in w_V$. If V' is another finite subset of Δ then clearly $w_V = w_{V \cup V'} = w_{V'}$. Thus there is a unique $w \in S(R/A)_{\max}$ with $w = w_V$ for every finite $V \subset \Delta$. It follows that $w \geq v$ for every $v \in \Delta$. □

We summarize what we have gained about coirreducible overrings in the case, that $A \subset R$ is connective and $S(R/A)$ has enough maximal elements.

Scholium 8.14. Assume that $A \subset R$ is connective and $S(R/A)$ has enough maximal elements.

a) The minimal coirreducible overrings F correspond bijectively with the elements w of $S(R/A)_{\max}$ via $F = C(\Lambda_w)$. Also $S(R/F) = \{v \in S(R/A) \mid v \leq w\}$.
b) The coirreducible overrings D are precisely all overrings D, for which there exists some $u \in S(R/A)$ (or $u \in S(R/A)_{\max}$), such that $v \leq u$ for every $v \in S(R/D)$. □

Nothing about irreducible overrings is in sight for us of similar nature.

We write down some permanence properties for connective Prüfer extensions.

Proposition 8.15. *As before, $A \subset R$ is a Prüfer extension.*

i) Let B be an overring. If $A \subset R$ is connective, the extension $B \subset R$ is connective.

ii) Let $A \subset C$ be a ws extension. If $A \subset R$ is connective, the Prüfer extension $C \subset RC$ is connective. {Here $RC = R \otimes_A C$, and we regard R and C as subrings of $R \otimes C$ in the obvious way.} Conversely, if $C \subset RC$ is connective, then $R \cap C \subset R$ is connective.

iii) Let $(B_\lambda \mid \lambda \in \Lambda)$ be a family of overrings with $\sum_{\lambda \in \Lambda} B_\lambda = R$ and $B_\lambda \cap B_\mu = A$ for $\lambda \neq \mu$. Then $A \subset R$ is connective iff $A \subset B_\lambda$ is connective for every $\lambda \in \Lambda$.

iv) Let $(B_\lambda \mid \lambda \in \Lambda)$ be a family of overrings with finite avoidance in R. Assume that $B_\lambda \subset R$ is connective for every $\lambda \in \Lambda$, and $B_\lambda + B_\mu = R$ for $\lambda \neq \mu$. Then $\bigcap_{\lambda \in \Lambda} B_\lambda \subset R$ is connective.

Proof. i): This is obvious from Definition 1 above.

ii): If $A \subset R$ is connective, then $R \cap C \subset R$ is connective. We now assume without loss of generality that $A = R \cap C$. As proved in [Vol. I, Chap. II], we have an isomorphism $D \mapsto DC$ from the lattice of overrings of A in R onto the lattice of overrings of C in RC, the inverse isomorphism being given by $D' \mapsto R \cap D'$. It follows that the coirreducible overrings D of A in R correspond uniquely with the coirreducible overrings D' of C in RC via $D' = DC$ and $D = R \cap D'$. For every such D the posets $S(R/D)$ and $S(RC/DC)$ are isomorphic, cf. Scholium 3.7. Thus $S(R/D)$ is connected iff $S(RC/DC)$ is connected. It follows that $A \subset R$ is connective iff $C \subset RC$ is connective.

iii): We rely on Proposition 6.10. For every $\lambda \in \Lambda$ let $B^\lambda := \sum_{\mu \neq \lambda} B_\mu$. Then $R = B_\lambda \times_A B^\lambda$. By the proved claim (ii) it follows that $A \subset B_\lambda$ is connective iff $B^\lambda \subset R$ is connective.

Now assume that $A \subset R$ is connective. Then $B^\lambda \subset R$ is connective, hence $A \subset B_\lambda$ is connective for every $\lambda \in \Lambda$. Conversely, if this holds, then $B^\lambda \subset R$ is connective for every $\lambda \in \Lambda$. Proposition 6.16(iv) tells us that every coirreducible overring D of A in R contains some ring B^λ. Thus $S(R/D)$ is connected. The extension $A \subset R$ is connective.

iv): Proposition 6.12(i) tells us that every coirreducible overring of $\bigcap_{\mu \in \Lambda} B_\mu$ in R contains some ring B_λ. The claim follows. □

We now use some propositions from above to improve the transfer principle for irreducible and coirreducible factors from Sect. 7 (Proposition 7.15) in the case of Prüfer-extensions which are connective and completely irreducible. Notice that this case includes PF-extensions. Indeed, we know from above that PF-extensions are connective, and from Sect. 7 that they are completely reducible (Corollary 7.6).

Theorem 8.16. *Assume that $A \subset R$ is a connective and completely reducible Prüfer extension and that $A \subset C$ is a ws extension. We regard both R and C as subrings of $R \otimes_A C = RC$. We choose an indexing $(E_i \mid i \in I)$ of* $\mathrm{Ir}(R/A)_{\max}$ *and an indexing of $(F_i \mid i \in I)$ of* $\mathrm{Coir}(R/A)_{\min}$*, both without repetitions, such that*

$F_i = E_i^\circ$, hence $R = E_i \times_A F_i$ for every $i \in I$. Let J denote the set of all $i \in I$ with $E_i \not\subset C$.

a) $E_iC \times_C F_iC = RC$ for every $i \in I$.
b) The family $(E_iC \mid i \in J)$ runs through the whole set $\text{Ir}(RC/C)_{\text{max}}$ without repetitions.
c) J is the set of all $i \in I$ with $F_iC \neq RC$, and the family $(F_iC \mid i \in J)$ runs through the whole set $\text{Coir}(RC/C)_{\text{min}}$ without repetitions.
d) The Prüfer extension $C \subset RC$ is again connective and completely reducible.

Proof. a): This is clear from [Vol. I, Proposition II.7.15]).
b): If $J = \emptyset$, then $E_i \subset C$ for every $i \in I$, hence $R \subset C$, and $RC = C$. In this case the claim is trivially true. Assume now that $J \neq \emptyset$. Then we conclude from $R = \sum_{i\in I} E_i$ that $RC = \sum_{i\in J} E_iC$. Proposition 15(iii) tells us that, for every $i \in I$, the extension $A \subset E_i$ is connective. Since this extension is irreducible, the poset $S(E_i/A)$ is connected. By Proposition 6 the poset $S(E_i/R \cap C)$ is connected. Since this poset is isomorphic to $S(E_iC/C)$, also $S(E_iC/C)$ is connected for every $i \in I$. {It is empty for $i \in I \setminus J$.} It follows by Proposition 2 that E_iC is irreducible over C for every $i \in J$. Let $B := R \cap C$. We have $E_iB \cap E_kB = (E_i + B) \cap (E_k + B) = (E_i \cap E_k) + B = B$ for indices $i \neq k$ in J, and this implies $E_iC \cap E_kC = C$ for $i \neq k$ (cf. [Vol. I, Proposition II.6.9]). Now Proposition 7.2 tells us that $(E_iC \mid i \in J)$ is the family of all maximal irreducible coverings of C in RC, and that RC is completely reducible over C.
c): Let $i \in I$ be given. The factorization $RC = (E_iC) \times_C (F_iC)$ implies that $E_iC = C$ iff $F_iC = RC$. Since J is the set of all $i \in I$ with $E_iC \neq C$, it is also the set of all $i \in I$ with $F_iC \neq RC$. The ring F_iC is the polar of E_iC in RC over C for every $i \in I$. If the index i runs through J, then E_iC runs through the whole set $\text{Ir}(RC/C)_{\text{max}}$ without repetitions, and thus F_iC runs through $\text{Coir}(RC/C)_{\text{min}}$ without repetitions (cf. Theorem 7.3).
d): We have already proved that the extension $C \subset RC$ is completely reducible. Proposition 15(ii) tells us that this extension is also connective. □

9 Integral Extensions

In this section we are given an integral ring extension $R \subset R'$ (i.e. a ring extension $R \subset R'$, such that every element of R' is integral over R). If A is a subring of R, let $\tilde{A}$ denote the integral closure of A in R'. We want to obtain insight in the relation between the restricted PM-spectra $S(R/A)$ and $S(R'/\tilde{A})$, and to draw some conclusions from that.

A major input for our study will be [Vol. I, Theorem I.5.9] stating that, if A is Prüfer in R, then $\tilde{A}$ is Prüfer in R' and $R' = R\tilde{A}$.

We start with some very general observations. Assume that a valuation v' on R' is given. Let v denote its restriction to R, $v = v'|R$, and let $\mathfrak{q}'$, $\mathfrak{q}$ denote the supports of v' and v respectively. $\mathfrak{q}'$ is a prime ideal of R', and $\mathfrak{q} = \mathfrak{q}' \cap R$. Our valuations

v' and v give us Krull valuations $\widehat{v'}$ and $\hat{v}$ on the residue class fields $k(\mathfrak{q}')$ and $k(\mathfrak{q})$ respectively (cf. [Vol. I, Chap. I §1]). Since R' is integral over R, the field extension $k(\mathfrak{q}) \subset k(\mathfrak{q}')$ is algebraic.

As is well known from the valuation theory of fields, the value group $\Gamma_v = \Gamma_{\hat{v}}$ is a subgroup of $\Gamma_{v'} = \Gamma_{\widehat{v'}}$ with $\Gamma_{v'}/\Gamma_v$ a torsion group [Bo, VI §8, Prop. 1]. It follows that the convex hull of Γ_v in $\Gamma_{v'}$ is the full group $\Gamma_{v'}$. {If $x \in \Gamma_{v'}$, say $x \geq 0$, and $nx \in \Gamma_v$, then $0 \leq x \leq nx$.} From this we conclude that *v' is special iff v is special.* Indeed, v is special iff the convex subgroup $c_v(\Gamma_v)$ of Γ_v generated by $\{v(x) \mid x \in R \setminus \mathfrak{q},\ v(x) \leq 0\}$ coincides with Γ_v. Then the convex subgroup generated by this set in $\Gamma_{v'}$ is the whole of $\Gamma_{v'}$, and a fortiori we have $c_{v'}(\Gamma_{v'}) = \Gamma_{v'}$. Conversely it is easy to check that $c_{v'}(\Gamma_{v'}) = \Gamma_{v'}$ implies $c_v(\Gamma_v) = \Gamma_v$.

Lemma 9.1. *Let v' be a valuation on R' and let v denote its restriction to R. Let further A be a subring of R.*

a) v is trivial iff v' is trivial.
b) A_v contains the ring A iff $A_{v'}$ contains $\tilde{A}$.
c) If v is PM, then v' is PM.
d) Assume that $A \subset A_v$ and that R is convenient over A (cf. [Vol. I, Definition 2 in I §6]). Assume further that v' is PM. Then v is PM.

Proof. a): We have seen that the convex closure of Γ_v in $\Gamma_{v'}$ is the whole group $\Gamma_{v'}$. Thus $\Gamma_v = 1$ iff $\Gamma_{v'} = 1$.

b): If $A \subset A_v$, then $A \subset A_{v'}$. Since $A_{v'}$ is integrally closed in R', it follows that $\tilde{A} \subset A_{v'}$. Conversely, if $\tilde{A} \subset A_{v'}$ then $A \subset \tilde{A} \cap R \subset A_{v'} \cap R = A_v$.

c): The extension $\tilde{A}_v \subset R'$ is Prüfer since $A_v \subset R$ is Prüfer. We have $\tilde{A}_v \subset A_{v'} \subset R'$ (as stated in (b)). The valuation v' is special, since v is special. It follows that v' is PM (cf. [Vol. I, Proposition I.5.1.iii]).

d): We may assume that $A_v \neq R$. The set $R \setminus A_v$ is closed under multiplication. This implies that the extension $A_v \subset R$ is PM, since A is convenient in R. The valuation v is special, since v' is special. Thus v is "the" PM-valuation corresponding to the PM-extension $A_v \subset R$. □

If the extension $A \subset R$ is convenient, Lemma 1 tells us that we have a well defined restriction map between PM-spectra

$$p\colon \operatorname{pm}(R'/\tilde{A}) \longrightarrow \operatorname{pm}(R/A), \quad v' \longmapsto v'|R.$$

Of course, this map is compatible with the orderings of the posets $\operatorname{pm}(R'/\tilde{A})$ and $\operatorname{pm}(R/A)$: If $v_1' \leq v_2'$ then $p(v_1') \leq p(v_2')$. Lemma 1 also tells us that $p^{-1}(S(R/A)) = S(R'/\tilde{A})$. In particular p restricts to a map

$$\pi\colon S(R'/\tilde{A}) \longrightarrow S(R/A).$$

We will focus attention on π instead of p.

Without further work we can establish a valuation theoretic description of the integral closures of R-overrings of A in R'.

Lemma 9.2. *Assume that A is a subring of R and R is convenient over A. Let C be an R-overring of A. Then*

$$S(R'/\tilde{C}) = \pi^{-1}(S(R/C)).$$

Proof. This follows from Lemma 1(b). □

Proposition 9.3. *Assume that A is a Prüfer subring of R.*

i) If C is any R-overring of A, then $\tilde{C} = C\tilde{A}$.
ii) If $(B_\lambda \mid \lambda \in \Lambda)$ is a family of R-overrings of A having finite avoidance in R, then the family $(\tilde{B}_\lambda \mid \lambda \in \Lambda)$ has finite avoidance in $\tilde{R}$, and

$$\left(\bigcap_{\lambda\in\Lambda} B_\lambda\right)^{\sim} = \bigcap_{\lambda\in\Lambda} \tilde{B}_\lambda.$$

Proof. i): This follows from [Vol. I, Theorem I.5.9], applied to the extensions $A \subset C$, instead of $A \subset R$ there, and $C \subset \tilde{C}$, instead of $R \subset R'$ there.
ii): Let $x \in R'$ be given. Since $R' = R\tilde{A}$, we have an equation $x = \sum_{i=1}^{n} a_i x_i$ with $a_i \in \tilde{A}$, $x_i \in R$. For every $i \in \{1,\dots,n\}$ there exists a finite subset U_i of Λ such that $x_i \in B_\lambda$ for every $\lambda \in \Lambda \setminus U$. The set $U := U_1 \cup \dots \cup U_n$ is again finite and $x \in \tilde{B}_\lambda$ for every $\lambda \in \Lambda \setminus U$. Thus $(\tilde{B}_\lambda | \lambda \in \Lambda)$ has finite avoidance in R'.

Let $B := \bigcap_{\lambda\in\Lambda} B_\lambda$ and $D := \bigcap_{\lambda\in\Lambda} \tilde{B}_\lambda$. We read off from Corollary 2.7 that $S(R/B) = \bigcup_{\lambda\in\Lambda} S(R/B_\lambda)$ and $S(R'/D) = \bigcup_{\lambda\in\Lambda} S(R'/\tilde{B}_\lambda)$. By the preceding lemma we conclude that

$$S(R'/\tilde{B}) = \pi^{-1}(S(R/B)) = \bigcup_{\lambda\in\Lambda} \pi^{-1}(S(R/B_\lambda)) = \bigcup_{\lambda\in\Lambda} S(R'/\tilde{B}_\lambda) = S(R'/D).$$

This implies $\tilde{B} = D$. □

Remark 9.4. Assume again that A is Prüfer in R. It follows from part (i) of Proposition 3 that

$$\left(\sum_{\lambda\in\Lambda} B_\lambda\right)^{\sim} = \sum_{\lambda\in\Lambda} \tilde{B}_\lambda$$

for *any* family $(B_\lambda \mid \lambda \in \Lambda)$ of R-overrings. □

Theorem 9.5. *Let A be a subring of R such that R is convenient over A.*

a) The restriction map $\pi \colon S(R'/\tilde{A}) \to S(R/A)$ is surjective.
b) For every $v \in S(R/A)$ the fiber $\pi^{-1}(v)$ consists of elements which are pairwise incomparable.

c) *If the extension $R \subset R'$ is finite, every fiber $\pi^{-1}(v)$ is finite. More precisely, if R' can be generated by n elements as an R-module, every fiber has at most n elements.*

d) *The extension $\tilde{A} \subset R'$ is convenient.*

Proof. Let $v \in S(R/A)$ be given, and let $\mathfrak{q}$ denote the support of v. By the well known "lying over theorem" ([Bo, V §2, Th. 1]) there exists a prime ideal $\mathfrak{q}'$ of R' with $\mathfrak{q}' \cap R = \mathfrak{q}$. By one of the most basic theorem about the extension of valuations in the case of fields ([Bo, VI §1, Th. 2]) there exists a valuation u on $k(\mathfrak{q}')$ with $u|k(\mathfrak{q}) = \hat{v}$. Given u, we have a unique valuation v' on R' with supp $v' = \mathfrak{q}'$ and $\widehat{v'} = u$. It follows that $v'|R = v$. Lemma 1 tells us that v' is PM and $A_{v'} \supset \tilde{A}$. Thus $\pi^{-1}(v)$ is certainly not empty.

Assume now that R' can be generated by n elements as an R-module. Then the same holds for the artinian ring $R'/\mathfrak{q}R' \otimes_{R/\mathfrak{q}} k(\mathfrak{q})$, considered as a vector space over $k(\mathfrak{q})$. It follows, that there exists only finitely many prime ideals $\mathfrak{q}'_1, \dots, \mathfrak{q}'_r$ of R' lying over $\mathfrak{q}$, and $\sum_{i=1}^{r} [k(\mathfrak{q}'_i) : k(\mathfrak{q})] \le n$. The extension theory for valuations on fields tells us that $\hat{v}$ has at most $[k(\mathfrak{q}'_i) : k(\mathfrak{q})]$ extensions to the field $k(\mathfrak{q}'_i)(1 \le i \le r)$ (cf. [Bo, VI §8, Prop. 2]). Thus v has at most n extensions to R'. This proves claim c) of the theorem.

We also know that there is no inclusion relation between the prime ideals $\mathfrak{q}'_1, \dots, \mathfrak{q}'_r$ (cf. [Bo, V §2]), and that, for a given $i \in \{1, \dots, r\}$, the extensions of $\hat{v}$ to $k(\mathfrak{q}'_i)$ are pairwise incomparable ([Bo, VI §8, Th. 1]). Thus the elements of $\pi^{-1}(v)$ are pairwise incomparable. This is claim (b) in the case that R' is finite over R.

In general we have a directed family $(R'_\lambda \mid \lambda \in \Lambda)$ of subrings R'_λ of R', all containing R, such that each extension $R \subset R'_\lambda$ is finite and R' is the union of the R'_λ. If v', w' are two different extension of v to R', then $v'|R'_\lambda$ and $w'|R'_\lambda$ are incomparable or equal for each $\lambda \in \Lambda$, and thus v' and w' are incomparable.

It remains to be proved that $\tilde{A}$ is convenient in R'. Assume that C is a subring of R' with $\tilde{A} \subset C$ and $R' \setminus C$ closed under multiplication. We have to verify that C is PM in R'. The set $R \setminus (R \cap C)$ is closed under multiplication, and $R \cap C$ is a subring of R containing A. Since A is convenient in R, there exists a PM-valuation v on R with $A_v = R \cap C$. [Vol. I, Theorem I.2.1] tells us that C is integrally closed in R'. Since $A_v \subset C$, it follows that $\tilde{A}_v \subset C$. By our key result [Vol. I, Theorem I.5.9] the extension $\tilde{A}_v \subset R'$ is Prüfer, hence convenient, and this implies that C is PM in R'. □

Theorem 9.6. *Assume again that A is a subring of R and R is convenient over A. Let $v' \in S(R'/\tilde{A})$ be given and $v := v'|R$. Then $v' \in \omega(R'/\tilde{A})$ iff $v \in \omega(R/A)$.*[8]

Proof. a) Assume that $v \in \omega(R/A)$. Let $u' \in S(R'/\tilde{A})$ be given with $u' \le v'$. This implies $(u'|R) \le v$, hence $(u'|R) = v$ due to the minimality of v. Thus both u' and v' are elements of the fiber $\pi^{-1}(v)$. Since $u' \le v'$, it follows from Theorem 5(b) that $u' = v'$. This proves the minimality of v' in $S(R'/\tilde{A})$.

[8]Recall that $\omega(R/A)$ denotes the set of minimal elements in the poset $S(R/A)$.

b) Assume that $v' \in \omega(R'/\tilde{A})$. Let $u \in S(R/A)$ be given with $u \le v$. The ring $B := A_u$ is PM in R and $B \subset A_v$. It follows that $\tilde{B} \subset A_{v'}$, and $\tilde{B}$ is Prüfer in R' (by [Vol. I, Theorem I.5.9]). Let $\mathfrak{p}$ denote the center of v on B and $\mathfrak{P}$ denote the center of v' on $\tilde{B}$. Then $\mathfrak{P} \cap B = \mathfrak{p}$. We have $v' \in \omega(R/\tilde{B})$. This means that $\mathfrak{P}$ is a maximal ideal of $\tilde{B}$. Since $\tilde{B}$ is integral over B, it follows that $\mathfrak{p}$ is a maximal ideal of B [Bo, V §2, Prop. 1], which means that $v \in \omega(R/B)$. Now $u \in S(R/B)$ and $u \le v$. Thus $u = v$. We have proved that $v \in \omega(R/A)$. □

Corollary 9.7. *If v is a PM-valuation on R, then $\omega(R'/\tilde{A}_v)$ is the set of all PM-valuations v' on R' with $v'|R = v$. In particular, the ring $\tilde{A}_v$ is the intersection of the rings $A_{v'}$ given by these valuations v'.*

Proof. We apply the theorem with $A = A_v$. We have $\omega(R/A_v) = \{v\}$, hence $\omega(R'/\tilde{A}_v) = \pi^{-1}(v)$. □

Corollary 9.8. *Assume that the extension $R \subset R'$ is finite. Let A be a Prüfer subring of R. The extension $\tilde{A} \subset R'$ is PF (resp. PM-finite) iff the extension $A \subset R$ has this property.*

Proof. We know by [Vol. I, Theorem I.5.9] that the extension $\tilde{A} \subset R'$ is Prüfer and $R' = R\tilde{A}$. We further know by Theorems 5 and 6 that $\pi(\omega(R'/\tilde{A}) = \omega(R/A)$, $\pi^{-1}(\omega(R/A)) = \omega(R'/\tilde{A})$, and the fibers of π are finite. Thus the set $\omega(R'/\tilde{A})$ is finite iff $\omega(R/A)$ is finite. Thus our claim about the property "PM-finite" is proved.

Assume now that R' is PF over $\tilde{A}$. Let $x \in R$ be given. We have a finite set $U \subset \omega(R'/\tilde{A})$ such that $x \in A_{v'}$ for every $v' \in \omega(R'/\tilde{A}) \setminus U$. The set $\pi(U)$ is again finite and $x \in A_v$ for every $v \in \omega(R/A) \setminus \pi(U)$. This proves that R is PF over A.

Assume finally that R is PF over A. Let $x \in R'$ be given. Since $R' = R\tilde{A}$, we have an equation $x = \sum_{i=1}^{n} a_i x_i$ with $a_i \in \tilde{A}$ and $x_i \in R$. There exists a finite set $U \subset \omega(R/A)$ such that $x_i \in A_v$ for every $1 \in \{1, \dots, n\}$ and $v \in \omega(R/A) \setminus U$. The set $\pi^{-1}(U)$ is again finite. For every $v' \in \omega(R'/\tilde{A}) \setminus \pi^{-1}(U)$, we have $v'(x) \ge 0$. This proves that $\tilde{A} \subset R'$ is PF. □

Example 9.9. Assume that R' is not finite over R and v is a nontrivial PM-valuation on R which has infinitely many extensions to R'. This situation can be encountered already in very classical areas. For example take for R a number field or a function field in one variable over $\mathbb{C}$, and take for R' the algebraic closure of R. Then any nontrivial valuation on R will do.

The extension $\tilde{A}_v \subset R'$ *does not have finite avoidance.* Indeed, choose some $x \in R$ with $v(x) < 0$. Then $v'(x) < 0$ for every $v' \in S(R'/\tilde{A}_v)$, and these are infinitely many. □

Theorem 9.10. *Let v be a nontrivial PM-valuation on R and v' a valuation on R' extending v. {N.B. v' is again PM by Lemma 1.} The nontrivial coarsenings w of v correspond bijectively with the nontrivial coarsenings w' of v' (up to equivalence) via $w = w'|R$. We have $A_{w'} = A_w A_{v'} = \tilde{A}_w A_{v'} = \tilde{A}_w + A_{v'}$.*

Proof. a) Let $A := A_v$. Then $v \in S(R/A)$, $v' \in S(R'/\tilde{A})$, and we can work with the restriction map $\pi\colon S(R'/\tilde{A}) \to S(R/A)$ as above. We introduce the chains $U := \{w \in S(R/A) \mid w \geq v\}$, $U' := \{w' \in S(R'/\tilde{A}) \mid w' \geq v'\}$. We have to verify that the restriction $\pi|U'$ is a bijection from U' to U.

If G is any totally ordered abelian group and M a subset of G, we denote the convex hull of M in G by $\mathrm{conv}_G(M)$. Let $\Gamma := \Gamma_v$, $\Gamma' := \Gamma_{v'}$. Then $\Gamma' = \mathrm{conv}_{\Gamma'}(\Gamma)$. The elements $w' \in U'$ (resp. $w \in U$) correspond bijectively with the convex subgroups $H' \neq \Gamma'$ of Γ' (resp. convex subgroups $H \neq \Gamma$ of Γ) via $w' = v'/H'$, $w = v/H$, and

$$A_{w'} = \{x \in R' \mid \exists h \in H',\ w'(x) \geq h\}, \quad A_w = \{x \in R \mid \exists h \in H,\ w(x) \geq h\} \tag{$*$}$$

On the other hand, since $\mathrm{conv}_{\Gamma'}(\Gamma) = \Gamma'$, the convex subgroups $H' \neq \Gamma'$ of Γ' correspond bijectively with the convex subgroups $H \neq \Gamma$ of Γ via $H = H' \cap \Gamma$, $H' = \mathrm{conv}_{\Gamma'}(H)$. It is also evident from ($*$), that $A_w = A_{w'} \cap H$ for the valuations $w' \in U'$, $w \in U$, corresponding to H' and H. This means $w = w'|R$. Thus the theorem is proved up to the claims about A_w and $A_{w'}$.

b) Let $w \in U$, $w' \in U'$ be given with $w'|R = w$. Then $\tilde{A}_w \subset A_{w'}$ and $A_{v'} \subset A_{w'}$, hence $A_{v'} \subset \tilde{A}_w A_{v'} \subset A_{w'}$. Since $A_{v'}$ is PM in R', we have a unique PM-valuation u' on R' with $\tilde{A}_w A_{v'} = A_{u'}$, and we conclude that $v' \leq u' \leq w'$. Now $A_w = \tilde{A}_w \cap R \subset A_{u'} \cap R \subset A_{w'} \cap R = A_w$. Thus $A_{u'} \cap R = A_w$, i.e. $u'|R = w = w'|R$. Since π is injective on U', as already proved, it follows that $u' = w'$. We have verified that $A_{w'} = \tilde{A}_w A_{v'}$.

Both $A_{v'}$ and $\tilde{A}_w$ are R'-overrings of $\tilde{A}$ and $\tilde{A}$ is Prüfer in R'. Thus we have $\tilde{A}_w A_{v'} = \tilde{A}_w + A_{v'}$. Lemma 2 tells us that $\tilde{A}_w = A_w \tilde{A}_v$. It follows that $\tilde{A}_w A_{v'} = A_w A_{v'}$. □

Scholium 9.11. Let v and w be nontrivial PM-valuations on R with $v \leq w$.

a) If v' is a valuation on R' with $v'|R = v$, there exists *a unique* (PM-)valuation w' on R' with $v' \leq w'$ and $w'|R = w$ ("Going up"). This is evident from Theorem 10 (and Lemma 1(c)).
b) If w' is a valuation on R' with $w'|R = w$, there exists *some* PM-valuation v' on R' with $v' \leq w'$ and $v'|R = v$ ("Going down"). Indeed, it follows from Theorem 6 that the v' with $v'|R = v$ are the elements of $\omega(R'/\tilde{A}_v)$. Of course, there exists a minimal element v' of $S(R/\tilde{A}_v)$ with $v' \leq w'$. □

Theorem 9.12. *Assume that R is convenient over A.*

a) If Δ' is a connected component of $S(R'/\tilde{A})$, then $\pi(\Delta')$ is a connected component of $S(R/A)$.
b) The preimage $\pi^{-1}(\Delta)$ of any connected component Δ of $S(R/A)$ is a union of connected components of $S(R'/\tilde{A})$. If R' is a finite extension of R, which as an R-module can be generated by n elements, then the number of connected components in $\pi^{-1}(\Delta)$ is at most n.

Proof. a): It is obvious, that the poset $\pi(\Delta')$ is connected. Thus there exists a connected component Δ of $S(R/A)$ with $\pi(\Delta') \subset \Delta$. Let $u \in \Delta$ be given. We choose some $v' \in \Delta'$. By Lemma 8.5 there exists some $w \in \Delta$ such that $\pi(v') \leq w$ and $u \leq w$. Scholium 11 now tells us that there is a unique $w' \in \pi^{-1}(w)$ with $v' \leq w'$ and some $u' \in \pi^{-1}(u)$ with $u' \leq w'$. Since Δ' is a connected component of $S(R'/\tilde{A})$, we have $w' \in \Delta'$ and $u' \in \Delta'$. Thus $u \in \pi(\Delta')$. This proves that $\pi(\Delta') = \Delta$.

b): We choose some $v \in \Delta$. Let $v' \in \pi^{-1}(v)$ and let Δ' denote the connected component of v' in $S(R'/\tilde{A})$. As just proved, $\pi(\Delta')$ is a connected component of $S(R/A)$. Since $v \in \pi(\Delta')$, this forces $\pi(\Delta') = \Delta$. It is now evident, that $\pi^{-1}(\Delta)$ is the union of the connected components of all the points $v' \in \pi^{-1}(v)$ in $S(R'/\tilde{A})$. If R' is generated as an R-module by n elements, the fiber $\pi^{-1}(v)$ consists of at most n points (cf. Theorem 5(c)). It follows that $\pi^{-1}(\Delta)$ has at most n connected components. □

Lemma 9.13. *Let D be an overring of A in R, and $R = U \times_D V$ a factorization of R over D. Then $R' = \tilde{U} \times_{\tilde{D}} \tilde{V}$.*

Proof. Proposition 3 tells us that $\tilde{U} = U\tilde{D}$, $\tilde{V} = V\tilde{D}$, and $\tilde{U} \cap \tilde{V} = \tilde{D}$. Also $R' = R\tilde{D} = (U + V)\tilde{D} = U\tilde{D} + V\tilde{D}$. □

From now on, up to the end of the section, *we assume that $A \subset R$ is a Prüfer extension*. It follows that $A' \subset \tilde{R}$ is a Prüfer extension. We turn to a study of the relations between the coirreducible overrings of A in R and of $\tilde{A}$ in R'.

Proposition 9.14. *Let D' be a coirreducible overring of $\tilde{A}$ in R'. Then $D := R \cap D'$ is a coirreducible overring of A in R.*

Proof. We have $\tilde{D} \subset D'$. Let $R = U \times_D V$ be a factorization of R over D. By the preceding lemma, $R' = \tilde{U} \times_{\tilde{D}} \tilde{V}$. This implies $R' = \tilde{U}D' \times_{D'} \tilde{V}D'$ (cf. [Vol. I, Proposition II.7.15]). Since R' is irreducible over D', this factorization is trivial, say $\tilde{U}D' = D'$ (and $\tilde{V}D' = R'$). We have $\tilde{U} \subset D'$. Intersecting with R, we obtain $U \subset D$, i.e. $U = D$. □

Proposition 9.15. *Assume that the Prüfer extension $A \subset R$ is connective (cf. Sect. 8, Definition 1). Let F' be a* minimal *coirreducible overring of $\tilde{A}$ in R'. Then $F' \cap R$ is a* minimal *coirreducible overring of A in R, and $\pi(S(R'/F')) = S(R/F' \cap R)$.*

Proof. By the preceding Proposition 14 we know that $R \cap F'$ is coirreducible in R. Let F denote the coirreducible core of $R \cap F'$ over A, $F = C(R \cap F')$. This is the minimal coirreducible overring of A in R contained in $R \cap F'$ (cf. Sect. 6). We verify that $\pi(S(R'/F')) = S(R/F)$. It then will follow that $F' \cap R = F$, and we will be done. Indeed, $A_{\pi(u')} = R \cap A_{u'}$ for every $u' \in S(R/F')$, and F' is the intersection of the rings $A_{u'}$, while F is the intersection of the rings A_u with $u \in S(R/F)$.

Of course, $S(R/F)$ contains $\pi(S(R'/F'))$. Let $u \in S(R/F)$ be given. We choose some $v' \in S(R/F')$. {Notice that $S(R'/F')$ is not empty.} The poset

$S(R/F)$ is connected, since $A \subset R$ is assumed to be connective. Thus there exists some $w \in S(R/F)$ with $u \leq w$ and $\pi(v') \leq w$. By Scholium 11 there exists a (unique) element w' in $\pi^{-1}(w)$ with $v' \leq w'$ and some $u' \in \pi^{-1}(u)$ with $u' \leq w'$. We have $F' \subset A_{v'} \subset A_{w'}$ and $A_{u'} \subset A_{w'}$. Thus $F' = C(A_{v'}) = C(A_{w'}) = C(A_{u'})$. It follows that $F' \subset A_{u'}$, hence $u' \in S(R/F')$. We have $\pi(u') = u$. □

We now are able to exhibit connective Prüfer extensions which are not necessarily PF, albeit still of rather special nature.

Theorem 9.16. *Assume that the Prüfer extension $A \subset R$ is PF and that the poset $S(R/A)$ has enough maximal elements (cf. Sect. 3, Definition 7). Then the extension $\tilde{A} \subset R'$ is connective.*

Remark. If the integral extension $R \subset R'$ is not finite, we cannot expect that the Prüfer extension $\tilde{A} \subset R'$ is PF, cf. Example 9 above. There exist many examples of PF-extensions $A \subset R$ such that $S(R/A)$ has enough maximal elements, as is clear from 8.12.

Proof of Theorem 9.16. We choose a direct system $(R_\alpha \mid \alpha \in I)$ of finite subextensions R_α of R in R', with $R_\alpha \subset R_\beta$ if $\alpha \leq \beta$, such that $R' = \bigcup_{\alpha \in I} R_\alpha$. Let A_α denote the integral closure of A in R_α. We have $A_\alpha = R_\alpha \cap A'$ and $A_\alpha = R_\alpha \cap A_\beta$ for $\alpha \leq \beta$. Every extension $A_\alpha \subset R_\alpha$ is PF by Corollary 8 above. It follows from Scholium 11(a) that each poset $S(R_\alpha/A_\alpha)$ and also $S(R'/\tilde{A})$ has enough maximal elements. Also, if $\alpha \leq \beta$, and w is a maximal element of $S(R_\beta/A_\beta)$, the restriction $w|R_\alpha$ is a maximal element of $S(R_\alpha/A_\alpha)$.

Let now D' be a coirreducible overring of $\tilde{A}$ in R'. We have to prove that $S(R'/D')$ is connected. Let two elements v_1', v_2' of $S(R'/D')$ be given. We are looking for an element $w' \in S(R'/D')$ such that $v_1' \leq w'$ and $v_2' \leq w'$.

For every $\alpha \in I$ let $v_{1\alpha} := v_1'|R_\alpha$ and $v_{2\alpha} = v_2'|R_\alpha$. These are elements of $S(R_\alpha/D_\alpha)$ with $D_\alpha := R_\alpha \cap D'$. Every ring D_α is coirreducible in R_α by Proposition 14. The Prüfer extension $A_\alpha \subset R_\alpha$ is PF by Corollary 8 above, hence connective (cf. Theorem 8.3). We conclude that $S(R_\alpha/D_\alpha)$ is connected for every $\alpha \in I$. Let w_α denote the unique maximal element of $S(R_\alpha/A_\alpha)$ with $v_{1\alpha} \leq w_\alpha$. It now is clear from Lemma 8.5 that also $v_{2\alpha} \leq w_\alpha$. If $\alpha \leq \beta$, we know by Scholium 11(a) that $w_\beta|R_\alpha = w_\alpha$, hence $A_{w_\alpha} = R_\alpha \cap A_{w_\beta}$. Let B denote the union of all the rings A_{w_α}, $\alpha \in I$. We have $\tilde{A} \subset B \subset R'$. Since all the extensions $A_{w_\alpha} \subset R_\alpha$ are PM, the set $R' \setminus B$ is certainly closed under multiplication. It follows that B is PM in R', hence $B = A_{w'}$ for some $w' \in S(R'/\tilde{A})$. Now $A_{v_1'}$ is the union of the rings $A_{v_{1\alpha}}$, $\alpha \in I$. This implies that $A_{v_1'} \subset A_{w'}$. For the same reason $A_{v_2'} \subset A_{w'}$. Thus $v_1' \leq w'$ and $v_2' \leq w'$, and we are done. {w' is the (unique) maximal element of $S(R'/\tilde{A})$ dominating v_1' and v_2', as is pretty obvious now.} □

Chapter 2
Approximation Theorems

Summary. In this chapter we embed the important work of Gräter (cf. [Gr], [Gr_1] and [Gr_2]) on approximation theorems in the book. Approximation theorems are a well-known and important topic in classical valuation theory of fields (cf. [E] and [Rib]). The question is to decide for given valuations $v_1, \ldots, v_n$ of a field, elements $a_1, \ldots, a_n$ in the field and $\alpha_1, \ldots, \alpha_n$ in the value groups whether there is an element x in the field such that

$$v_i(x - a_i) \geq \alpha_i \text{ resp. } v_i(x - a_i) = \alpha_i$$

for all i; i.e. if the elements a_i can be approximated by some x up to a certain degree. The approximation theorems were then generalized to certain classes of rings as "rings of Krull type" (cf. [G_3]).

Gräter elaborated various approximation theorems in our general setting of R-Prüfer rings and has found deep connections, to be reflected below.

We consider three types of approximation theorems: the approximation theorem in the neighbourhood of zero, the general approximation theorem and the reinforced approximation theorem. The first concerns the condition $v_i(x) = \alpha_i$, the second $v_i(x - a_i) \geq \alpha_i$ and the last one $v_i(x - a_i) = \alpha_i$. The reinforced approximation theorem was formulated by Gräter. He perceived the important connection with the intersection ring of the v_i to be Prüfer. He did this also in the case of families with finite avoidance (in his terminology "with finite character", cf. also [G_4]). The approximation theorem in the neighbourhood of zero (going already back to Manis [M], cf. also [Gr_1]) was stated in [Al], but only for finitely many valuations. Our conception of the general approximation theorem appears to be more natural then the formulation of Gräter. We give it also in the case of families having finite avoidance.

The approximation theorems are treated in Sect. 5–7. After the basic Sect. 1 we give in Sect. 2–4 the important notions of dependence, inverse property and essential valuation which will be used widely for the approximation theorems.

In this chapter R, A denote commutative rings with 1.

M. Knebusch and T. Kaiser, *Manis Valuations and Prüfer Extensions II*,
Lecture Notes in Mathematics 2103, DOI 10.1007/978-3-319-03212-2_2,

1 Coarsening of Valuations

Already in Volume I coarsenings of a given valuation $v : R \to \Gamma \cup \{\infty\}$ (cf. [Vol. I, Definition 9 in I §1]) played a major role at many places, and then again here in Chap. 1. In the present section we collect basic facts about coarsenings, mostly in the case that v is Manis. Section 1 should be regarded as a tool box for the approximation theorems on families of valuations, to be studied later.

All these facts are explicitly or virtually contained in the previous chapters, or very easy consequences of results there. We recommend that the reader does not bother much about the (hints of) proofs in Sect. 1. Instead, whenever he has doubts about an assertion in Sect. 1, he should first try to prove it as exercise.

Definition 1 (cf. [Vol. I, Definition 9 in I §1]). Let v, w be valuations on R.

a) Then w is called *coarser* than v, or v is called *finer* than w, if there exists a homomorphism of ordered monoids $f : \Gamma_v \cup \{\infty\} \to \Gamma_w \cup \{\infty\}$[1] such that $w(x) = f(v(x))$ for all $x \in R$.[2] We write $v \leq w$.
b) If $v \leq w$ or $w \leq v$, the valuations v and w are called *comparable*. Otherwise they are called *incomparable*.

It is obvious that $A_v \subset A_w$, $\mathfrak{p}_v \supset \mathfrak{p}_w$ and $\operatorname{supp} v = \operatorname{supp} w$ if $v \leq w$. The latter shows that in the case of PM-valuations the above relation is not the same as the relation in Definition 2 of Chap. 1, Sect.3. (In the case of non-trivial valuations they are the same as we will see later.) Note also that we follow the definition of [Gr$_1$] and not the definition of [Al-M] where the opposite ordering is used.

Remark 1.1. The homomorphism in Definition 1 is uniquely determined and an epimorphism.

Proof. We work in the situation of Definition 1.

a) Let f and g be homomorphisms of ordered monoids fulfilling the condition. Let $\gamma \in \Gamma_v$. There are $x, y \in R \setminus \operatorname{supp} v$ such that $\gamma = v(x) - v(y)$. We obtain

$$\begin{aligned} f(\gamma) &= f(v(x) - v(y)) = f(v(x)) - f(v(y)) = w(x) - w(y) \\ &= g(v(x)) - g(v(y)) = g(v(x) - v(y)) = g(\gamma). \end{aligned}$$

b) Given $\delta \in \Gamma_w$ we have to find some $\gamma \in \Gamma_v$ such that $f(\gamma) = \delta$. There are $x, y \in R \setminus \operatorname{supp} w$ such that $\delta = w(x) - w(y)$. We obtain with $\gamma := v(x) - v(y) \in \Gamma_v$

$$f(\gamma) = f(v(x) - v(y)) = f(v(x)) - f(v(y)) = w(x) - w(y) = \delta.$$

□

[1]This means that $f(\alpha) \geq f(\beta)$ if $\alpha \geq \beta$ (cf. [Vol. I, p. 17]). Note that necessarily $f(\Gamma_v) \subset \Gamma_w$ and that $f|\Gamma_v : \Gamma_v \to \Gamma_w$ is a homomorphism of ordered groups.

[2]Note that then necessarily $f(\infty) = \infty$.

Proposition 1.2. *The set of valuations on R is partially ordered by the coarsening relation $\leq$ up to equivalence.*

Proof. It is clear that the relation $\leq$ is reflexive and transitive. For antisymmetry we have to show the following. Let v, w be valuations on R such that $v \leq w$ and $w \leq v$. Then v and w are equivalent. Let $f : \Gamma_v \cup \{\infty\} \to \Gamma_w \cup \{\infty\}$ and $g : \Gamma_w \cup \{\infty\} \to \Gamma_v \cup \{\infty\}$ be the homomorphisms of ordered monoids such that $w = f \circ v$ and $v = g \circ w$. We show that f is an isomorphism and then are done. By Remark 1 f is surjective. For the injectivity of f let $\gamma \in \Gamma_v$ be given such that $f(\gamma) = 0$. Let $x, y \in R \setminus \operatorname{supp} v$ such that $\gamma = v(x) - v(y)$. Then by the above

$$\gamma = v(x) - v(y) = g(w(x)) - g(w(y))$$

and therefore

$$0 = f(\gamma) = f(g(w(x))) - f(g(w(y))) = w(x) - w(y).$$

Hence

$$\gamma = g(w(x)) - g(w(y)) = g(w(x) - w(y)) = g(0) = 0$$

and we are done. □

Before looking at characterizations of coarsening we collect facts about the influence of coarsening on various properties and constructions introduced in Volume I.

Remarks 1.3. Let v, w be valuations on R such that $v \leq w$. Then the following holds.

(1) If v is trivial then w is trivial.
(2) If v is special then w is special.
(3) If v is Manis then w is Manis.
(4) If w is local then v is local.
(5) v is Manis and local iff w is Manis and local.
(6) v has maximal support iff w has maximal support.
(7) If v is Prüfer–Manis then w is Prüfer–Manis.
(8) If v is principal then w is principal.

Proof. (1), (2), (3) and (6) are clear by the definitions. To prove (4) let $x \in A_v \setminus \mathfrak{p}_v$. We have to show that x is a unit of A_v. Since $v(x) = 0$ also $w(x) = 0$, hence $x \in A_w \setminus \mathfrak{p}_w$. Therefore $x \in A_w^*$ by the assumption and we obtain some $y \in A_w$ such that $xy = 1$. Consequently $v(y) = 0$ and so $y \in A_v$ and $x \in A_v^*$. (5) follows from [Vol. I, Proposition I.1.3.ii]. (7) follows from (3) and [Vol. I, Corollary I.5.3]. (8) can be seen by [Vol. I, Proposition III.8.1.a] and Remark 1. □

Proposition 1.4. *Let v, w be valuations on R such that $v \leq w$. Let B be a subring of R. Then $v|_B \leq w|_B$ holds for the special restrictions of v resp. w to B.*

Proof. Let $u_1 := v|B : B \to \Gamma_v \cup \{\infty\}$ and $u_2 := w|B : B \to \Gamma_w \cup \{\infty\}$. Let $\Delta_1 := c_{u_1}(\Gamma_v)$ and $\Delta_2 := c_{u_2}(\Gamma_w)$. Let $f : \Gamma_v \cup \{\infty\} \to \Gamma_w \cup \{\infty\}$ be the homomorphism of ordered monoids such that $w = f \circ v$. We show that $f(\Delta_1) \subset \Delta_2$ and $f(\Gamma_v \setminus \Delta_1) \subset \Gamma_w \setminus \Delta_2$. For the first assertion let $\gamma \in \Delta_1$. Then there is some $x \in B$ with $v(x) \leq 0$ such that $v(x) \leq \gamma \leq -v(x)$. We obtain $w(x) = f(v(x)) \leq 0$ and

$$w(x) = f(v(x)) \leq f(\gamma) \leq f(-v(x)) = -f(v(x)) = -w(x).$$

Hence $f(\gamma) \in \Delta_2$. For the second assertion let $\delta \in \Gamma_v \setminus \Delta_1$. Clearly $\delta \neq 0$. We may assume that $\delta > 0$. Then $\delta > -v(x)$ for all $x \in B$. Assume that $f(\delta) \in \Delta_2$. Then there is some $x_0 \in B$ with $w(x_0) < 0$ such that $f(\delta) \leq -w(x_0)$. We obtain $\delta > -v(x_0^2)$ and $f(\delta) < -w(x_0^2)$. This contradicts the fact that f is order preserving. From this two observations and the definition of $v|_B = u_1|\Delta_1$ resp. $w|_B = u_2|\Delta_2$ we see that f induces a well-defined homomorphism of ordered monoids $g : \Delta_1 \cup \{\infty\} \to \Delta_2 \cup \{\infty\}$ such that $g \circ v|_B = w|_B$. □

Proposition 1.5. *Let v, w be valuations on R with $\operatorname{supp} v = \operatorname{supp} w =: \mathfrak{q}$. Let S be a multiplicative subset of R with $S \cap \mathfrak{q} = \emptyset$. We consider the valuations $v_S : S^{-1}R \to \Gamma_v \cup \{\infty\}, w_S : S^{-1}R \to \Gamma_w \cup \{\infty\}$ (cf. [Vol. I, Chap. I §1]). The following are equivalent:*

(1) $v \leq w$.
(2) $v_S \leq w_S$.

Proof. (1) ⇒ (2): Let $f : \Gamma_v \cup \{\infty\} \to \Gamma_w \cup \{\infty\}$ be the homomorphism of ordered monoids such that $f \circ v = w$. Let $\tilde{v} := v_S$ and $\tilde{w} := w_S$. We have $\Gamma_{\tilde{v}} = \Gamma_v$ and $\Gamma_{\tilde{w}} = \Gamma_w$. Hence f is a homomorphism from $\Gamma_{\tilde{v}} \cup \{\infty\}$ to $\Gamma_{\tilde{w}} \cup \{\infty\}$. For $a/s \in S^{-1}R$ we have

$$f(\tilde{v}(\frac{a}{s})) = f(v(a) - v(s)) = f(v(a)) - f(v(s)) = w(a) - w(s) = \tilde{w}(\frac{a}{s}).$$

(2) ⇒ (1): Let $f : \Gamma_v \cup \{\infty\} \to \Gamma_w \cup \{\infty\}$ be the homomorphism of ordered monoids such that $f \circ v_S = w_S$. Then obviously $f \circ v = w$. We are done. □

Proposition 1.6. *Let v, w be valuations on R with $\operatorname{supp} v = \operatorname{supp} w := \mathfrak{q}$. We consider the valuations $\overline{v} : R/\mathfrak{q} \to \Gamma_v \cup \{\infty\}, \overline{w} : R/\mathfrak{q} \to \Gamma_w \cup \{\infty\}$ and $\hat{v} : k(\mathfrak{q}) \to \Gamma_w \cup \{\infty\}, \hat{w} : k(\mathfrak{q}) \to \Gamma_w \cup \{\infty\}$ (cf. [Vol. I, Chap. I §1]). The following are equivalent:*

(1) $v \leq w$.
(2) $\overline{v} \leq \overline{w}$.
(3) $\hat{v} \leq \hat{w}$.

Proof. The equivalence of (1) and (2) is clear. The equivalence of (2) and (3) follows from Proposition 5. □

Note that in the situation above $\hat{v} \leq \hat{w}$ iff $\mathfrak{o}_v \subset \mathfrak{o}_w$ by classical valuation theory (cf. [Vol. I, Theorem I.2.6]).

Scholium 1.7 (cf. [Vol. I, Remarks I.1.12]). Let Γ be a totally ordered abelian group. Given a convex subgroup H of Γ, the quotient Γ/H can be made naturally into an ordered abelian group such that the canonical map $\Gamma \to \Gamma/H$ is an order preserving homomorphism. If $v : R \to \Gamma \cup \{\infty\}$ is a valuation we obtain a coarsening $w : R \to \Gamma/H \cup \{\infty\}$ by setting $w(x) := v(x) + H$ for $x \in R$. This latter valuation is denoted by v/H.

Assume that $\Gamma = \Gamma_v$ (for the given valuation v on R). The coarsenings w of v correspond, up to equivalence, uniquely with the convex subgroups H of Γ via $w = v/H$.

Assume that v is Manis. Let H be a convex subgroup of Γ. Then the following holds for the coarsening $w := v/H$ of v (cf. [Vol. I, Scholium I.1.18]):

$$A_w = A_H := \{x \in R \mid v(x) \geq h \text{ for some } h \in H\}$$

$$\mathfrak{p}_w = \mathfrak{p}_H := \{x \in R \mid v(x) > h \text{ for all } h \in H\}.$$

In the case of Manis valuations we establish more criteria for coarsening.

Proposition 1.8. *Let v, w be non-trivial Manis valuations on R. The following are equivalent:*

(1) $v \leq w$.
(2) $\mathfrak{p}_w \subset \mathfrak{p}_v \subset A_v \subset A_w$,
(3) $\mathfrak{p}_w$ *is an ideal of* A_v *contained in* $\mathfrak{p}_v$.
(4) $\mathfrak{p}_w$ *is a proper* v*-convex ideal of* A_v.

Proof. The equivalence of (1)–(3) was established in [Vol. I, Theorem I.2.6.i].

(3) $\Rightarrow$ (4): Note that $\mathfrak{p}_w$ is a prime ideal of A_v since $A_v \subset A_w$ by (2) and $\mathfrak{p}_w$ is a prime ideal of A_w. We show that $\operatorname{supp} v \subset \mathfrak{p}_w$ and then will be done by [Vol. I, Proposition I.1.10]. Since v and w are both Manis and non-trivial and since $A_v \subset A_w$ we obtain $\operatorname{supp} v = [A_v : R] \subset [A_w : R] = \operatorname{supp} w \subset \mathfrak{p}_w$.

(4) $\Rightarrow$ (3): Assume that $\mathfrak{p}_w \not\subset \mathfrak{p}_v$. Then there is some $x \in \mathfrak{p}_w$ with $v(x) = 0$. Since $\mathfrak{p}_w$ is v-convex we obtain $A_v \subset \mathfrak{p}_w$, contradiction. □

Remark 1.9. In the situation of Proposition 8 we have that $\mathfrak{p}_w$ is a prime ideal of A_v.

Scholium 1.10 (cf. [Vol. I, Corollary I.2.7]). Let v be a Manis valuation on R. The coarsenings w of v correspond uniquely, up to equivalence, with the prime ideals $\mathfrak{p}$ of $A := A_v$ between $\operatorname{supp} v$ and $\mathfrak{p}_v$ via $\mathfrak{p} = \mathfrak{p}_w$. Also $A_w = A_{[\mathfrak{p}]}$.

Remark 1.11. Let v be a Manis valuation on R and let $\mathfrak{p}$ be a prime ideal of $A := A_v$ with $\operatorname{supp} v \subset \mathfrak{p} \subset \mathfrak{p}_v$. Then the following holds.

i) $(A_{[\mathfrak{p}]}, \mathfrak{p})$ is a Manis pair in R.
ii) $A_{[\mathfrak{p}]} = [\mathfrak{p} : \mathfrak{p}] = \{x \in R \mid x\mathfrak{p} \subset \mathfrak{p}\}$.
iii) $(A_{[\mathfrak{p}]})_{[\mathfrak{p}]} = A_{[\mathfrak{p}]}$ and $\mathfrak{p}_{[\mathfrak{p}]} = \mathfrak{p}$.

Proof. By Scholium 10 there is a coarsening w of v such that $A_w = A_{[\mathfrak{p}]}$ and $\mathfrak{p}_w = \mathfrak{p}$.

i): This follows since w is Manis by (3) of Remarks 3.
ii): This has been proved in [Vol. I, Theorem I.2.6.ii].
iii): This follows from [Vol. I, Lemma III.1.0]. □

Definition 2. Let v be a Manis valuation on R and let $\mathfrak{p}$ be a prime ideal of A_v with $\operatorname{supp} v \subset \mathfrak{p} \subset \mathfrak{p}_v$. Then the corresponding coarsening of v is denoted by $v^{\mathfrak{p}}$.

Proposition 1.12. *Let v be a Manis valuation on R and let $\mathfrak{p}$ be a prime ideal of A_v with* $\operatorname{supp} v \subset \mathfrak{p} \subset \mathfrak{p}_v$. *Then the following are equivalent:*

(1) $v^{\mathfrak{p}}$ *is non-trivial.*
(2) $\operatorname{supp} v \subsetneqq \mathfrak{p}$.

Proof. (1) ⇒ (2): If $\operatorname{supp} v = \mathfrak{p}$ we have $A_{[\mathfrak{p}]} = [\mathfrak{p} : \mathfrak{p}] = R$ by Remark 11(ii). Hence $v^{\mathfrak{p}}$ is trivial.

(2) ⇒ (1): Let $\operatorname{supp} v \subsetneqq \mathfrak{p}$. Then there is some $x \in \mathfrak{p}$ with $v(x) \neq \infty$. Since v is Manis there is some $y \in R$ with $v(y) = -v(x)$. By Remark 11(ii) we get $A_{v^{\mathfrak{p}}} = [\mathfrak{p} : \mathfrak{p}] \subset [\mathfrak{p}_v : \mathfrak{p}]$. But $y \notin [\mathfrak{p}_v : \mathfrak{p}]$ since $yx \in A_v \setminus \mathfrak{p}_v$. So $A_{v^{\mathfrak{p}}} \neq R$. □

Combining Scholium 7 and Scholium 10 we obtain

Remark 1.13. Let v be a Manis valuation on R.

a) Let H be a convex subgroup of Γ_v. Then $v/H = v^{\mathfrak{p}}$ where

$$\mathfrak{p} = \mathfrak{p}_H = \{x \in R \mid v(x) > h \text{ for all } h \in H\}.$$

b) Let $\mathfrak{p}$ be a prime ideal of A_v with $\operatorname{supp} v \subset \mathfrak{p} \subset \mathfrak{p}_v$. Then $v^{\mathfrak{p}} = v/H$ where

$$H = \{\gamma \in \Gamma_v \mid v(z) > \gamma > -v(z) \text{ for all } z \in \mathfrak{p}\} = \{\pm v(x) \mid x \in A_v \setminus \mathfrak{p}\}.$$

Theorem 1.14. *Let v and w be non-trivial Manis valuations on R. Assume that v is Prüfer–Manis. Then the following are equivalent.*

(1) $v \leq w$,
(2) $A_v \subset A_w$,

Proof. (1) ⇒ (2): This follows from Proposition 8.

(2) ⇒ (1): Since v is PM we have that A_v is Prüfer in R. By [Vol. I, Theorem III.1.2] there is an R-regular prime ideal $\mathfrak{p}$ of $A := A_v$ such that $A_w = A_{[\mathfrak{p}]}$, namely $\mathfrak{p} = \mathfrak{p}_{A_w} \cap A$ where $\mathfrak{p}_{A_w} = \{x \in A_w \mid \exists s \in R \setminus A_w \text{ such that } sx \in A_w\}$ (see [Vol. I, Definition 2 in I §2]). Note that $\mathfrak{p}_{A_w} = \mathfrak{p}_w$ by [Vol. I, Proposition I.2.3]. By [Vol. I,Theorem III.1.3] we get $\mathfrak{p} \subset \mathfrak{p}_v$ and $\mathfrak{p} = \mathfrak{p}_{[\mathfrak{p}]}$. But $\mathfrak{p}_{[\mathfrak{p}]} = \mathfrak{p}_w$ by the proof of [Vol. I, Theorem III.1.2]. Hence $\mathfrak{p}_w \subset \mathfrak{p}_v$. By Proposition 8 we obtain $v \leq w$. □

Theorem 1.15. *Let v be a non-trivial Prüfer–Manis valuation on R. Let B be a proper R-overring of A. Then there is up to equivalence a unique valuation w on R such that $B = A_w$. Moreover, $v \leq w$.*

Proof. For the existence we can copy the proof of Theorem 14 (2) $\Rightarrow$ (1): Since v is PM we have by [Vol. I, Corollary III.3.2] that B is Prüfer–Manis in R. Let w be a Manis valuation such that $A_w = B$. By Theorem 14 we have $v \leq w$ for the corresponding Manis valuation.

For the uniqueness let w_1, w_2 be Manis valuations on R such that $A_{w_1} = A_{w_2} = B$. Applying Theorem 14 we get $w_1 \leq w_2$ and $w_2 \leq w_1$. By Proposition 2 we get that w_1 and w_2 are equivalent. □

Scholium 1.16. Let v be a Prüfer–Manis valuation on R.

a) The coarsenings of v correspond uniquely, up to equivalence, with the R-overrings of A_v.
b) The non-trivial coarsenings of v correspond uniquely, up to equivalence, with the R-regular prime ideals of A_v.

Proof. a): This is a consequence of Theorem 15 (the unique trivial valuation being coarser than v corresponds of course with R itself).
b): This is a consequence of Scholium 10, Proposition 12 and [Vol. I, Theorem III.2.5]. □

Corollary 1.17. *Let v and w be non-trivial Manis valuations in R. Let A be a Prüfer subring of R such that $A \subset A_v \cap A_w$. The following are equivalent.*

(1) $v \leq w$,
(2) $A_v \subset A_w$,
(3) $\mathfrak{p}_w \subset \mathfrak{p}_v$.
(4) $\mathfrak{p}_w \cap A \subset \mathfrak{p}_v \cap A$.

Proof. Since v is PM by [Vol. I, Corollary I.5.3] we can apply Theorem 14. This gives the equivalence of (1) and (2). The implication (1) $\Rightarrow$ (3) follows from Proposition 8.
(3) $\Rightarrow$ (4): This is trivial.
(4) $\Rightarrow$ (2): Let $\mathfrak{p} := \mathfrak{p}_v \cap A$ and $\mathfrak{p}' := \mathfrak{p}_w \cap A$. We have $A_v = A_{[\mathfrak{p}]}$ and $A_w = A_{[\mathfrak{p}']}$ by [Vol. I, Theorem III.1.2]. This gives $A_v \subset A_w$. □

2 Dependent Families of Manis Valuations

In this section the classical notion of dependence resp. independence (cf. [E]) is formulated in the setting of Manis valuations (cf. [Gr$_1$], [Gr$_2$]). Special attention is paid to the case of valuations over a Prüfer subring.

Definition 1. Let $(v_i \mid i \in I)$ be a family of Manis valuations on R. If there is a non-trivial Manis valuation u on R with $v_i \le u$ for $i \in I$ the family is called *dependent*; otherwise it is called *independent*.

We then often say that *the valuations $v_i, i \in I$, are dependent resp. independent.*

N.B. If $(v_i \mid i \in I)$ is a family of dependent Manis valuations on R then v_i is non-trivial for $i \in I$ and $\operatorname{supp} v_i = \operatorname{supp} v_j$ for $i, j \in I$.

Proposition 2.1. *Let $(v_i \mid i \in I)$ be a family of Manis valuations on R. The following are equivalent.*

(1) $(v_i \mid i \in I)$ is dependent.
(2) There is a subset $\mathfrak{p}$ of R such that $\mathfrak{p}$ is a prime ideal of A_{v_i} with $\operatorname{supp} v_i \subsetneqq \mathfrak{p} \subset \mathfrak{p}_{v_i}$ for all $i \in I$.

Proof. (1) $\Rightarrow$ (2): Let u be a non-trivial Manis valuation on R with $v_i \le u$ for $i \in I$. Let $\mathfrak{p} := \mathfrak{p}_u$. Then $\operatorname{supp} v_i = \operatorname{supp} u$ and $\operatorname{supp} u \subsetneqq \mathfrak{p}_u$ since u is non-trivial. By Proposition 1.8 (and Remark 1.9) we see that $\mathfrak{p} \subset \mathfrak{p}_{v_i}$ and that $\mathfrak{p}$ is a prime ideal of A_{v_i} for all $i \in I$.

(2) $\Rightarrow$ (1): By Scholium 1.10 the valuation u corresponding to $\mathfrak{p}$ is a coarsening of v_i for $i \in I$. Since $\operatorname{supp} u = \operatorname{supp} v_i \subsetneqq \mathfrak{p}$ for some (resp. all) $i \in I$ we get that u is non-trivial. Hence $(v_i \mid i \in I)$ is dependent. □

Definition 2. Assume that $(B_i \mid i \in I)$ is a family of subrings of R. Then we denote the set of subsets $\mathfrak{p}$ of R such that $\mathfrak{p}$ is a prime ideal of every B_i by $\bigcap_{i \in I} \operatorname{Spec} B_i$.

Remark 2.2. Let $\mathfrak{V} = (v_i \mid i \in I)$ be a family of Manis valuations on R. By Proposition 1 we know that $\mathfrak{V}$ is dependent iff

$$X_{\mathfrak{V}} := \{\mathfrak{p} \in \bigcap_{i \in I} \operatorname{Spec} A_{v_i} \mid \operatorname{supp} v_i \subsetneqq \mathfrak{p} \subset \mathfrak{p}_{v_i} \text{ for } i \in I\}$$

is non-empty. By [Vol. I, Proposition I.1.10] every $\mathfrak{p} \in X_{\mathfrak{V}}$ is v_i-convex for all $i \in I$. Hence the elements of $X_{\mathfrak{V}}$ are ordered by inclusion, and $X_{\mathfrak{V}}$ has the largest element

$$\mathfrak{p}_{\mathfrak{V}} := \bigcup_{\mathfrak{p} \in X_{\mathfrak{V}}} \mathfrak{p}$$

if $\mathfrak{V}$ is dependent.

Corollary 2.3. *Let $\mathfrak{V} = (v_i \mid i \in I)$ be a dependent family of Manis valuations on R. Then there is a finest (non-trivial) Manis valuation which is coarser than v_i for every $i \in I$. It is (up to equivalence) the unique Manis valuation u on R with $\mathfrak{p}_u = \mathfrak{p}_{\mathfrak{V}}$.*

Definition 3. Let $(v_i \mid i \in I)$ be a family of dependent Manis valuations on R. The finest Manis valuation which is coarser than v_i for all $i \in I$ is denoted by $\bigvee_{i \in I} v_i$.

We use the results of Sect. 1 to establish criteria for dependence and to describe $\bigvee v_i$.

Remark 2.4. Let $\mathfrak{V} = (v_i \mid i \in I)$ be a dependent family of Manis valuations on R.

i) We have $\mathfrak{p}_{\bigvee v_i} = \mathfrak{p}_{\mathfrak{V}}$.
ii) Let u be a Manis valuation such that $v_i \leq u$ for $i \in I$. Then $\bigvee_{i \in I} v_i \leq u$.

Definition 4. Assume that $\mathfrak{V} = (v_i \mid i \in I)$ is a family of Manis valuations on R. If $\mathfrak{V}$ is dependent and $i \in I$, let $H^i_{\mathfrak{V}}$ denote the convex subgroup of Γ_{v_i} generated by $v_i(A_{v_i} \setminus \mathfrak{p}_{\bigvee v_j})$ with j running through I. If $\mathfrak{V}$ is independent, we set $H^i_{\mathfrak{V}} = \Gamma_{v_i}$ for $i \in I$.

Remark 2.5 (cf. Remarks 1.13). If $\mathfrak{V} = (v_i \mid i \in I)$ is dependent then for $i \in I$

$$\Gamma_{\bigvee v_j} \cong \Gamma_{v_i}/H^i_{\mathfrak{V}} \text{ and } \bigvee v_j = v_i/H^i_{\mathfrak{V}}.$$

Remark 2.6. Let $(v_i \mid i \in I)$ be a family of dependent Manis valuations on R. Then $A_{\bigvee v_i} \supset \prod_{i \in I} A_{v_i}$ and $\mathfrak{p}_{\bigvee v_i} \subset [\bigcap_{i \in I} \mathfrak{p}_{v_i} : \prod_{i \in I} A_{v_i}]$.

Proposition 2.7. *Let $(v_i \mid i \in I)$ be a family of PM-valuations on R. The following are equivalent.*

(1) $(v_i \mid i \in I)$ is dependent.
(2) There is a proper subring of R containing A_{v_i} for all $i \in I$.
(3) $\prod_{i \in I} A_{v_i} \neq R$.

Proof. The equivalence of (1) and (2) is a consequence of Scholium 1.16(a). The equivalence of (2) and (3) is obvious. □

Corollary 2.8. *Let $(v_i \mid i \in I)$ be a family of Manis valuations on R such that $\bigcap_{i \in I} A_{v_i}$ is Prüfer in R. The following are equivalent.*

(1) $(v_i \mid i \in I)$ is dependent.
(2) There is a proper subring of R containing A_{v_i} for all $i \in I$.
(3) $\prod_{i \in I} A_{v_i} \neq R$.

Proposition 2.9. *Let $(v_i \mid i \in I)$ be a family of Manis valuations on R. Let A be a Prüfer subring of R with $A \subset \bigcap_{i \in I} A_{v_i}$. The following are equivalent.*

(1) $(v_i \mid i \in I)$ is dependent.
(2) There is an R-regular prime ideal $\mathfrak{p}$ of A contained in $\mathfrak{p}_{v_i} \cap A$ for all $i \in I$.

Proof. (1) ⇒ (2): Let $u := \bigvee v_i$ and $\mathfrak{p} := \mathfrak{p}_u \cap A$. Then $\mathfrak{p} \subset \mathfrak{p}_{v_i} \cap A$ for $i \in I$ by Corollary 1.17. Since u is non-trivial we have $A_u = A_{[\mathfrak{p}]}$ by [Vol. I, Theorem III.1.2] and [Vol. I, Proposition I.2.3]. By [Vol. I, Lemma III.1.1], $\mathfrak{p}$ is an R-regular prime ideal of A.

(2) $\Rightarrow$ (1): Let $\mathfrak{p}$ be an R-regular prime ideal $\mathfrak{p}$ of A contained in $\mathfrak{p}_{v_i} \cap A$ for $i \in I$. Then again by [Vol. I, Theorem III.1.2] $A_{[\mathfrak{p}]} \neq R$ and $A_{[\mathfrak{p}]}$ is an R-overring of $A_{[\mathfrak{p}_{v_i} \cap A]} = A_{v_i}$ for $i \in I$. We get the claim by Proposition 7 or Corollary 8. □

Proposition 2.10. *Let $(v_i \mid i \in I)$ be a dependent family of PM-valuations on R. Then $A_{\bigvee v_i} = \prod_{i \in I} A_{v_i}$ and $\mathfrak{p}_{\bigvee v_i} = [\bigcap_{i \in I} \mathfrak{p}_{v_i} : \prod_{i \in I} A_{v_i}]$.*

Proof. Let $u := \bigvee_{i \in I} v_i$.

a) We have $A_u \supset \prod_{i \in I} A_{v_i}$ by Remark 6. On the other hand there is by Scholium 1.16(a) a Manis valuation u' on R such that $A_{u'} = \prod_{i \in I} A_{v_i}$ and $v_i \leq u'$ for $i \in I$. Since $u \leq u'$ we get $A_u \subset A_{u'}$.
b) Let $\mathfrak{p} := [\bigcap_{i \in I} \mathfrak{p}_{v_i} : \prod_{i \in I} A_{v_i}]$. Then $\mathfrak{p}$ is a proper ideal of A_u containing $\mathfrak{p}_u$ by Remark 6. Since u is non-trivial, the ideal $\mathfrak{p}_u$ is R-regular by [Vol. I, Theorem III.2.5]. Since $\mathfrak{p}$ contains $\mathfrak{p}_u$ it is clearly also R-regular. By [Vol. I, Theorem III.3.10] we get that $\mathfrak{p}$ is contained in $\mathfrak{p}_u$. □

Corollary 2.11. *Let $(v_i \mid i \in I)$ be a dependent family of Manis valuations on R. If $\bigcap_{i \in I} A_{v_i}$ is Prüfer in R, then $A_{\bigvee v_i} = \prod_{i \in I} A_{v_i}$ and $\mathfrak{p}_{\bigvee v_i} = [\bigcap_{i \in I} \mathfrak{p}_{v_i} : \prod_{i \in I} A_{v_i}]$.*

We collect basic facts about dependent and independent families.

Remark 2.12. Let $(v_i \mid i \in I)$ be a family of Manis valuations on R and let $J \subset I$. If $(v_i \mid i \in I)$ is dependent then $(v_i \mid i \in J)$ is dependent and $\bigvee_{i \in J} v_i \leq \bigvee_{i \in I} v_i$.

Proposition 2.13. *Let $(v_i \mid i \in I), (w_i \mid i \in I)$ be families of Manis valuations on R such that $v_i \leq w_i$ for all $i \in I$. If $(w_i \mid i \in I)$ is dependent then $(v_i \mid i \in I)$ is dependent and $\bigvee_{i \in I} v_i \leq \bigvee_{i \in I} w_i$.*

Proof. By the transitivity of $\leq$ we get $v_i \leq \bigvee_{j \in I} w_j$ for all $i \in I$. Hence $(v_i \mid i \in I)$ is dependent and $\bigvee_{i \in I} v_i \leq \bigvee_{i \in I} w_j$. □

Definition 5. Let $(v_i \mid i \in I)$ be an independent family of Manis valuations on R. If $\operatorname{supp} v_i = \operatorname{supp} v_j$ for all $i, j \in I$ we denote by $\bigvee_{i \in I} v_i$ the trivial valuation with $\operatorname{supp} \bigvee_{i \in I} v_i = \operatorname{supp} v_j$ for $j \in I$. Otherwise let $\bigvee_{i \in I} v_i$ denote the map $R \to \{0\}$. Notice that in this case $\bigvee_{i \in I} v_i$ is not a valuation.

Remark 2.14. Let $(v_i \mid i \in I)$ be a family of Manis valuations on R with $\operatorname{supp} v_i = \operatorname{supp} v_j$ for $i, j \in I$. Then $\bigvee_{j \in I} v_j$ is a coarsening of v_i for $i \in I$. We have

$$A_{\bigvee v_j} = (A_{v_i})_{[\mathfrak{p}_{\bigvee v_j}]}$$

for $i \in I$ (see Scholium 1.10).

Proposition 2.15. *Let $\mathfrak{V} = (v_i \mid i \in I)$ be a family of Manis valuations on R with $\operatorname{supp} v_i = \operatorname{supp} v_j$ for all $i, j \in I$. Let $w_i := v_i|_{A_{\bigvee v_j}}$ for $i \in I$. Then w_i is a Manis valuation on $A_{\bigvee v_j}$ with $\Gamma_{w_i} = H^i_{\mathfrak{V}}$ and $\operatorname{supp} w_i = \mathfrak{p}_{\bigvee v_j}$ for $i \in I$ (cf. Definitions 4 and 5 above).*

Proof. a) If $\mathfrak{V}$ is independent then $A_{\bigvee v_j} = R$ and nothing is to show. So we assume that $\mathfrak{V}$ is dependent. Let $i \in I$. By [Vol. I, Proposition I.1.17] and Remark 6 w_i is a Manis valuation on A_u where $u := \bigvee v_j$.

b) We show that $\Gamma_{w_i} = H^i_{\mathfrak{V}}$. Let $\gamma \in \Gamma_{w_i}$. Then there is some $x \in A_u$ with $w_i(x) = \gamma$. Since $w_i(x) < \infty$ we have $w_i(x) = v_i(x)$ and find some $y \in A_u$ with $v_i(y) \leq 0$ and $v_i(y) \leq v_i(x) \leq -v_i(y)$ by the definition of the special restriction. Since $v_i \leq u$ and $y \in A_u$ we get $u(y) = u(x) = 0$ and therefore $\gamma \in H^i_{\mathfrak{V}}$. Let $\delta \in H^i_{\mathfrak{V}} \subset \Gamma_{v_i}$. Let $x \in R$ with $v_i(x) = \delta$. Then $u(x) = 0$ and therefore $x \in A_u$. Let $x' \in R$ with $v_i(x') = -\delta$. By the same argument we get $x' \in A_u$. By the definition of the special restriction we get $w_i(x) = \delta$ and therefore $\delta \in \Gamma_{w_i}$.

c) We show that $\operatorname{supp} w_i = \mathfrak{p}_u$. Let $x \in \operatorname{supp} w_i \subset A_u$. Assume that $x \notin \mathfrak{p}_u$. Then $u(x) = 0$ and therefore $v(x) \in H^i_{\mathfrak{V}}$ and $w_i(x) = v_i(x) \neq \infty$ by the argument in b), contradiction. Let $x \in \mathfrak{p}_u \subset A_u$. Assume that $w_i(x) \neq \infty$. Then by the definition of the special restriction there is some $y \in A_u$ with $v_i(y) \leq 0$ and $v_i(y) \leq v_i(x) \leq -v_i(y)$. Since $v_i \leq u$ we obtain $u(y) \leq u(x) \leq -u(y)$. This gives $u(y) < 0$, contradiction to $y \in A_u$. □

3 The Inverse Property

The inverse property is a substitute for the inverse element in the case of fields (cf. [M]). We prove various inequalities to be used later and investigate the connection between dependence and the inverse property (see also [Gr$_1$], [Al] and [Al$_1$]).

Definition 1. Let $(v_i \mid i \in I)$ be a family of Manis valuations on R.

i) The family has the *inverse property* if for every $x \in R$, there is some $x' \in R$ such that $v_i(xx') = 0$ for all $i \in I$ with $v_i(x) \neq \infty$.

ii) The family has the *finite inverse property* if every finite subfamily has the inverse property.

We then often say that the valuations $v_i, i \in I$, have the inverse (resp. finite inverse) property.

Remarks 3.1. a) Let v be a Manis valuation on R. Then v has the inverse property.

b) If a family of Manis valuations on R has the inverse property then it has also the finite inverse property.

c) Let $(v_i \mid i \in I)$ be a family of Manis valuations on R having the inverse (resp. finite inverse) property. Then for $J \subset I$, the subfamily $(v_i \mid i \in J)$ has the inverse (resp. finite inverse) property.

d) Any family of valuations on a field has the inverse property.

Proof. a): The inverse property for a single valuation is equivalent with the Manis property.

b), c): This is obvious.

d): Let R be a field. Given $x \in R^*$ take x^{-1}. □

Remark 3.2. A family $(v_i \mid i \in I)$ of Manis valuations on R has the inverse property iff for every $x \in R$ there is some $x' \in R$ such that $v_i(x^2x') = v_i(x)$ for all $i \in I$.

Remark 3.3. Let $(v_i \mid i \in I_1)$ be a family of Manis valuations on R with the (finite) inverse property. Let $(v_i \mid i \in I_2)$ be a family of trivial Manis valuations on R such that for every $i \in I_2$ there is an $i' \in I_1$ with $\operatorname{supp} v_{i'} = \operatorname{supp} v_i$. Then the family $(v_i \mid i \in I_1 \cup I_2)$ has the (finite) inverse property.

Remark 3.4. Let $(v_i \mid i \in I)$ be a family of Manis valuations on R. Let I be an ideal of R such that $I \subset \operatorname{supp} v_i$ for all $i \in I$. For $i \in I$ we denote by $\overline{v}_i$ the corresponding Manis valuation of v_i on R/I. The following are equivalent.

(1) $(v_i \mid i \in I)$ has the inverse (resp. finite inverse) property.
(2) $(\overline{v}_i \mid i \in I)$ has the inverse (resp. finite inverse) property.

Before exploiting the inverse property we prove useful inequalities for a finite set of Manis valuations.

Lemma 3.5 (cf. [Vol. I, Lemma I.6.9]). *Let k be a subring of R and let $v_1, \ldots, v_n$ be valuations on R with $A_{v_i} \supset k$ for all $1 \leq i \leq n$. Let $m \in \mathbb{N}$. Given an element x of R, there exists a monic polynomial $F(T) \in k[T]$ with $F(0) = 0$ and the following property:*

If $G(T) \in k[T]$ is any monic polynomial of degree ≥ 1 with absolute term $G(0) \in k^*$, then $v_i(G(F(x))) = 0$ if $v_i(x) \geq 0$ and $v_i(G(F(x))) \leq mv_i(x)$ if $v_i(x) < 0$ for $1 \leq i \leq n$.

Proof. We take $F(T) := T^m F_1(t) \ldots F_n(T)$ in the proof of [Vol. I, Lemma I.6.9]. □

Proposition 3.6. *Let $v_1, \ldots, v_n$ be valuations on R and let $x \in R$. Let $m \in \mathbb{N}$. Then there exists $y \in R$ such that $v_i(y) = 0$ if $v_i(x) \geq 0$ and $v_i(y) \leq mv_i(x)$ if $v_i(x) < 0$ for all $1 \leq i \leq n$.*

Proof. In the previous Lemma 5 we take $k := \mathbb{Z} \cdot 1_R$. Let $F(T) \in k[T]$ be a monic polynomial with $F(0) = 0$ and the above property. We take $G(T) := 1 + T$. Then $y := 1 + F(x)$ fulfills the requirements. □

Lemma 3.7. *Let $v_1, \ldots, v_n, w_1, \ldots, w_m$ be valuations on R. Let $x_1, \ldots, x_k \in R$ such that for every $i \in \{1, \ldots, n\}$ there is at most one $l \in \{1, \ldots, k\}$ with $v_i(x_l) = 0$. Then there are $y_1, \ldots, y_k \in R$ such that the following properties hold.*

i) If $u \in \{v_1, \ldots, v_n, w_1, \ldots, w_m\}$, $l \in \{1, \ldots, k\}$ and $u(x_l) < \infty$, then $u(y_l) < \infty$.
ii) If $u \in \{v_1, \ldots, v_n, w_1, \ldots, w_m\}$, $l \in \{1, \ldots, k\}$ and $u(x_l)\sigma 0$, then $u(y_l)\sigma 0$ where $\sigma \in \{<, =, >\}$.
iii) If $i \in \{1, \ldots, n\}$, $l_1, l_2 \in \{1, \ldots, k\}, l_1 \neq l_2$, and $v_i(x_{l_1}) < \infty$ or $v_i(x_{l_2}) < \infty$, then $v_i(y_{l_1}) \neq v_i(y_{l_2})$.

Proof. We do induction on k.
$k = 1$: There is nothing to show, we can take $y_1 := x_1$.
$k \to k + 1$: By the inductive hypothesis applied to $x_1, \ldots, x_k$ we find $y_1, \ldots, y_k \in R$ with the above properties.

Claim: Let $1 \leq i \leq n$. Let $1 \leq l \leq k$ such that $v_i(x_{k+1}) < \infty$ or $v_i(y_l) < \infty$. Then there is at most one $m_{i,l} \in \mathbb{N}$ such that $v_i(x_{k+1}^{m_{i,l}}) = v_i(y_l)$.

Proof of the Claim: Let $1 \leq l \leq k$. By the assumption of the lemma and the inductive hypotheses we have $v_i(x_{k+1}) \neq 0$ or $v_i(y_l) \neq 0$. This gives the claim.

We choose now $m \in \mathbb{N}$ such that $m > m_{i,l}$ for all such $m_{i,l}$ above. Then we take $y_{k+1} := x_{k+1}^m$. □

Proposition 3.8. *Let $v_1, \ldots, v_n, w$ be valuations on R such that $A_w \not\subset A_{v_i}$ for all $1 \leq i \leq n$. Then there is some $x \in R$ such that $w(x) = 0$ and $v_i(x) < 0$ for all $1 \leq i \leq n$.*

Proof. We do induction on n.
$n = 1$: Since $A_w \not\subset A_{v_1}$ there is some $x' \in R$ such that $w(x') \geq 0$ and $v_1(x') < 0$. By Proposition 6 there is some $x \in R$ such that $w(x) = 0$ and $v_1(x) < 0$.
$n \to n + 1$: By the inductive hypothesis and the case $n = 1$ there are $x', x'' \in R$ such that

$$w(x') = 0, v_1(x') < 0, \ldots, v_n(x') < 0,$$
$$w(x'') = 0, v_{n+1}(x'') < 0.$$

If $v_{n+1}(x') < 0$ we take $x := x'$. So we assume that $v_{n+1}(x') \geq 0$. Applying Lemma 7 we can assume that $v_i(x') \neq v_i(x'')$ for all $1 \leq i \leq n + 1$. Let $\hat{x} := x' + x''$. Then $w(\hat{x}) \geq 0$ and $v_i(\hat{x}) < 0$ for all $1 \leq i \leq n + 1$. By Proposition 6 there is some $x \in R$ with $w(x) = 0$ and $v_i(x) < 0$ for all $1 \leq i \leq n$. □

Corollary 3.9. *Let $v_1, \ldots, v_n, w_1, \ldots, w_m$ be valuations on R such that the following properties hold.*

a) $A_{w_1} \not\subset A_{v_i}$ for all $1 \leq i \leq n$.
b) $A_{w_1} \subset A_{w_j}$ for all $1 \leq j \leq m$.

Then there is some $x \in R$ such that $w_j(x) = 0$ for all $1 \leq j \leq m$ and $v_i(x) < 0$ for all $1 \leq i \leq n$.

Proof. We do induction on m.
$m = 1$: This is covered by Proposition 8.
$m \to m + 1$: By the inductive hypothesis there is some $x' \in R$, such that $w_j(x') = 0$ for all $1 \leq j \leq m$, and $v_i(x') < 0$ for all $1 \leq i \leq n$. By assumption b) we have $w_{m+1}(x') \geq 0$. By Proposition 6 we find some $x \in R$ such that $w_j(x) = 0$ for all $1 \leq j \leq m + 1$ and $v_i(x) \leq v_i(x') < 0$ for $1 \leq i \leq n$. □

Remark 3.10. Assume that v is a non-trivial special valuation on R. Then for every $\alpha \in \Gamma_v$ there exists some $x \in R$ with $v(x) < \alpha$.

Proof. Let $\alpha \in \Gamma_v$. Then there are $y, z \in R \setminus \operatorname{supp} v$ with $\alpha = v(y) - v(z)$. Since v is special and non-trivial there is some $z' \in R$ with $v(z') < -v(z)$. We take $x := yz'$. □

Proposition 3.11. *Let $v_1, \dots, v_n$ be non-trivial special valuations on R. Let $\alpha_i \in \Gamma_{v_i}$ for $1 \le i \le n$. Then there is some x in R such that $v_i(x) < \alpha_i$ for all $1 \le i \le n$.*

Proof. We do induction on n.

$n = 1$: This is clear by Remark 10.

$n \to n+1$: Since the valuations $v_1, \dots, v_{n+1}$ are non-trivial we may assume that $\alpha_i < 0$ for all $1 \le i \le n+1$. By the inductive hypothesis there are $x', x'' \in R$ such that $v_i(x') < \alpha_i$ for $1 \le i \le n$ and $v_i(x'') < \alpha_i$ for $2 \le i \le n+1$. By Lemma 7 we can assume that $v_i(x') \neq v_i(x'')$ for all $1 \le i \le n+1$. If $v_{n+1}(x') < \alpha_{n+1}$ or $v_1(x'') < \alpha_1$ we are done. Otherwise let $x := x' + x''$. For $2 \le i \le n$ we have $v_i(x) = \min\{v_i(x'), v_i(x'')\} < \alpha_i$. Also, $v_1(x) = v_1(x') < \alpha_1$ and $v_{n+1}(x) = v_{n+1}(x'') < \alpha_{n+1}$. □

Proposition 3.12. *Let $v_1, \dots, v_n$ be non-trivial special valuations on R and let $w_1, \dots, w_m$ be trivial valuations on R. Let $\alpha_i \in \Gamma_{v_i}$ for $1 \le i \le n$. Then there is some $x \in R$ such that $v_i(x) < \alpha_i$ for $1 \le i \le n$ and $w_j(x) = 0$ for $1 \le j \le m$.*

Proof. We do induction on m.

$m = 1$: We may assume that $\alpha_i < 0$ for $1 \le i \le n$. By Proposition 11 there is some $x' \in R$ such that $v_i(x') < \alpha_i$ for $1 \le i \le n$. If $w_1(x') = 0$ we take $x := x'$. If $w_1(x') = \infty$ we take $x = 1 + x'$. Then $v_i(x) = v_i(x') < \alpha_i$ for $1 \le i \le n$ and $w_1(x) = 0$.

$m \to m+1$: Again we may assume that $\alpha_i < 0$ for $1 \le i \le n$. By the inductive hypothesis there is for $1 \le j \le m+1$ some $x_j \in R$ such that $v_i(x_j) < \alpha_i$ for $1 \le i \le n$ and $w_k(x_j) = 0$ for $k \in \{1, \dots, m+1\} \setminus \{j\}$. If there is some $j \in \{1, \dots, m+1\}$ such that $w_j(x_j) = 0$ we are done. Otherwise let $y_j := \prod_{k \neq j} x_k$ for $1 \le j \le m+1$. Then

$$v_i(y_j) = \sum_{k \neq j} v_i(x_k) < m\alpha_i < \alpha_i$$

for $1 \le i \le n$. Moreover, $w_j(y_j) = 0$ and $w_k(y_j) = \infty$ for $k \in \{1, \dots, m+1\} \setminus \{j\}$. By Lemma 7 we may assume that $v_i(y_{j_1}) \neq v_i(y_{j_2})$ for all $1 \le i \le n$ and $j_1 \neq j_2$. We set $x := y_1 + \ldots + y_{m+1}$. Then $v_i(x) = \min\{v_i(y_1), \dots, v_i(y_{m+1})\} < \alpha_i$ for $1 \le i \le n$ and $w_j(x) = 0$ for $1 \le j \le m+1$. □

Corollary 3.13. *Let $v_1, \dots, v_n$ be special valuations on R. Let $\alpha_i \in \Gamma_{v_i}$ for $1 \le i \le n$. Then there is some $x \in R$ such that $v_i(x) \le \alpha_i$ for all $1 \le i \le n$.*

Proof. Without restriction we may assume that there is some $k \in \{0, \dots, n\}$ such that v_i is non-trivial for $1 \le i \le k$ and trivial for $k + 1 \le i \le n$. Then $\alpha_i = 0$ for $k + 1 \le i \le n$. We get the claim by Proposition 12. □

Now the inverse property comes into the game. We start by exploiting the results above.

Corollary 3.14. *Let $v_1, \dots, v_n, w$ be Manis valuations on R having the inverse property such that $A_w \not\subset A_{v_i}$ for all $1 \le i \le n$. Then there is some $x \in R$ such that $w(x) = 0$ and $0 < v_i(x) < \infty$ for all $1 \le i \le n$.*

Proof. By Proposition 8 there is some $x' \in R$ such that $w(x') = 0$ and $v_i(x') < 0$ for all $1 \le i \le n$. Since $v_1, \dots, v_n, w$ have the inverse property there is some $x \in R$ such that $w(x) = w(x') = 0$ and $v_i(x) = -v_i(x') > 0$ for all $1 \le i \le n$. □

Corollary 3.15. *Let $v_1, \dots, v_n, w_1, \dots, w_m$ be Manis valuations on R having the inverse property such that the following hold.*

a) $A_{w_1} \not\subset A_{v_i}$ for all $1 \le i \le n$.
b) $A_{w_1} \subset A_{w_j}$ for all $1 \le j \le m$.

Then there is some $x \in R$ such that $w_j(x) = 0$ for all $1 \le j \le m$ and $0 < v_i(x) < \infty$ for all $1 \le i \le n$.

Proof. This follows from Corollary 9 and the inverse property. □

Corollary 3.16. *Let $v_1, \dots, v_n$ be non-trivial Manis valuations on R having the inverse property. Let $\alpha_i \in \Gamma_{v_i}$ for $1 \le i \le n$. Then there is some x in R such that $\alpha_i < v_i(x) < \infty$ for all $1 \le i \le n$.*

Proof. By Proposition 11 there is some $x' \in R$ such that $v_i(x') < -\alpha_i$ for all $1 \le i \le n$. The statement follows from the inverse property. □

Corollary 3.17. *Let $v_1, \dots, v_n, w_1, \dots, w_m$ be Manis valuations on R having the inverse property such that $v_1, \dots, v_n$ are non-trivial and $w_1, \dots, w_m$ are trivial. Let $\alpha_i \in \Gamma_{v_i}$ for $1 \le i \le n$. Then there is some $x \in R$ such that $\alpha_i < v_i(x) < \infty$ for $1 \le i \le n$ and $w_j(x) = 0$ for $1 \le j \le m$.*

Proof. This follows from Proposition 12 and the inverse property. □

Corollary 3.18. *Let $v_1, \dots, v_n$ be Manis valuations on R having the inverse property. Let $\alpha_i \in \Gamma_{v_i}$ for $1 \le i \le n$. Then there is some $x \in R$ such that $\alpha_i \le v_i(x) < \infty$ for all $1 \le i \le n$.*

Proof. This follows from Corollary 13 and the inverse property. □

Proposition 3.19. *Let $v_1, \dots, v_n$ be Manis valuations on R. The following are equivalent.*

(1) $v_1, \dots, v_n$ have the inverse property.
(2) For any $x \in R$ there is an element $y \in R$ such that, for all $1 \le i \le n$, $v_i(y) = v_i(x)$ if $v_i(x) \ge 0$, and $-v_i(x) \le v_i(y) < \infty$ if $v_i(x) < 0$.

(3) *For any* $x \in R$ *there is an element* $y \in R$ *such that, for all* $1 \leq i \leq n$, $v_i(y) = v_i(x)$ *if* $v_i(x) \geq 0$, *and* $-v_i(x) \leq v_i(y)$ *if* $v_i(x) < 0$.
(4) *If* $1 \leq i \leq n$ *and* $x \in R$ *with* $v_i(x) < 0$ *and* $v_j(x) = 0$ *for* $j \neq i$, *then there is an element* $y \in R$ *such that* $0 \leq v_i(xy) < \infty$ *and* $v_j(y) = 0$ *for* $j \neq i$.
(5) *If* $1 \leq i \leq n$ *and* $x \in R$ *with* $v_i(x) < 0$ *and* $v_j(x) = 0$ *for* $j \neq i$, *then there is an element* $y \in R$ *such that* $0 \leq v_i(xy)$ *and* $v_j(y) = 0$ *for* $j \neq i$.

Proof. (1) $\Rightarrow$ (2): By Proposition 6 we find some $y' \in R$ such that for all $1 \leq i \leq n$ $v_i(y') = 0$ if $v_i(x) \geq 0$ and $v_i(y') \leq 2v_i(x)$ if $v_i(x) < 0$. Since $v_1, \dots, v_n$ have the inverse property there is some $y'' \in R$ such that for all $1 \leq i \leq n$ $v_i(y'') = 0$ if $v_i(x) \geq 0$ and $-2v_i(x) \leq v_i(y'') < \infty$ if $v_i(x) < 0$. We set $y := xy''$. Let $1 \leq i \leq n$. If $v_i(x) \geq 0$ then $v_i(y) = v_i(x) + v_i(y'') = v_i(x)$. If $v_i(x) < 0$ then $v_i(y) = v_i(x) + v_i(y'') \geq v_i(x) - 2v_i(x) = -v_i(x)$ and $v_i(y) \neq \infty$.
(2) $\Rightarrow$ (3): This is obvious.
(2) $\Rightarrow$ (4): Take y from (2).
(3) $\Rightarrow$ (5): Take y from (3).
(4) $\Rightarrow$ (5): This is obvious.
(5) $\Rightarrow$ (1): We do induction on n.
$n = 1$: $\{v_1\}$ has the inverse property since v_1 is Manis (cf. Remarks 1(a)).
$\leq n \rightarrow n + 1$: Let $x \in R$. By the inductive hypothesis we may assume that $v_i(x) \neq \infty$ for $1 \leq i \leq n+1$. Also by the inductive hypothesis there is some $y_1 \in R$ such that $v_i(xy_1) = 0$ for $2 \leq i \leq n + 1$. By (5) we may assume that $v_1(xy_1) \geq 0$. (Otherwise, $v_1(xy_1) < 0$. By (5) there is some $z_1 \in R$ with $v_i(xy_1z_1) = 0$ for $2 \leq i \leq n + 1$, and $v_1(xy_1z_1) \geq 0$. Replacing y_1 by y_1z_1, we are done.) In the same way there are $y_2, \dots, y_{n+1} \in R$ such that for $1 \leq i \leq n + 1$ $v_j(xy_i) = 0$ for $j \neq i$ and $v_i(xy_i) \geq 0$. If there is some $1 \leq i \leq n + 1$ such that $v_i(xy_i) = 0$ we take $x' := y_i$ and are done. Otherwise, $v_i(xy_i) > 0$ for all $1 \leq i \leq n + 1$. We set $x'_i := x^{n-1} \prod_{j \neq i} y_j$. Then

$$v_i(xx'_i) = v_i(x^n \prod_{j \neq i} y_j) = \sum_{j \neq i} v_i(xy_j) = 0$$

for $1 \leq i \leq n + 1$ and

$$v_j(xx'_i) = v_j(x^n \prod_{k \neq i} y_k) = \sum_{k \neq i} v_j(xy_k) = v_j(xy_j) > 0$$

for all $j \neq i$. Let $x' := x'_1 + \ldots + x'_{n+1}$. Then $v_i(xx') = v_i(xx'_i) = 0$ for all $1 \leq i \leq n + 1$.

□

Corollary 3.20. *Let* $v_1, \dots, v_n$ *be Manis valuations on* R *having the inverse property. Let* B *be an* R*-overring of* $\prod_{i=1}^n A_{v_i}$. *Then the special restrictions* $v_1|_B, \dots, v_n|_B$ *of* $v_1, \dots, v_n$ *to* B *have the inverse property.*

Proof. By [Vol. I, Proposition I.1.17], $v_i|_B$ is a Manis valuation for $1 \leq i \leq n$. We show that condition (5) of Proposition 19 holds. Without restriction we show it for $i = 1$. Let $x \in B$ with $v_1|_B(x) < 0$ and $v_j|_B(x) = 0$ for $2 \le j \le n$. Then $v_i|_B(x) = v_i(x)$ for all $1 \le i \le n$. By Proposition 19(5) applied to $v_1, \dots, v_n$ there is some $y \in R$ such that $v_1(xy) \ge 0$ and $v_j(y) = 0$ for $2 \le j \le n$. We see that $y \in \bigcap_{i=1}^n A_{v_i} \subset B$. We get $v_1|_B(xy) \ge 0$ and $v_j|_B(y) = 0$ for $2 \le j \le n$. □

We show in Proposition 4.19 below that the statement of Corollary 20 holds even for overrings of $\bigcap_{1 \le i \le n} A_{v_i}$.

Now we investigate the connection of the inverse property with coarsening and independence.

Proposition 3.21. *Let $(v_i \mid i \in I), (w_j \mid j \in J)$ be families of Manis valuations on R such that the following holds.*

a) $(v_i \mid i \in I)$ has the inverse (resp. finite inverse) property.
b) For every $j \in J$ there is some $i \in I$ such that $v_i \le w_j$.

Then $(w_j \mid j \in J)$ has the inverse (resp. finite inverse) property.

Proof. We may concentrate on the inverse property. Let $x \in R$. Since $(v_i \mid i \in I)$ has the inverse property there is by Remark 2 some $x' \in R$ such that $v_i(x^2x') = v_i(x)$ for all $i \in I$. For $j \in J$ let $i_j \in I$ with $v_{i_j} \le w_j$. Let $f_{i_j} : \Gamma_{v_{i_j}} \cup \{\infty\} \to \Gamma_w \cup \{\infty\}$ be the homomorphism of ordered monoids such that $w_j = f_{i_j} \circ v_{i_j}$. We obtain

$$w_j(x^2x') = f_{i_j}(v_{i_j}(x^2x')) = f_{i_j}(v_{i_j}(x)) = w_j(x)$$

for all $j \in J$. By Remark 2 we get the claim. □

Proposition 3.22. *Let v, w be two Manis valuations on R with w non-trivial. The following are equivalent.*

(1) $v \le w$.
(2) v, w have the inverse property and $A_v \subset A_w$.
(3) v, w have the inverse property and $\operatorname{supp} v \subsetneqq \mathfrak{p}_w \subset \mathfrak{p}_v$.

Proof. (1) $\Rightarrow$ (2): The valuations v, w have the inverse property by Remarks 1(a) and Proposition 21. The valuation v is non-trivial by Remarks 1.3(1). We get by Proposition 1.8 that $A_v \subset A_w$.

(2) $\Rightarrow$ (3): We show that $\mathfrak{p}_w \subset \mathfrak{p}_v$. Assume that $\mathfrak{p}_w \not\subset \mathfrak{p}_v$. Let $x \in \mathfrak{p}_w \setminus \mathfrak{p}_v$. Then $w(x) > 0$ and $v(x) \le 0$.

Case 1: $w(x) < \infty$. Since v, w have the inverse property there is some $x' \in R$ such that $v(x') = -v(x) \ge 0$ and $w(x') = -w(x) < 0$, i.e. $x' \in A_v \setminus A_w$, contradiction.

Case 2: $w(x) = \infty$. Since w is non-trivial there is some $y \in R$ such that $0 < w(y) < \infty$. By Case 1 we get $y \in \mathfrak{p}_v$, i.e. $v(y) > 0$. Let $x' := x + y$. Then $v(x') = v(x) \le 0$ and $0 < w(x') = w(y) < \infty$, contradiction to Case 1.

It remains to show that $\operatorname{supp} v \subsetneqq \mathfrak{p}_w$. Since w is non-trivial $A_w \ne R$. Hence $A_v \ne R$ and therefore v is also non-trivial. We obtain

$$\operatorname{supp} v = [A_v : R] \subset [A_w : R] = \operatorname{supp} w \subsetneq \mathfrak{p}_w.$$

(3) $\Rightarrow$ (1): Since $\operatorname{supp} v \subsetneq \mathfrak{p}_v$ the valuation v is non-trivial. By Proposition 1.8 it suffices to show that $A_v \subset A_w$. Assume that this does not hold. By Corollary 14 we find some $x \in R$ with $v(x) = 0$ and $w(x) > 0$, hence $x \in \mathfrak{p}_w \setminus \mathfrak{p}_v$, contradiction to (3). □

Corollary 3.23. *Let $v_1, \dots, v_n, w$ be Manis valuations on R having the inverse property such that $v_1, \dots, v_n$ are non-trivial and $w \not\leq v_i$ for all $1 \leq i \leq n$. Then there is some $x \in R$ such that $w(x) = 0$ and $0 < v_i(x) < \infty$ for all $1 \leq i \leq n$.*

Proof. By Proposition 22 we see that $A_w \not\subset A_{v_i}$ for all $1 \leq i \leq n$. We get the claim by Corollary 14. □

Corollary 3.24. *Let $v_1, \dots, v_n, w_1, \dots, w_m$ be Manis valuations on R having the inverse property such that $v_1, \dots, v_n$ are non-trivial and the following hold.*

a) $w_1 \not\leq v_i$ *for all* $1 \leq i \leq n$.
b) $w_1 \leq w_j$ *for all* $1 \leq j \leq m$.

Then there is some $x \in R$ such that $w_j(x) = 0$ for all $1 \leq j \leq m$ and $0 < v_i(x) < \infty$ for all $1 \leq i \leq n$.

Proof. By Proposition 22 we get that $A_{w_j} \not\subset A_{v_i}$ for all $1 \leq j \leq m$ and $1 \leq i \leq n$. Clearly, $A_{w_1} \subset A_{w_j}$ for all $2 \leq j \leq m$. We get the claim by Corollary 15. □

Let v, w be Manis valuations with $\operatorname{supp} v = \operatorname{supp} w$ having the inverse property. Our next goal is to describe $v \vee w$ in this situation.

Lemma 3.25. *Let v be a valuation on R and let B be a subring of R. Then $\mathfrak{p} := [\mathfrak{p}_v : B]$ is a v-convex prime ideal of A_v with* $\operatorname{supp} v \subset \mathfrak{p} \subset \mathfrak{p}_v$.

Proof. Since $\mathfrak{p}_v$ is an ideal of A_v, clearly $\mathfrak{p}$ is an ideal of A_v. Since $1 \in B$ we have $\mathfrak{p} \subset \mathfrak{p}_v$. Obviously $\mathfrak{p}$ is a v-convex ideal with $\operatorname{supp} v \subset \mathfrak{p}$. It remains to show that $\mathfrak{p}$ is prime. Let $a, b \in A_v \setminus \mathfrak{p}$. There are $x, y \in B$ such that $v(ax) \leq 0$ and $v(by) \leq 0$. Hence $v(abxy) \leq 0$ and therefore $ab \notin \mathfrak{p}$ since $xy \in B$. □

Remark 3.26. Let v be a valuation on R and let B be a subring of R. Then $[\mathfrak{p}_v : B] = [\mathfrak{p}_v : \operatorname{conv}(B)]$, with $\operatorname{conv}(B)$ denoting the v-convex hull of B.

Proposition 3.27. *Let v be a Manis valuation on R and let B_1, B_2 be subrings of R. The following are equivalent.*

(1) $[\mathfrak{p}_v : B_1] = [\mathfrak{p}_v : B_2]$,
(2) $\operatorname{conv}(B_1) = \operatorname{conv}(B_2)$.

Proof. (2) $\Rightarrow$ (1): This follows with Remark 26.

(1) $\Rightarrow$ (2): By Remark 26 we can assume that B_1 and B_2 are v-convex and have to show that $B_1 = B_2$. Assume that there is some $x \in B_1 \setminus B_2$. Then $v(x) < v(y)$

for all $y \in B_2$. Let $z \in R$ with $v(z) = -v(x)$. Then $v(zy) > 0$ for all $y \in B_2$, hence $z \in [\mathfrak{p}_v : B_2]$. But $v(zx) = 0$, hence $z \notin [\mathfrak{p}_v : B_1]$, contradiction. □

Proposition 3.28. *Let v, w be Manis valuations on R having the inverse property such that* $\operatorname{supp} v = \operatorname{supp} w$. *Let $\mathfrak{p} := [\mathfrak{p}_v : A_w]$. Then $\mathfrak{p}$ is a w-convex prime ideal of A_w with* $\operatorname{supp} w \subset \mathfrak{p} \subset \mathfrak{p}_w$.

Proof. a) Since $\operatorname{supp} w = \operatorname{supp} v$ we get that $\operatorname{supp} w \subset \mathfrak{p}$ by Lemma 25.

b) We show that $\mathfrak{p} \subset \mathfrak{p}_w$. Suppose that $\mathfrak{p} \not\subset \mathfrak{p}_w$. Let $x \in \mathfrak{p} \setminus \mathfrak{p}_w$. Then $x \notin \operatorname{supp} w = \operatorname{supp} v$. Since v, w have the inverse property there is some $x' \in R$ with $v(xx') = w(xx') = 0$. Since $x \notin \mathfrak{p}_w$ we get $x' \in A_w$. From $x \in \mathfrak{p} = [\mathfrak{p}_v : A_w]$ we see that $xx' \in \mathfrak{p}_v$, contradiction to $v(xx') = 0$.

c) We show that $\mathfrak{p}$ is an ideal of A_w. Given $x \in \mathfrak{p}$ and $a \in A_w$ we get $xaA_w \subset xA_w \subset \mathfrak{p}_v$, hence $xa \in \mathfrak{p}$.

d) Finally we show that $\mathfrak{p}$ is a prime ideal of A_w. Let $x, y \in A_w$ with $xy \in \mathfrak{p}$ and $y \notin \mathfrak{p}$. Then there is some $a \in A_w$ with $ya \notin \mathfrak{p}_v$ and therefore $v(ya) \leq 0$. For any $b \in A_w$ we get

$$v(xb) \geq v(xb) + v(ya) = v(xyab).$$

Since $xy \in \mathfrak{p}, ab \in A_w$ and $\mathfrak{p}A_w \subset \mathfrak{p}_v$ by the definition of $\mathfrak{p}$, we have $v(xyab) > 0$. So $v(xb) > 0$ for all $b \in B$ and hence $x \in \mathfrak{p}$.

By [Vol. I, Proposition I.1.10] $\mathfrak{p}$ is w-convex. □

Proposition 3.29. *Let v, w be Manis valuations on R having the inverse property such that* $\operatorname{supp} v = \operatorname{supp} w$. *Then $\mathfrak{p}_{v \vee w} = [\mathfrak{p}_v : A_w]$.*

Proof. Let $\mathfrak{p} := [\mathfrak{p}_v : A_w]$. By Lemma 25 and Proposition 28, $\mathfrak{p}$ is a prime ideal of A_v and A_w such that $\operatorname{supp} v, \operatorname{supp} w \subset \mathfrak{p} \subset \mathfrak{p}_v \cap \mathfrak{p}_w$.

Case 1: v and w are dependent. By Remark 2.2 and Corollary 2.3 we have to show the following. Let $\mathfrak{r}$ be a prime ideal of A_v and A_w with $\mathfrak{r} \subset \mathfrak{p}_v \cap \mathfrak{p}_w$. Then $\mathfrak{r} \subset \mathfrak{p}$. Let $x \in \mathfrak{r}$ and let $a \in A_w$ be arbitrary. Then $xa \in \mathfrak{r} \subset \mathfrak{p}_v$ and therefore $x \in \mathfrak{p}$ by the definition of $\mathfrak{p}$.

Case 2: v and w are independent. Again by Remark 2.2 we get that $\mathfrak{p} = \operatorname{supp} v = \operatorname{supp} w$ and therefore $\mathfrak{p} = \mathfrak{p}_{v \vee w} = \operatorname{supp}(v \vee w)$ according to Definition 5 in Sect. 2. □

Corollary 3.30. *Let v, w be Manis valuations on R having the inverse property such that* $\operatorname{supp} v = \operatorname{supp} w$. *Then $[\mathfrak{p}_v : A_w] = [\mathfrak{p}_w : A_v]$.*

Proof. By Proposition 29 we get

$$[\mathfrak{p}_v : A_w] = \mathfrak{p}_{v \vee w} = \mathfrak{p}_{w \vee v} = [\mathfrak{p}_w : A_v].$$

□

Proposition 3.31. *Let v, w be Manis valuations on R having the inverse property such that* $\operatorname{supp} v = \operatorname{supp} w$. *Then* $[\mathfrak{p}_v : A_w] = [\mathfrak{p}_v \cap \mathfrak{p}_w : A_v A_w] = [\mathfrak{p}_w : A_v]$.

Proof. Let $\mathfrak{p} := [\mathfrak{p}_v : A_w]$. We show that $[\mathfrak{p}_v \cap \mathfrak{p}_w : A_v A_w] = \mathfrak{p}$ and then are done by Corollary 30. By Lemma 25 and Proposition 28, $\mathfrak{p}$ is a prime ideal of A_v and A_w with $\mathfrak{p} \subset \mathfrak{p}_v \cap \mathfrak{p}_w$. Hence $\mathfrak{p} \subset [\mathfrak{p}_v \cap \mathfrak{p}_w : A_v A_w]$. The other inclusion is trivial. □

Corollary 3.32. *Let v, w be Manis valuations on R having the inverse property such that* $\operatorname{supp} v = \operatorname{supp} w$. *Then* $\mathfrak{p}_{v \vee w} = [\mathfrak{p}_v \cap \mathfrak{p}_w : A_v A_w]$.

Proof. This follows from Propositions 29 and 31. □

We extend the above results to finitely many Manis valuations. Let $(v_i \mid i \in I)$ be a family of Manis valuations on R. For $J \subset I$ we set $v_J := \bigvee_{i \in J} v_i$, $A_J := \prod_{i \in J} A_{v_i}$ and $\mathfrak{p}_J := \mathfrak{p}_{v_J}$. For J finite we often omit the brackets in the index, writing $A_{1,2}$ instead of $A_{\{1,2\}}$ etc.

Proposition 3.33. *Let I be a finite set and let $(v_i \mid i \in I)$ be a family of Manis valuations on R having the inverse property such that* $\operatorname{supp} v_i = \operatorname{supp} v_j$ *for all $i, j \in I$. Given a non-empty subset J of I we have* $\mathfrak{p}_I = [\mathfrak{p}_J : A_K]$ *for every non-empty subset K of I with $J \cup K = I$.*

Proof. Let $\emptyset \neq J \subset I$ and let $\emptyset \neq K \subset I$ with $J \cup K = I$.

Case 1: $J = I$. Clearly $\mathfrak{p}_I = [\mathfrak{p}_I : A_{v_I}]$. By Remark 2.6 and Definition 5 in Sect. 2 $A_{v_I} \supset A_K$. Hence $\mathfrak{p}_I \subset [\mathfrak{p}_I : A_K]$. Since $1 \in A_K$, we have equality.

Case 2: $J \subsetneqq I$. Without restriction we assume that $I = \{1, \dots, n\}$ for some $n \in \mathbb{N}$. We do induction on n.

$n = 1$: Nothing is to show.

$n \to n + 1$: Without restriction we may assume that $n + 1 \notin J$. We have $v_{1,\dots,n+1} = v_{1,\dots,n} \vee v_{n+1}$. By Proposition 21 the valuations $v_{1,\dots,n}, v_{n+1}$ have the inverse property. By Proposition 29 we obtain $\mathfrak{p}_{1,\dots,n+1} = [\mathfrak{p}_{1,\dots,n} : A_{n+1}]$. By the inductive hypothesis we get $\mathfrak{p}_{1,\dots,n+1} = [[\mathfrak{p}_J : A_{K \setminus \{n+1\}}] : A_{n+1}]$. But the latter set clearly coincides with $[\mathfrak{p}_J : A_K]$. □

Proposition 3.34. *Let I be a finite set and let $(v_i \mid i \in I)$ be a family of Manis valuations on R having the inverse property such that* $\operatorname{supp} v_i = \operatorname{supp} v_j$ *for all $i, j \in I$. For every non-empty subset J of I we have* $\mathfrak{p}_I = [\bigcap_{i \in J} \mathfrak{p}_i : A_I]$.

Proof. We may assume that $I = \{1, \dots, n\}$ for some $n \in \mathbb{N}$.

Special Case: $J = I$. We do induction on n.

$n = 1$: Clearly $\mathfrak{p}_v = [\mathfrak{p}_v : A_v]$ for any valuation v on R.

$n \to n + 1$: We have $v_{1,\dots,n+1} = v_{1,\dots,n} \vee v_{n+1}$. By Proposition 21 the valuations $v_{1,\dots,n}, v_{n+1}$ have the inverse property. By Corollary 32 we obtain

$$\mathfrak{p}_I = [\mathfrak{p}_{1,\dots,n} \cap \mathfrak{p}_{n+1} : A_{1,\dots,n} A_{n+1}] = [\mathfrak{p}_{1,\dots,n} \cap \mathfrak{p}_{n+1} : A_I].$$

By the inductive hypothesis we get

$$\mathfrak{p}_I = [[\bigcap_{i=1}^{n} \mathfrak{p}_i : A_{1,\dots,n}] \cap \mathfrak{p}_{n+1} : A_I].$$

But the latter set clearly coincides with $[\bigcap_{i \in I} \mathfrak{p}_i : A_I]$.

General Case. By Proposition 33 we have $\mathfrak{p}_I = [\mathfrak{p}_J : A_I]$. By the special case we obtain

$$\mathfrak{p}_I = [[\bigcap_{i \in J} \mathfrak{p}_i : A_J] : A_I].$$

But the latter set clearly coincides with $[\bigcap_{i \in J} \mathfrak{p}_i : A_I]$. □

Proposition 3.35. *Let* $v_1, \dots, v_n$ *be Manis valuations on R having the inverse property such that* $\operatorname{supp} v_i = \operatorname{supp} v_j$ *for all* $1 \le i, j \le n$. *Then* $v_1|_{A_{\bigvee v_j}}, \dots, v_n|_{A_{\bigvee v_j}}$ *are independent Manis valuations on* $A_{\bigvee v_j}$ *having the inverse property.*

Proof. If $v_1, \dots, v_n$ are independent then $A_{\bigvee v_j} = R$ and nothing is to show. So we assume that $v_1, \dots, v_n$ are dependent. Let $u := \bigvee_{1 \le j \le n} v_j$. For $1 \le i \le n$ let $w_i := v_i|_{A_u}$. Suppose that $w_1, \dots, w_n$ are dependent. Let $u' := \bigvee_{1 \le i \le n} w_i$. By Remark 2.6, $A_u \supset \prod_{i=1}^n A_{v_i}$. By Corollary 20 the valuations $w_1, \dots, w_n$ have the inverse property. By Remark 2.2 we find some $x \in \mathfrak{p}_{u'} \setminus \operatorname{supp} w_1$. Then $x \notin \mathfrak{p}_u$ by Proposition 2.15. By Proposition 33 we get $x \notin [\mathfrak{p}_{v_1} : B]$ where $B := \prod_{i=2}^n A_{v_i}$. Hence we find some $a \in B$ such that $xa \notin \mathfrak{p}_{v_1}$ resp. $v_1(ax) \le 0$. By the above $B \subset A_u$. Hence $ax \in A_u$ and therefore $w_1(ax) = v_1(ax) \le 0$. But $ax \in \mathfrak{p}_{u'} \subset \mathfrak{p}_{w_1}$ by Remark 2.2. So $w_1(ax) > 0$, contradiction. □

4 Essential Valuations

The notion of essentiality in the field case (cf. [E]) and for rings of Krull type (cf. [G4], [Al-O]) is generalized to arbitrary Manis valuations: A valuation over a subring A is called A-essential if v can be recovered from the trace of v on A. If the intersection ring A of a family of valuations is Prüfer then every valuation in this family is A-essential. We want to establish criteria for the converse. For this we introduce the important notion of a family of valuations having finite avoidance (compare with Definition 2 in Chap. 1, Sect. 1). (Compare the notion of "endlichem Typ" in the work of Gräter and of "finite character" in the work of Griffin and Alajbegović.) For finitely many valuations the inverse property is equivalent with essentiality. We establish this result for families having finite avoidance by defining the finite avoidance inverse property.

Definition 1. Let v be a valuation on R and let A be a subring of R.

a) If $A \subset A_v$ we say that v *is a valuation over* A.
b) Let v be a valuation over A. Then the prime ideal $\mathfrak{p}_v \cap A$ of A is called the *center* of v on A and is denoted by $\text{cent}_A(v)$.
c) Let v be a valuation over A. We say that v is A-*essential* if $A_v = A_{[\text{cent}_A(v)]}$.

Example 4.1. Let v be any valuation on R. Then v is A_v-essential by [Vol. I, Lemma III.1.0].

Remarks 4.2. Let v be any valuation on R.

i) $\mathfrak{p}_v$ is the center of v on A_v (compare with [Vol. I, p. 11]).
ii) Let A be a subring of R such that v is a valuation over A. Then $A_{[\text{cent}_A(v)]} \subset A_v$.

Proposition 4.3. *Let A be a subring of R and let v be an A-essential valuation on R. Let B be an R-overring of A such that v is a valuation over B. Then v is B-essential.*

Proof. Let $\mathfrak{r}_1 := \text{cent}_A(v)$ and $\mathfrak{r}_2 := \text{cent}_B(v)$. By Remark 2(ii) it is enough to show $A_v \subset B_{[\mathfrak{r}_2]}$. Let $x \in A_v$. Since v is A-essential there is some $y \in A \setminus \mathfrak{r}_1$ such that $xy \in A$. Then clearly $y \in B \setminus \mathfrak{r}_2$ and $xy \in B$. Hence $x \in B_{[\mathfrak{r}_2]}$. □

Proposition 4.4. *Let A be a subring of R and let v be an A-essential valuation on R. Then $\mathfrak{p}_v = \text{cent}_A(v)_{[\text{cent}_A(v)]}$.*

Proof. Let $\mathfrak{r} := \text{cent}_A(v)$. Obviously $\mathfrak{p}_v \supset \mathfrak{r}_{[\mathfrak{r}]}$. Let $x \in \mathfrak{p}_v \subset A_v$. Then since v is A-essential there is some $s \in A \setminus \mathfrak{r}$ with $xs \in A$. We have $v(s) = 0$ and therefore $v(xs) = v(x) > 0$. So $xs \in \mathfrak{p}_v \cap A = \mathfrak{r}$ and therefore $x \in \mathfrak{r}_{[\mathfrak{r}]}$. □

Proposition 4.5. *Let v, w be Manis valuations on R with $v \leq w$. Let A be a subring of R. If v is A-essential then w is A-essential.*

Proof. We have $\mathfrak{p}_w \subset \mathfrak{p}_v \subset A_v \subset A_w$ by Sect. 1. In particular w is a valuation over A. Let $\mathfrak{r}_v := \text{cent}_A(v)$ and $\mathfrak{r}_w := \text{cent}_A(w)$. Then $\mathfrak{r}_w \subset \mathfrak{r}_v$. By Remark 2(ii) it is enough to show $A_w \subset A_{[\mathfrak{r}_w]}$. Let $x \in A_w$.

Case 1: $x \in A_v$. Since $A_v = A_{[\mathfrak{r}_v]}$ by assumption there is some $s \in A \setminus \mathfrak{r}_v$ such that $xs \in A$. Since $A \setminus \mathfrak{r}_v \subset A \setminus \mathfrak{r}_w$ we are done.

Case 2: $x \in A_w \setminus A_v$. Then $x \in A_w \setminus \mathfrak{p}_w$ and therefore $w(x) = 0$. Since $x \notin A_v$ there is some $y \in \mathfrak{p}_v$ with $xy \in A_v \setminus \mathfrak{p}_v$. Therefore $xy \in A_w \setminus \mathfrak{p}_w$ and hence $w(xy) = 0$. We obtain $w(y) = 0$ and so $y \in \mathfrak{p}_v \setminus \mathfrak{p}_w$. From $A_v = A_{[\mathfrak{r}_v]}$ we get $s, t \in A \setminus \mathfrak{r}_v$ such that $xys \in A$ and $yt \in A$. Then $xyst \in A$. We have $s \in A \setminus \mathfrak{r}_w$, so we show that $yt \in A \setminus \mathfrak{r}_w$ and are done. But $y \in \mathfrak{p}_v \setminus \mathfrak{p}_w \subset A_w \setminus \mathfrak{p}_w$ and $t \in A \setminus \mathfrak{r}_w$, hence $yt \notin \mathfrak{p}_w$. □

Corollary 4.6. *Let v, w be a Manis valuations on R such that $v \leq w$. Then w is B-essential for every ring B with $A_v \subset B \subset A_w$.*

Proof. This follows immediately from Example 1, Propositions 3 and 5. □

In particular we obtain in the above situation $A_w = (A_v)_{[\mathrm{cent}_{A_v}(w)]}$ and $\mathfrak{p}_w = \mathrm{cent}_{A_v}(w)$ (cf. Scholium 1.10).

Theorem 4.7. *Let $A \subset B$ be subrings of R and let v be an A-essential Manis valuation on R. Then the special restriction $v|_B$ is a Manis valuations on B that is A-essential.*

Proof. Let $\mathfrak{r} := \mathrm{cent}_A(v)$ and $w := v|_B$.

a) We show that w is A-essential. Since v is A-essential we have

$$A_v = A^R_{[\mathfrak{r}]} = \{x \in R \mid \exists s \in A \setminus \mathfrak{r} \text{ such that } xs \in A\}$$

and

$$\mathfrak{p}_v = \mathfrak{r}^R_{[\mathfrak{r}]} = \{x \in R \mid \exists s \in A \setminus \mathfrak{r} \text{ such that } xs \in \mathfrak{r}\}.$$

Since $A_w = A_v \cap B$ and $\mathfrak{p}_w = \mathfrak{p}_v \cap B$ we get that

$$A_w = A^B_{[\mathfrak{r}]} = \{x \in B \mid \exists s \in A \setminus \mathfrak{r} \text{ such that } xs \in A\}$$

and

$$\mathfrak{p}_w = \mathfrak{r}^B_{[\mathfrak{r}]} = \{x \in B \mid \exists s \in A \setminus \mathfrak{r} \text{ such that } xs \in \mathfrak{r}\}.$$

This shows that w is A-essential.

We show that w is Manis. Let $x \in B$ with $w(x) < \infty$. We need to find some $x' \in B$ with $w(xx') = 0$.

Case 1: $w(x) \leq 0$. Then $w(x) = v(x) < \infty$. Since v is Manis there is some $y \in R$ such that $v(y) = -v(x) \geq 0$. By the above we have some $z \in A$ and $s \in A \setminus \mathfrak{r}$ with $ys = z$, hence $v(z) = v(y) = -v(x)$. Then $x' := z \in B$ and $w(xx') = 0$.

Case 2: $w(x) > 0$. Since w is special, there exists some $y \in B$ with $w(y) \leq -w(x)$, hence $w(xy) \leq 0$. Then by Case 1 we obtain some $z \in B$ with $w(xyz) = 0$. So $x' := yz \in B$ does the job. □

Theorem 4.8. *Let A be Prüfer in R and let v be a Manis valuation over A. Then v is A-essential.*

Proof. Let $\mathfrak{r} := \mathrm{cent}_A(v)$. Since A is Prüfer the pair $(A_{[\mathfrak{r}]}, \mathfrak{r}_{[\mathfrak{r}]})$ is a Manis pair. By Remarks 2(ii) we know that $A_{[\mathfrak{r}]} \subset A_v$. We show that $\mathfrak{p}_v \cap A_{[\mathfrak{r}]} = \mathfrak{r}_{[\mathfrak{r}]}$ and obtain by [Vol. I, Theorem I.2.4 i) $\Rightarrow$ ii)] that $A_{[\mathfrak{r}]} = A_v$.

Let $x \in \mathfrak{p}_v \cap A_{[\mathfrak{r}]}$. Then $v(x) > 0$ and there is some $s \in A \setminus \mathfrak{r}$ such that $xs \in A$. Since $v(s) = 0$ we get $v(xs) > 0$ and hence $xs \in \mathfrak{r}$. This shows that $x \in \mathfrak{r}_{[\mathfrak{r}]}$. Let

$x' \in \mathfrak{r}_{[\mathfrak{r}]}$. Then there is some $s' \in A \setminus \mathfrak{r}$ with $x's' \in \mathfrak{r}$. So $v(x') = v(x's') > 0$ and therefore $x' \in \mathfrak{p}_v \cap A_{[\mathfrak{r}]}$. □

Corollary 4.9. *Let $(v_i \mid i \in I)$ be a family of Manis valuations on R. If $A := \bigcap_{i \in I} A_{v_i}$ is Prüfer in R then every v_i is A-essential.*

Corollary 4.10. *Let A be Prüfer in R and let v be a Manis valuation over A. Let B be an R-overring of A. Then $v|_B$ is a PM-valuation on B.*

Proof. Let $w := v|_B$. By Theorem 7 w is a Manis valuation on B. By the definition of a special restriction (cf. [Vol. I, Definition 11 in I §1]) we have $A_w = B \cap A_v \supset A$. By [Vol. I, Corollary I.5.3] A_w is Prüfer in B. Hence w is a PM-valuation. □

Corollary 4.11. *Let $(v_i \mid i \in I)$ be a family of Manis valuations on R. Then the following are equivalent.*

(1) $\bigcap_{i \in I} A_{v_i}$ is Prüfer in R.
(2) For all $i \in I$ we have that $\bigcap_{j \neq i} A_{v_j}$ is Prüfer in R and $v_i|_{\bigcap_{j \neq i} A_{v_j}}$ is a PM-valuation on $\bigcap_{j \neq i} A_{v_j}$.
(3) There is some $i \in I$ such that $\bigcap_{j \neq i} A_{v_j}$ is Prüfer in R and $v_i|_{\bigcap_{j \neq i} A_{v_j}}$ is a PM-valuation on $\bigcap_{j \neq i} A_{v_j}$.

Proof. Let $A := \bigcap_{i \in I} A_{v_i}$. For $i \in I$ let $B_i := \bigcap_{j \neq i} A_{v_i}$ and $w_i := v_i|_{B_i}$. By the very definition of special restriction we see that $A_{w_i} = A_{v_i} \cap B_i = A$.

(1) ⇒ (2): For $i \in I$ v_i is a valuation over A. By Corollary 10 we obtain that w_i is a PM-valuation on B_i.

(2) ⇒ (3): This is trivial.

(3) ⇒ (1): Let $i \in I$ such that the property in (3) holds. Since w_i is a PM-valuation by assumption we get that $A = A_{w_i}$ is Prüfer in B_i. Since B_i is Prüfer in R by assumption we get that A is Prüfer in R by [Vol. I, Theorem I.5.6]. □

Corollary 4.12. *Let v, w be PM-valuations on R. The following are equivalent.*

(1) $A_v \cap A_w$ is Prüfer in R.
(2) $v|_{A_w}$ is a PM-valuation on A_w.
(3) $w|_{A_v}$ is a PM-valuation on A_v.

Proposition 4.13. *Let $v_1, \dots, v_n$ be Manis valuations on R and let $A \subset A_{v_1} \cap \dots \cap A_{v_n}$ be Prüfer in R. Let B be an R-overring of A. If $v_1|_B, \dots, v_n|_B$ are dependent then $v_1, \dots, v_n$ are dependent and then $\bigvee v_i|_B = (\bigvee v_i)|_B$.*

Proof. Let $w_i := v_i|_B$ for $1 \leq i \leq n$. By Corollary 10 these valuations are Manis on B. We have

$$\bigcap_{i \leq i \leq n} A_{w_i} = \bigcap_{1 \leq i \leq n} A_{v_i} \cap B \supset A.$$

By [Vol. I, Corollary I.5.3] $\bigcap_{1\le i\le n} A_{w_i}$ is Prüfer in B. We assume that $w_1,\dots,w_n$ are dependent.

a) We show that $v_1,\dots,v_n$ are dependent. We have $\prod_{i=1}^n A_{w_i} \neq B$ by Corollary 2.8 and $A_{\bigvee w_i} = \prod_{i=1}^n A_{w_i}$ by Corollary 2.11. Again by Corollary 2.8 we have to show that $\prod_{i=1}^n A_{v_i} \neq R$. Assume this does not hold. Since $\prod_{i=1}^n A_{w_i} \neq B$ we find some $a \in B$ with $(\bigvee w_i)(a) < 0$. Clearly $a \in R = \prod_{i=1}^n A_{v_i}$. Hence there is some $q \in \mathbb{N}$ and $b_{i,1},\dots,b_{i,q} \in A_{v_i}$ for $1 \le i \le n$ such that $a = \sum_{p=1}^q \prod_{i=1}^n b_{i,p}$. The valuations $v_1,\dots,v_n$ are A-essential by Theorem 8. Hence there are $s_{i,p} \in A\setminus \mathrm{cent}_A(v_i)$ such that $b_{i,p}s_{i,p} \in A$ for $1 \le i \le n$ and $1 \le p \le q$. By Proposition 1.8 we have for $1 \le i \le n$

$$\mathfrak{p}_{\bigvee w_j} \subset \mathfrak{p}_{w_i} = \mathrm{cent}_B(v_i) \subset \mathfrak{p}_{v_i}.$$

Hence $(\bigvee w_j)(s_{i,p}) = 0$ for all $1 \le i \le n$ and $1 \le p \le q$. Let $x := \prod_{i=1}^n \prod_{p=1}^q s_{i,p}$. For $1 \le p \le q$ we get

$$(\bigvee w_j)\left(\prod_{i=1}^n b_{i,p}x\right) = (\bigvee w_j)\left(\prod_{i=1}^n b_{i,p}s_{i,p}\right) \ge 0$$

since $b_{i,p}s_{i,p} \in A \subset A_{\bigvee w_j}$. Hence

$$(\bigvee w_j)(ax) = (\bigvee w_j)\left(\sum_{p=1}^q (\prod_{i=1}^n b_{i,p}x)\right)$$

$$\ge \min\{(\bigvee w_j)\left(\prod_{i=1}^n b_{i,p}x\right) \mid 1 \le p \le q\} \ge 0.$$

But $(\bigvee w_j)(ax) = (\bigvee w_j)(a) < 0$, contradiction.

b) We have

$$A_{w_1} \cap \ldots \cap A_{w_n} = A_{v_1} \cap \ldots \cap A_{v_n} \cap B = A.$$

By [Vol. I, Corollary I.5.3] A is Prüfer in B. By the definition of special restriction and Scholium 1.16 we have to show that $A_{\bigvee w_i} = A_{\bigvee v_i} \cap B$. By a) $v_1,\dots,v_n$ are dependent. The proof has shown that $(\prod_{i=1}^n A_{v_i}) \cap B \subset \prod_{i=1}^n A_{w_i}$. Since

$$\prod_{i=1}^n A_{w_i} = \prod_{i=1}^n (A_{v_i} \cap B) \subset (\prod_{i=1}^n A_{v_i}) \cap B$$

we get that $\prod_{i=1}^n A_{w_i} = (\prod_{i=1}^n A_{v_i}) \cap B$. Hence $A_{\bigvee w_i} = A_{\bigvee v_i} \cap B$ by Corollary 2.11. □

We introduce the important notion of a family of Manis valuations with finite avoidance. Many theorems for the finite case can be extended to this more general situation.

Definition 2. A family of Manis valuations $(v_i \mid i \in I)$ on R has *finite avoidance* if the family $(A_{v_i} \mid i \in I)$ has finite avoidance in the sense of Definition 2 in Chap. 1, Sect. 1; i.e. for every $x \in R$ the set of indices $i \in I$ such that $v_i(x) < 0$ is finite.

The following will be useful for later.

Remark 4.14. Let $(v_i \mid i \in I)$ be a family of Manis valuations on R having finite avoidance. Let $i \in I$ such that v_i is non-trivial. Then the set $\{j \in I \mid v_j \leq v_i\}$ is finite.

Proof. Let J be the above set. Since v_i is non-trivial there is some $x \in R$ such that $v_i(x) < 0$. By the finite avoidance property there is a finite set $\tilde{J} \subset I$ such that $v_k(x) \geq 0$ for all $k \in I \setminus \tilde{J}$. Hence $J \subset \tilde{J}$ and we are done. □

Proposition 4.15. *Let $(v_i \mid i \in I)$ be a family of Manis valuations on R having finite avoidance. Let $A := \bigcap_{i \in I} A_{v_i}$. The following are equivalent.*

(1) A is Prüfer in R.
(2) $A \subset R$ is a PF-extension (cf. Definition 1 in Chap. 1, Sect. 4).

Proof. (1) ⇒ (2): We can assume that all v_i are non-trivial. We can also assume that the valuations v_i are pairwise incomparable (by taking only the minimal elements with respect to $\leq$, using Remark 14). By Theorem 1.4.1 we get that A is a PF-extension.

(2) ⇒ (1): This is obvious by the definition of a PF-extension. □

Theorem 4.16. *Let $(v_i \mid i \in I)$ be a family of Manis valuations on R having finite avoidance such that $A := \bigcap_{i \in I} A_{v_i}$ is Prüfer in R. Let $\mathfrak{m}$ be a maximal ideal of A such that $\mathfrak{m} \neq \mathrm{cent}_A(v_i)$ for all $i \in I$ and let w be a non-trivial Manis valuation on R such that $v_i \leq w$ for some $i \in I$. Then $\mathrm{cent}_A(w) \not\subset \mathfrak{m}$.*

Proof. It clearly suffices to do the proof in the case that all v_i are non-trivial. Moreover, we can assume that the valuations are pairwise incomparable. Then $A_{v_i} \not\subset A_{v_j}$ for $i \neq j$ by Corollary 1.17. Let $\mathfrak{r}_i := \mathrm{cent}_A(v_i)$ for $i \in I$. Then $A_{v_i} = A_{[\mathfrak{r}_i]}$ by Theorem 8. Hence $\mathfrak{r}_i \not\subset \mathfrak{r}_j$ for $i \neq j$. By [Vol. I, Lemma III.1.1] we have that $\mathfrak{r}_i$ is an R-regular prime ideal of A for all $i \in I$. Similarly to Proposition 1.1.7 we see that $\{\mathfrak{r}_i \mid i \in I\}$ is the set of all maximal R-regular prime ideals of A. So if $\mathfrak{m}$ is a maximal ideal of A with $\mathfrak{m} \neq \mathfrak{r}_i$ for all $i \in I$ then $\mathfrak{m}$ is not R-regular.

Let now w be a non-trivial Manis valuation on R such that $v_i \leq w$ for some $i \in I$. By Proposition 5 we get that w is A-essential, so $A_{[\mathrm{cent}_A(w)]} = A_w \neq R$. Therefore again by [Vol. I, Lemma III.1.1] the ideal $\mathrm{cent}_A(w)$ of A is R-regular and so $\mathrm{cent}_A(w) \not\subset \mathfrak{m}$, simply by [Vol. I, Definition 1 in II §1]. □

We show a result converse to Corollary 9 (in the case of finite avoidance) and Theorem 16 for PM-valuations.

Theorem 4.17. *Let* $(v_i \mid i \in I)$ *be a family of PM-valuations on* R *having finite avoidance. Let* $A := \bigcap_{i \in I} A_{v_i}$. *Assume that the following properties hold.*

i) Every v_i *is* A*-essential.*
ii) If $\mathfrak{m}$ *is a maximal ideal of* A *such that* $\mathfrak{m} \neq \mathrm{cent}_A(v_i)$ *for all* $i \in I$ *and if* w *is a non-trivial Manis valuation on* R *such that* $v_i \leq w$ *for some* $i \in I$*, then* $\mathrm{cent}_A(w) \not\subset \mathfrak{m}$.

Then A *is Prüfer in* R.

Proof. Let $\mathfrak{m}$ be a maximal ideal of A. We show that $(A_{[\mathfrak{m}]}, \mathfrak{m}_{[\mathfrak{m}]})$ is a Manis pair and are done.

Case 1: $\mathfrak{m} = \mathrm{cent}_A(v_i)$ for some $i \in I$. Since v_i is A-essential by (i) we have $A_{[\mathfrak{m}]} = A_{v_i}$. By Proposition 4 we get $\mathfrak{m}_{[\mathfrak{m}]} = \mathfrak{p}_v$. Since v_i is Manis by assumption we are done.

Case 2: $\mathfrak{m} \neq \mathrm{cent}_A(v_i)$ for all $i \in I$. It is enough to show $A_{[\mathfrak{m}]} = R$. Suppose that $A_{[\mathfrak{m}]} \neq R$. Let $x \in R \setminus A_{[\mathfrak{m}]}$. Then

$$(A :_A x) = \{a \in A \mid xa \in A\} \subset \mathfrak{m}.$$

We have $(A :_A x) = \bigcap_{i \in I}(A_{v_i} :_A x)$. Since the family $(v_i \mid i \in I)$ has finite avoidance there are $i_1, \ldots, i_n$ in I such that $v_i(x) \geq 0$ for all $i \notin J := \{i_1, \ldots, i_n\}$. Hence $A = (A_{v_i} :_A x)$ for all $i \notin J$. This gives $\bigcap_{1 \leq l \leq n}(A_{v_{i_l}} :_A x) \subset \mathfrak{m}$ and therefore $(A_{v_{i_k}} :_A x) \subset \mathfrak{m}$ for some $1 \leq k \leq n$. We write $j := i_k$.

Case 2.1: v_j is trivial. Then $(A_{v_j} :_{A_{v_j}} x) = R$ and hence

$$(A_{v_j} :_A x) = (A_{v_j} :_{A_{v_j}} x) \cap A = A.$$

We get $\mathfrak{m} = A$, contradiction.

Case 2.2: v_j is non-trivial. From $(A_{v_j} :_A x) \subset \mathfrak{m}$ we conclude $\sqrt{(A_{v_j} :_A x)} \subset \mathfrak{m}$. Obviously the ideals $(A_{v_j} :_{A_{v_j}} x)$ and $\sqrt{(A_{v_j} :_{A_{v_j}} x)}$ of A_{v_j} are v_j-convex. Therefore $\sqrt{(A_{v_j} :_{A_{v_j}} x)}$ is a prime ideal of A_{v_j}. Since v_j is Manis we have $(A_{v_j} :_{A_{v_j}} x) \supsetneq \mathrm{supp}\, v_j$. With [Vol. I, Theorem III.2.5] we get that $\mathfrak{n} := \sqrt{(A_{v_j} :_{A_{v_j}} x)}$ is an R-regular prime ideal of A_{v_i} contained in $\mathfrak{p}_{v_j}$. By Scholium 1.10 and Proposition 1.12 we obtain a non-trivial coarsening w of v_j with $A_w = (A_{v_j})_{[\mathfrak{n}]}$ and $\mathfrak{p}_w = \mathfrak{n}$. Since $\sqrt{(A_{v_j} :_{A_{v_j}} x)} \cap A = \sqrt{(A_{v_j} :_A x)}$ we get $\mathfrak{p}_w \cap A \subset \mathfrak{m}$, contradiction to (ii). □

The next results establish the connection between the inverse property and essential valuations.

Theorem 4.18. *Let $v_1, \ldots, v_n$ be Manis valuations on R and let $A := A_{v_1} \cap \ldots \cap A_{v_n}$. The following are equivalent.*

(1) $v_1, \ldots, v_n$ have the inverse property.
(2) Every v_i is A-essential.

Proof. Let $\mathfrak{r}_i := \operatorname{cent}_A(v_i)$ for $1 \leq i \leq n$.

(1) $\Rightarrow$ (2): Let $i \in \{1, \ldots, n\}$. Clearly v_i is a valuation over A. By Remarks 2(ii) it is enough to show $A_{v_i} \subset A_{[\mathfrak{r}_i]}$. Let $x \in A_{v_i}$. By Proposition 3.6 there is some $y \in R$ such that for $1 \leq j \leq n$ $v_j(y) = 0$ if $v_j(x) \geq 0$ and $v_j(y) \leq v_j(x)$ if $v_j(x) \leq 0$. Since $v_1, \ldots, v_n$ have the inverse property we get some $s \in R$ such that $v_j(s) = -v_j(y)$ for all $1 \leq j \leq n$. Then $v_j(s) \geq 0$ for all $1 \leq j \leq n$ and $v_i(s) = 0$. This gives $s \in A \setminus \mathfrak{r}_i$. We have $v_j(xs) = v_j(x) - v_j(y) \geq 0$ for all $1 \leq j \leq n$. Hence $xs \in A$. This shows that $x \in A_{[\mathfrak{r}_i]}$.

(2) $\Rightarrow$ (1): By Proposition 3.21 we can assume that v_i, v_j are incomparable for all $i \neq j$.

Special Case: We assume that every v_i is non-trivial. We do induction on n.

$n = 1$: $\{v_1\}$ has the inverse property since v_1 is Manis (cf. Remark 3.1(a)).

$n \to n + 1$: Let $x \in R$. By the inductive hypothesis and Proposition 3 we may assume that $v_i(x) \neq \infty$ for all $1 \leq j \leq n + 1$ (otherwise we can omit those v_i with $v_i(x) = \infty$). We have to find some $x' \in R$ with $v_i(x') = -v_i(x)$ for all $1 \leq i \leq n+1$. Let $B := \bigcap_{i=1}^{n} A_{v_i}$. By Proposition 3 $v_1, \ldots, v_n$ are B-essential. By the inductive hypothesis the valuations $v_1, \ldots, v_n$ have the inverse property.

Claim 1: There is some $y \in R$ such that

$$v_1(y) = 0, v_2(y) > 0, \ldots, v_{n+1}(y) > 0.$$

Proof of Claim 1: By Corollary 3.23 there is some $y' \in R$ with $v_1(y') = 0$ and $0 < v_i(y') < \infty$ for all $2 \leq i \leq n$. We may assume that $v_{n+1}(y') \geq 0$. Otherwise, since $A_{v_1} = A_{[\mathfrak{r}_1]}$ there is some $a \in A \setminus \mathfrak{r}_1$ with $y'a \in A$. Then we replace y' by $y'a$. Applied the same arguments to the valuations $v_1, v_3, \ldots, v_{n+1}$ we find some $y'' \in R$ with $v_1(y'') = 0, v_2(y'') \geq 0$ and $v_i(y'') > 0$ for all $3 \leq i \leq n+1$. With $y := y'y''$ we get $v_1(y) = 0$ and $v_i(y) > 0$ for all $2 \leq i \leq n + 1$.

By above $v_1, \ldots, v_n$ have the inverse property. Hence there is some $z_1 \in R$ with $v_1(xz_1) = 0$ and $v_i(xz_1) \geq 0$ for all $2 \leq i \leq n$. As in the proof of Claim 1 we can assume that $v_{n+1}(xz_1) \geq 0$. Let $y_1 \in R$ be as in Claim 1. Then $v_1(xy_1z_1) = 0$ and $v_i(xy_1z_1) > 0$ for all $2 \leq i \leq n + 1$. In the same way we obtain $y_2, \ldots, y_{n+1} \in R$ and $z_2, \ldots, z_{n+1} \in R$ such that $v_i(xy_iz_i) = 0$ and $v_j(xy_iz_i) > 0$ for all $j \neq i$. Let $x' := y_1z_1 + \ldots + y_nz_n$. Then $v(xx') = 0$ for all $1 \leq i \leq n$.

General Case: We may assume that $v_1, \ldots, v_k$ are trivial and $v_{k+1}, \ldots, v_{n+1}$ are non-trivial for some $0 \leq k \leq n + 1$. We do induction on k.

$k = 0$: This is covered by the special case.

$k \to k + 1$: We may assume that $\operatorname{supp} v_1$ is minimal in $\{\operatorname{supp} v_i \mid 1 \leq i \leq k + 1\}$ with respect to inclusion.

Claim 2: Let $\varepsilon_i \in \Gamma_{v_i}$ for $k+2 \leq i \leq n$. Then there is some $z \in R$ such that

$$v_1(z) = 0, v_2(z) = \infty, \ldots, v_{k+1}(z) = \infty, v_{k+2}(z) > \varepsilon_{k+2}, \ldots, v_n(z) > \varepsilon_n.$$

Proof of Claim 2: We can assume that $\varepsilon_i > 0$ for $k+2 \leq i \leq n$. By Proposition 3.11 there is some $z' \in R$ with $v_i(z') < -\varepsilon_i$ for all $k+2 \leq i \leq n$. We have $A_{v_1} = R = A_{[\mathfrak{r}_1]}$ by assumption. Hence there is some $s \in A \setminus \mathfrak{r}_1$ with $z's \in A$, We get that $v_1(s) = 0$ and $v_i(s) > \varepsilon_i$ for $k+1 \leq i \leq n$. By assumption $\operatorname{supp} v_1$ is minimal in $\{\operatorname{supp} v_i \mid 1 \leq i \leq k+1\}$. Since v_i and v_j are incomparable for $i \neq j$ we have $\operatorname{supp} v_1 \neq \operatorname{supp} v_i$ for $2 \leq i \leq k+1$. Hence there is for $2 \leq i \leq k+1$ some $b_i \in \operatorname{supp} v_i \setminus \operatorname{supp} v_1$. Let $b := sb_2 \cdot \ldots \cdot b_{k+1}$. Then $v_1(b) = 0$ and $v_i(b) = \infty$ for $2 \leq i \leq k+1$. Since $R = A_{[\mathfrak{r}_1]}$ there is some $t \in A \setminus \mathfrak{r}_1$ such that $c := bt \in A$. We obtain

$$v_1(c) = 0, v_2(c) = \infty, \ldots, v_{k+1}(c) = \infty, v_{k+2}(c) \geq 0, \ldots, v_n(c) \geq 0.$$

Let $z := sc$. Then z fulfills the requirements of the claim.

Let now $x \in R$. We have to find some $x' \in R$ with $v_i(x') = -v_i(x)$ for all $1 \leq i \leq n$ with $v_i(x) < \infty$. By the inductive hypothesis there is some $y \in R$ such that $v_i(y) = -v_i(x)$ for all $2 \leq i \leq n$ with $v_i(x) < \infty$. If $v_1(x) = \infty$ nothing is to show. So we assume that $v_1(x) = 0$. If $v_1(y) = 0$ we are done. Otherwise $v_1(y) = \infty$. By the claim there is some $z \in R$ such that $v_1(z) = 0$, $v_i(z) = \infty$ for $2 \leq i \leq k+1$ and $v_j(z) > v_j(y)$ for all $k+1 \leq j \leq n$ with $v_j(y) < \infty$. Let $x' := y + z$. Then $v_i(x') = -v_i(x)$ for all $1 \leq i \leq n$ such that $v_i(x) \neq \infty$. □

We want to extend Theorem 18 to the case of finite avoidance. For that reason we introduce the following technical definition.

Definition 3. Let $(v_i \mid i \in I)$ be a family of Manis valuations on R. We say that it has the *finite avoidance inverse property* if for all $i_1, \ldots, i_n \in I$ and all R-overrings B of $\bigcap_{i \in I} A_{v_i}$ the special restrictions $v_{i_1}|_B, \ldots, v_{i_n}|_B$ are Manis valuations on B having the inverse property.

Proposition 4.19. *Let $v_1, \ldots, v_n$ be finitely many Manis valuations on R. The following are equivalent.*

(1) $v_1, \ldots, v_n$ have the inverse property.
(2) $v_1, \ldots, v_n$ have the finite avoidance inverse property.

Proof. (1) ⇒ (2): Let $A := \bigcap_{1 \leq i \leq n} A_{v_i}$. Let $i_1, \ldots, i_q \in \{1, \ldots, n\}$ and let B be an R-overring of A. We have to show that $v_{i_1}|_B, \ldots, v_{i_q}|_B$ are Manis valuations on B having the inverse property. We write $w_p := v_{i_p}|_B$ for $1 \leq p \leq q$. By Theorems 18 and 7 $w_1, \ldots, w_q$ are Manis valuations on B and A-essential. Let $C := \bigcap_{1 \leq p \leq q} A_{w_i}$. Clearly $A \subset C \subset B$. By Proposition 3 we get that $w_1, \ldots, w_q$ are C-essential. By Theorem 18 we get that they have the inverse property.

(2) ⇒ (1): Apply (2) to $1, \ldots, n$ and $B = R$. □

Remark 4.20. Let $(v_i \mid i \in I)$ be a family of Manis valuations on R having the finite avoidance inverse property.

a) For $J \subset I$, the subfamily $(v_i \mid i \in J)$ has the finite avoidance inverse property.
b) The family $(v_i \mid i \in I)$ has the finite inverse property.

Proof. a): This is clear from the definition.
b): This follows from (a) and Proposition 19. □

Theorem 4.21. *Let $(v_i \mid i \in I)$ be a family of Manis valuations on R having finite avoidance and let $A := \bigcap_{i \in I} A_{v_i}$. The following are equivalent.*

(1) $(v_i \mid i \in I)$ has the finite avoidance inverse property.
(2) Every v_i is A-essential.

Proof. Let $\mathfrak{r}_i := \text{cent}_A(v_i)$ for $i \in I$.

(1) ⇒ (2): Let $i_0 \in I$. Clearly v_{i_0} is a valuation over A. By Remark 2(ii) it is enough to show $A_{v_{i_0}} \subset A_{[\mathfrak{r}_{i_0}]}$. Let $x \in A_{v_{i_0}}$. Since $(v_i \mid i \in I)$ has finite avoidance there is a finite subset $J = \{i_0, \dots, i_n\}$ of I such that $x \in A_{v_k}$ for all $k \notin J$. (Enlarging J if necessary we can assume that $i_0 \in J$.) Let $B := \bigcap_{k \notin J} A_{v_k}$. Then $x \in B$. Since $(v_i \mid i \in I)$ has the finite avoidance inverse property the valuations $v_{i_1}|_B, \dots, v_{i_n}|_B$ are Manis with the inverse property. Let $w_k := v_{i_k}|_B$ for $0 \le k \le n$. We have

$$\bigcap_{0 \le k \le n} A_{w_k} = \bigcap_{0 \le k \le n} (A_{v_{i_k}} \cap B) = \Big(\bigcap_{0 \le k \le n} A_{v_{i_k}}\Big) \cap B = A$$

and

$$\text{cent}_A(w_0) = \mathfrak{p}_{w_0} \cap A = \mathfrak{p}_{v_{i_0}} \cap B \cap A = \text{cent}_A(v_{i_0}) = \mathfrak{r}_{i_0}.$$

By Theorem 18, (1) ⇒ (2), we have that w_0 is A essential. Hence there is some $s \in A \setminus \mathfrak{r}_{i_0}$ such that $xs \in A$.

(2) ⇒ (1): Let $i_1, \dots, i_n \in I$ and let B be an R-overring of A. Let $w_k := v_{i_k}|_B$ for $1 \le k \le n$. By Theorem 7 the valuations $w_1, \dots, w_n$ are Manis on B and A-essential. We have

$$\bigcap_{1 \le k \le n} A_{w_k} = \bigcap_{1 \le k \le n} (A_{v_{i_k}} \cap B) = \Big(\bigcap_{1 \le k \le n} A_{v_{i_k}}\Big) \cap B \supset A.$$

By Proposition 3 $w_1, \dots, w_n$ are $\bigcap_{1 \le k \le n} A_{w_k}$-essential. By Theorem 18, (2) ⇒ (1), we get that $w_1, \dots, w_n$ have the inverse property. □.

Corollary 4.22. *Let $(v_i \mid i \in I)$ be a family of Manis valuation on R having finite avoidance. If $\bigcap_{i \in I} A_{v_i}$ is Prüfer in R then the family has the finite avoidance inverse property.*

Proof. Let $A := \bigcap_{i \in I} A_{v_i}$. By Corollary 9 v_i is A-essential for every $i \in I$. By Theorem 21 the family has the finite avoidance inverse property. □

5 The Approximation Theorem in the Neighbourhood of Zero

The approximation theorem in the neighbourhood of zero for finitely many valuations is equivalent to the inverse property (and essentiality). Therefore the approximation theorem holds if the intersection ring of the finitely many valuations is Prüfer. Extending the existing literature (cf. [Gr_1], [Al-O], [Al-M]) we introduce the notion of the approximation theorem in the neighbourhood of zero for arbitrary families. It implies the finite avoidance inverse property and is implied by the so-called strong finite avoidance inverse property and the Prüfer condition.

Proposition 5.1. *Let v, w be Manis valuations on R with the inverse property. Then the following are equivalent.*

(1) For any $\varepsilon \in \Gamma_v$ with $\varepsilon \geq 0$ there is some $x \in R$ with $v(x) \geq \varepsilon$ and $w(x) \leq 0$.
(2) For any $\varepsilon \in \Gamma_v$ with $\varepsilon \geq 0$ there is some $x \in R$ with $v(x) = \varepsilon$ and $w(x) \leq 0$.
(3) For any $\varepsilon \in \Gamma_v$ with $\varepsilon \geq 0$ there is some $x \in R$ with $v(x) \geq \varepsilon$ and $w(x) = 0$.
(4) For any $\varepsilon \in \Gamma_v$ with $\varepsilon \geq 0$ there is some $x \in R$ with $v(x) = \varepsilon$ and $w(x) = 0$.
(5) For any $\varepsilon \in \Gamma_v$ there is some $x \in R$ with $v(x) = \varepsilon$ and $w(x) = 0$.
(6) For any $\varepsilon \in \Gamma_v$ and $\eta \in \Gamma_w$ there is some $x \in R$ with $v(x) = \varepsilon$ and $w(x) = \eta$.

Proof. Clearly condition (6) implies all the other conditions. We verify (1) ⇒ (2) ⇒ (3) ⇒ (4) ⇒ (5) ⇒ (6), and then will be done.

(1) ⇒ (2): By (1), there is some $y \in R$ with $v(y) \geq \varepsilon$ and $w(y) \leq 0$. If $v(y) = \varepsilon$ we are done. We assume that $v(y) > \varepsilon$. Since v is Manis there is some $z \in R$ with $v(z) = \varepsilon$. If $w(z) \leq 0$ we are done. So we may assume that $w(z) > 0$. With $x := y + z$ we obtain $v(x) = v(y + z) = v(z) = \varepsilon$ and $w(x) = w(y) \leq 0$.

(2) ⇒ (3): For $\varepsilon = 0$ we can take $x = 1$. So we assume that $\varepsilon > 0$. By (2) there is some $y \in R$ such that $v(y) \geq \varepsilon$ and $w(y) \leq 0$. If $w(y) = 0$ we are done. Hence we assume that $w(y) < 0$. Then $v(1+y) = v(1) = 0$ and $w(1+y) = w(y) < 0$. Since v, w have the inverse property there is some $z \in R$ with $v(z) = -v(1 + y) = 0$ and $w(z) = -w(1 + y) = -w(y)$. With $x := yz$ we get $v(x) = v(y) \geq \varepsilon$ and $w(x) = w(y) + w(z) = w(y) - w(y) = 0$.

(3) ⇒ (4): For $\varepsilon = 0$ we can take $x = 1$. So we assume that $\varepsilon > 0$. By (3) there is some $y \in R$ with $v(y) \geq \varepsilon$ and $w(y) = 0$. If $v(y) = \varepsilon$ we are done. Hence we assume that $v(y) > \varepsilon$. Since v is Manis there is some $z \in R$ with $v(z) = \varepsilon$.

Case 1: $w(z) = 0$. Then we can take $x := z$.

Case 2: $w(z) > 0$. Let $x := y + z$. Then $v(x) = v(z) = \varepsilon$ and $w(x) = w(y) = 0$.

Case 3: $w(z) < 0$. Then $v(1+z) = v(1) = 0$ and $w(1+z) = w(z) < 0$. Since v, w have the inverse property there is some $z' \in R$ with $v(z') = -v(1+z) = 0$ and $w(z') = -w(1+z) = -w(z)$. With $x := zz'$ we get $v(x) = v(z) + v(z') = \varepsilon$ and $w(x) = w(z) + w(z') = w(z) - w(z) = 0$.

(4) $\Rightarrow$ (5): By (4), we may assume that $\varepsilon < 0$. Again by (4) there is some $y \in R$ with $v(y) = -\varepsilon$ and $w(y) = 0$. Since v, w have the inverse property there is some $x \in R$ such that $v(x) = -v(y) = \varepsilon$ and $w(x) = -w(y) = 0$.

(5) $\Rightarrow$ (6):

Case 1: $\eta = 0$. The claim follows with (5).

Case 2: $\eta < 0$. Since w is Manis there is some $y \in R$ with $w(y) = \eta$. We may assume that $v(y) \neq \infty$. (Otherwise, we replace y by $1 + y$.) By (5) there is some $z \in R$ with $v(z) = \varepsilon - v(y)$ and $w(z) = 0$. Let $x := yz$. Then $v(x) = \varepsilon$ and $w(x) = w(y) = \eta$.

Case 3: $\eta > 0$. By Case 2 there is an element $y \in R$ such that $v(y) = -\varepsilon$ and $w(y) = -\eta$. Since v and w have the inverse property the claim follows. □

Proposition 5.2. *Let $v_1, \ldots, v_n$ be Manis valuations on R with the inverse property. Let $\varepsilon_i \in \Gamma_{v_i}$ for $1 \leq i \leq n$. The following are equivalent.*

(1) There is some $x \in R$ with $v_1(x) = \varepsilon_1, \ldots, v_n(x) = \varepsilon_n$.
(2) For each pair (i, j) there is some $x \in R$ with $v_i(x) = \varepsilon_i$ and $v_j(x) = \varepsilon_j$.

Proof. (1) $\Rightarrow$ (2): This is obvious.

(2) $\Rightarrow$ (1):

Claim 1: If $v_1 \geq v_2$ it is enough to show (1) for $v_2, \ldots, v_n$.

Proof of Claim 1: Assume that there is some $x \in R$ such that $v_2(x) = \varepsilon_2, \ldots, v_n(x) = \varepsilon_n$. By (2) there is some $y \in R$ such that $v_1(y) = \varepsilon_1$ and $v_2(y) = \varepsilon_2$. Let $f : \Gamma_{v_2} \cup \{\infty\} \to \Gamma_{v_1} \cup \{\infty\}$ be the homomorphism of ordered monoids such that $v_1 = f \circ v_2$. Then

$$v_1(x) = f(v_2(x)) = f(\varepsilon_2) = f(v_2(y)) = v_1(y) = \varepsilon_1.$$

By Claim 1 we may assume that $v_i \not\leq v_j$ for all $i \neq j$. (Otherwise, if $v_i \leq v_j$ for some i, j, we may omit v_j.)

Special Case: $v_1, \ldots, v_n$ are non-trivial.

Claim 2: There is some $y \in R$ such that $v_1(y) = \varepsilon_1, v_2(y) > \varepsilon_2, \ldots, v_n(y) > \varepsilon_n$.

Proof of Claim 2: We prove Claim 2 by induction on n.

$n = 1$: This is obvious since v_1 is Manis.

$n \to n+1$: By the inductive hypothesis there is some $y' \in R$ such that

$$v_1(y') = \varepsilon_1, v_2(y') > \varepsilon_2, \ldots, v_n(y') > \varepsilon_n.$$

If $v_{n+1}(y') > \varepsilon_{n+1}$ we are done. Hence we assume that $v_{n+1}(y') \leq \varepsilon_{n+1}$. For the pair $(1, n+1)$ there is by (2) some $z \in R$ with $v_1(z) = \varepsilon_1$ and $v_{n+1}(z) = \varepsilon_{n+1}$. Since v_1 and v_{n+1} have the inverse property there is some $z' \in R$ such that $v_1(z') = -\varepsilon_1$ and $v_{n+1}(z') = -\varepsilon_{n+1}$. Hence $v_1(y'z') = 0$ and $\alpha := v_{n+1}(y'z') \leq 0$. By Proposition 3.19, (1) $\Rightarrow$ (3), applied to $y'z'$ there is some $a \in R$ such that

$$v_1(a) = 0, v_2(a) \geq 0, \ldots, v_n(a) \geq 0, v_{n+1}(a) \geq -\alpha.$$

Since $v_1 \not\leq v_{n+1}$ there is by Corollary 3.23 some $b \in R$ with $v_1(b) = 0$ and $v_{n+1}(b) > 0$. By Proposition 3.19, (1) $\Rightarrow$ (3), we may assume that

$$v_1(b) = 0, v_2(b) \geq 0, \ldots, v_n(b) \geq 0, v_{n+1}(b) > 0.$$

Hence

$$v_1(ab) = 0, v_2(ab) \geq 0, \ldots, v_n(ab) \geq 0, v_{n+1}(ab) > -\alpha.$$

We set $y := y'ab$. Then

$$\begin{aligned} v_1(y) &= v_1(y') + v_1(ab) = \varepsilon_1, \\ v_i(y) &= v_i(y') + v_i(ab) \geq v_i(y') > \varepsilon_i \end{aligned}$$

for $2 \leq i \leq n$ and

$$\begin{aligned} v_{n+1}(y) &= v_{n+1}(y') + v_{n+1}(ab) \\ &> v_{n+1}(y') - \alpha = v_{n+1}(y') - v_{n+1}(y'z') \\ &= -v_{n+1}(z') = v_{n+1}(z) = \varepsilon_{n+1}. \end{aligned}$$

This shows Claim 2.

By Claim 2 there is for each $1 \leq i \leq n$ some $x_i \in R$ such that $v_i(x_i) = \varepsilon_i$ and $v_j(x_i) > \varepsilon_j$ for all $j \neq i$ provided all v_i are non-trivial. Let $x := x_1 + \cdots + x_n$. Then $v_i(x) = \varepsilon_i$ for all $1 \leq i \leq n$.

General Case: We may assume that $v_1, \ldots, v_k$ are non-trivial and $v_{k+1}, \ldots, v_n$ are trivial for some $0 \leq k \leq n$. We can also assume that $v_{k+1}, \ldots, v_n$ are all different. Note that $\varepsilon_i = 0$ for all $k+1 \leq i \leq n$. We write $v_{k+1}, \ldots, v_n$ as

$$v_{11}, \ldots, v_{1l_1}, v_{21}, \ldots, v_{2l_2}, \ldots, v_{m1}, \ldots, v_{ml_m}$$

with $m \in \mathbb{N}_0$, $l_j \in \mathbb{N}$ (where $1 \leq j \leq m$ and $\sum_{j=1}^{m} l_j = n-k$) such that $\operatorname{supp} v_{j1}$ is maximal in $\{\operatorname{supp} v_i \mid k+1 \leq i \leq n\}$ and $\operatorname{supp} v_{j1} \supset \operatorname{supp} v_{jj'}$ for $1 \leq j \leq m$ and $1 \leq j' \leq l_j$. Note that after rewriting $\varepsilon_{jj'} = 0$ for all j, j'.

By the special case there is some $x \in R$ such that $v_i(x) = \varepsilon_i$ for all $1 \leq i \leq k$. Let

$$n(x) := \#\{1 \leq j \leq m \mid v_{j1}(x) = \infty\}.$$

Note that if $v_{j1}(x) = 0$ then $v_{jj'}(x) = 0$ for all $1 \leq j' \leq l_j$. Hence if $n(x) = 0$ we are done. So we may assume that $n(x) \geq 1$ and without restriction we assume that $v_{11}(x) = \infty$. Since $\operatorname{supp} v_{j1}$ is maximal and the trivial valuations are all different there is for $2 \leq j \leq m$ some $a_j \in R$ with $a_j \in \operatorname{supp} v_{j1} \setminus \operatorname{supp} v_{11}$. Let $a := a_2 \cdot \ldots \cdot a_m$. Then $v_{11}(a) = 0$ and $v_{j1}(a) = \infty$ for $2 \leq j \leq m$. By Proposition 3.19, (1) $\Rightarrow$ (3), we can assume that $v_1(a) \geq 0, \ldots, v_k(a) \geq 0$. By Corollary 3.17 we find some $b \in R$ with $v_{11}(b) = 0$ and $v_1(b) > \varepsilon_1, \ldots, v_k(b) > \varepsilon_k$. Then $v_{11}(ab) = 0$, $v_{j1}(ab) = \infty$ for $2 \leq j \leq n$ and $v_i(ab) = v_i(a) + v_i(b) > \varepsilon_i$ for $1 \leq i \leq k$. Let $x' := x + ab$. Then $v_i(x') = v_i(x) = \varepsilon_i$ for $1 \leq i \leq k$, $v_{11}(x') = v_{11}(ab) = 0$ and $v_{j1}(x') = v_{j1}(x)$ for $2 \leq j \leq m$. So $n(x') < n(x)$. Doing induction on $n(x)$ we obtain the claim. □

Proposition 5.3. *Let $v_1, \ldots, v_n$ be Manis valuations on R having the inverse property. Then the following are equivalent.*

(1) For every $i \neq j$ v_i and v_j are independent.
(2) For any $(\varepsilon_1, \ldots, \varepsilon_n) \in \Gamma_{v_1} \times \ldots \times \Gamma_{v_n}$ there is some $x \in R$ with $v_i(x) = \varepsilon_i$ for all $1 \leq i \leq n$.
(3) For every $i \neq j$ and $\varepsilon_i \in \Gamma_{v_i}$ there is some $x \in R$ with $v_i(x_i) = \varepsilon_i$ and $v_j(x) = 0$.

Proof. By Remark 3.1(c) and Proposition 2 it is enough to do the proof in the case $n = 2$. Let $v := v_1, w := v_2, \varepsilon := \varepsilon_1$ and $\eta := \varepsilon_2$

(1) $\Rightarrow$ (2):

Case 1: $\operatorname{supp} v \not\subset \operatorname{supp} w$. Let $y \in \operatorname{supp} v \setminus \operatorname{supp} w$. Since w is Manis there is some $z \in R$ with $w(yz) = 0$. For any $0 \leq \varepsilon' \in \Gamma_v$ we have $v(yz) = \infty \geq \varepsilon'$. By Proposition 1, (3) $\Rightarrow$ (6), we get some $x \in R$ such that $v(x) = \varepsilon$ and $w(x) = \eta$.

Case 2: $\operatorname{supp} w \not\subset \operatorname{supp} v$. We can copy the proof of Case 1.

Case 3: $\operatorname{supp} v = \operatorname{supp} w$. Let $\mathfrak{p} := [\mathfrak{p}_w : A_v]$. Then by Proposition 3.29 we have $\mathfrak{p} = \mathfrak{p}_{v \vee w}$. Since v and w are independent by assumption we get $\mathfrak{p} = \operatorname{supp} v$ (cf. Definition 5 in Sect. 2). Let $\varepsilon' \in \Gamma_v$ with $\varepsilon' \geq 0$. Let $y \in A_v$ with $v(y) = \varepsilon'$. Then $y \notin \operatorname{supp} v = \mathfrak{p}$. Hence there is some $a \in A_v$ such that $z := ay \notin \mathfrak{p}_w$. We obtain $v(z) = v(a) + v(y) \geq v(y) = \varepsilon'$ and $w(z) \leq 0$. By Proposition 1, (1) $\Rightarrow$ (6), we get some $x \in R$ such that $v(x) = \varepsilon$ and $w(x) = \eta$.

(2) $\Rightarrow$ (3): This is obvious.

(3) $\Rightarrow$ (1): Suppose that v and w are dependent. Then by Remark 2.2 there is a prime ideal $\mathfrak{p}$ of A_v and A_w such that

$$\operatorname{supp} v, \operatorname{supp} w \subsetneq \mathfrak{p} \subset \mathfrak{p}_v \cap \mathfrak{p}_w.$$

Let $y \in \mathfrak{p} \setminus \operatorname{supp} v$. Let $\varepsilon := v(y) < \infty$. We show that there is no $x \in R$ such that $v(x) = \varepsilon$ and $w(x) = 0$, contradiction. Hence suppose there is some $x \in R$ with $v(x) = \varepsilon$ and $w(x) = 0$. Since v, w have the inverse property and $v(y), w(y) \neq \infty$ there is some $y' \in R$ such that $v(yy') = 0$ and $w(yy') = 0$. We have

$$v(xy') = v(x) + v(y') = v(y) - v(y) = 0.$$

This shows that $xy' \in A_v$. Since $y \in \mathfrak{p}$ and $\mathfrak{p}$ is an ideal of A_v we get $xyy' \in \mathfrak{p}$. Since $\mathfrak{p} \subset \mathfrak{p}_w$ we get $w(xyy') > 0$. But

$$w(xyy') = w(x) + w(yy') = w(x) = 0,$$

contradiction. □

Definition 1. Let v_1, v_2 be Manis valuations on R. We write H_{v_1,v_2} for the subgroup $H^1_{(v_1,v_2)}$ of Γ_{v_1} introduced in Sect. 2, Definition 4. We set $\Gamma_{v_1,v_2} := \Gamma_{v_1}/H_{v_1,v_2}$ and write $f_{v_1,v_2} : \Gamma_{v_1} \to \Gamma_{v_1,v_2}$ for the canonical homomorphism of (ordered) groups.

Note that $\Gamma_{v_1,v_2} = \Gamma_{v_1 \vee v_2}$ if $\operatorname{supp} v_1 = \operatorname{supp} v_2$ (in particular if v_1, v_2 are dependent) and that $\Gamma_{v_1,v_2} = \{0\}$ if v_1, v_2 are independent. Let $\operatorname{supp} v_1 = \operatorname{supp} v_2$. Extending f_{v_1,v_2} by setting $f_{v_1,v_2}(\infty) = \infty$ the map $f_{v_1,v_2} : \Gamma_{v_1} \cup \{\infty\} \to \Gamma_{v_1 \vee v_2} \cup \{\infty\}$ is the homomorphism of ordered monoids such that $v_1 \vee v_2 = f_{v_1,v_2} \circ v_1$.

Corollary 5.4. *Let $v_1, \dots, v_n$ be Manis valuations on R with the inverse property. Let $\varepsilon_i \in \bigcap_{j \neq i} H_{v_i,v_j}$ for $1 \leq i \leq n$. Then there is some $x \in R$ with $v_1(x) = \varepsilon_1, \dots, v_n(x) = \varepsilon_n$.*

Proof. Let $i, j \in \{1, \dots, n\}$ with $i \neq j$. Since $\varepsilon_i \in H_{v_i,v_j}$ and $\varepsilon_j \in H_{v_j,v_i}$ it is enough by Proposition 2 to show the claim for v_i and v_j. Let $v := v_i$ and $w := v_j$. If v and w are independent the claim follows from Proposition 3. So we assume that v and w are dependent. Then $\operatorname{supp} v = \operatorname{supp} w$. By Proposition 3.35 $v' := v|_{A_{v \vee w}}$ and $w' := w|_{A_{v \vee w}}$ are independent Manis valuations on $A_{v \vee w}$ having the inverse property. By Propositions 2.15 and 3 we find some $x \in A_{v \vee w}$ such that $v'(x) = \varepsilon$ and $w'(x) = \eta$. We get $v(x) = v'(x) = \varepsilon$ and $w(x) = w'(x) = \eta$. □

Definition 2. i) Let v, w be Manis valuations on R. Let $(\alpha, \beta) \in \Gamma_v \times \Gamma_w$. Then (α, β) is called *compatible* if $f_{v,w}(\alpha) = f_{w,v}(\beta)$.

ii) Let $v_1, \dots, v_n$ be Manis valuations on R and let $(\alpha_1, \dots, \alpha_n) \in \Gamma_{v_1} \times \dots \times \Gamma_{v_n}$. The tuple $(\alpha_1, \dots, \alpha_n)$ is called *compatible* if (α_i, α_j) is compatible for every $1 \leq i, j \leq n$.

Remark 5.5. Let $v_1, \dots, v_n$ be Manis valuations on R.

a) Let $x \in R \setminus \bigcup_{1 \leq i \leq n} \operatorname{supp} v_i$. The tuple $(v_1(x), \dots, v_n(x)) \in \Gamma_{v_1} \times \dots \times \Gamma_{v_n}$ is compatible.

b) Let $(\alpha_1, \dots, \alpha_n) \in \Gamma_{v_1} \times \dots \times \Gamma_{v_n}$ be compatible. Then for every $1 \leq k \leq n$ the tuple $(\alpha_1, \dots, \alpha_k) \in \Gamma_{v_1} \times \dots \times \Gamma_{v_k}$ is compatible.

Remark 5.6. Let v, w be Manis valuation on R with $v \leq w$. Let $(\alpha, \beta) \in R$. Then (α, β) is compatible iff $\beta = f_{v,w}(\alpha)$.

Proposition 5.7. *Let $v_1, \ldots, v_n$ be Manis valuations on R. For $1 \leq i \leq n$ let $\Delta_i := \bigcap_{j \neq i} H_{v_i,v_j}$. Let $(\delta_1, \ldots, \delta_n) \in \Delta_1 \times \ldots \times \Delta_n$. Then $(\delta_1, \ldots, \delta_n)$ is compatible.*

Proof. Let $1 \leq i, j \leq n$. We may assume that $i \neq j$. Since $\delta_i \in H_{v_i,v_j}$ and $\delta_j \in H_{v_j,v_i}$ we have $f_{v_i,v_j}(\delta_i) = f_{v_j,v_i}(\delta_j) = 0$. □

Proposition 5.8. *Let $v_1, \ldots, v_n$ be Manis valuations on R. Then the following are equivalent.*

(1) For every $i \neq j$ v_i, v_j are independent.
(2) Every $(\alpha_1, \ldots, \alpha_n) \in \Gamma_{v_1} \times \ldots \times \Gamma_{v_n}$ is compatible.

Proof. It clearly suffices to do the proof in the case $n = 2$. We set $v := v_1, w := v_2, \alpha := \alpha_1$ and $\beta := \alpha_2$.

(1) ⇒ (2): This follows immediately from $\Gamma_{v,w} = \{0\}$ (cf. Definition 4 in Sect. 2).

(2) ⇒ (1): Assume that v, w are dependent. We choose $x \in R \setminus A_{v \vee w}$. Then $(v(x), 2w(x))$ is not compatible since

$$f_{v,w}(v(x)) = (v \vee w)(x) \neq 2(v \vee w)(x) = f_{w,v}(2w(x)),$$

contradiction. □

Let $v_1, \ldots, v_n$ be Manis valuations on R. Let $(\alpha_1, \ldots, \alpha_n) \in \Gamma_{v_1} \times \ldots \times \Gamma_{v_n}$. By Remark 5(a) it is necessary that $(\alpha_1, \ldots, \alpha_n)$ is compatible to find some $x \in R$ with $v_1(x) = \alpha_1, \ldots, v_n(x) = \alpha_n$.

Definition 3. Let $v_1, \ldots, v_n$ be Manis valuations on R. The *approximation theorem in the neighbourhood of zero* holds for $v_1, \ldots, v_n$ if for every compatible $(\alpha_1, \ldots, \alpha_n) \in \Gamma_{v_1} \times \ldots \times \Gamma_{v_n}$ there is some $x \in R$ such that $v_1(x) = \alpha_1, \ldots, v_n(x) = \alpha_n$.

Example 5.9. Let $v_1, \ldots, v_n$ be Manis valuations on R having the inverse property such that v_i and v_j are independent for every $i \neq j$. Then by Proposition 3 the approximation theorem in the neighbourhood of zero holds for $v_1, \ldots, v_n$.

Remark 5.10. Let $v_1, \ldots, v_n, w_1, \ldots, w_m$ be Manis valuations on R such that for every $1 \leq j \leq m$ there is some $1 \leq i \leq n$ with $v_i \leq w_j$. The following are equivalent.

(1) The approximation theorem in the neighbourhood of zero holds for $v_1, \ldots, v_n$.
(2) The approximation theorem in the neighbourhood of zero holds for $v_1, \ldots, v_n, w_1, \ldots, w_m$.

Proof. For $1 \leq j \leq m$ we choose $1 \leq i_j \leq n$ such that $v_{i_j} \leq w_j$.

(1) ⇒ (2): Let $(\alpha_1, \ldots, \alpha_n, \beta_1, \ldots, \beta_m) \in \Gamma_{v_1} \times \ldots \times \Gamma_{v_n} \times \Gamma_{w_1} \times \ldots \times \Gamma_{w_m}$ be compatible. Then $(\alpha_1, \ldots, \alpha_n) \in \Gamma_{v_1} \times \ldots \times \Gamma_{v_n}$ is compatible. Hence there is some $x \in R$ such that $v_i(x) = \alpha_i$ for all $1 \leq i \leq n$. Let $1 \leq j \leq m$. Then by Remark 6

$$w_j(x) = f_{v_{i_j},w_j}(v_{i_j}(x)) = f_{v_{i_j},w_j}(\alpha_{i_j}) = \beta_j.$$

(2) ⇒ (1): Let $(\alpha_1,\ldots,\alpha_n) \in \Gamma_{v_1} \times \ldots \times \Gamma_{v_n}$ be compatible. Let $\beta_j := f_{v_{i_j},w_j}(\alpha_{i_j})$ for $1 \le j \le m$. Then $(\alpha_1,\ldots,\alpha_n,\beta_1,\ldots,\beta_n) \in \Gamma_{v_1} \times \ldots \times \Gamma_{v_n} \times \Gamma_{w_1} \times \ldots \times \Gamma_{w_m}$ is compatible. To see this we distinguish three cases.

Claim 1: (α_k,α_l) is compatible for $1 \le k,l \le n$. This follows from the setting.

Claim 2: (α_k,β_l) is compatible for $1 \le k \le n$ and $1 \le l \le m$. We may assume that v_k and w_l are dependent. Then $v_k \vee v_{i_l} \le v_k \vee w_l$ by Proposition 2.13. Let $g : \Gamma_{v_k \vee v_{i_l}} \cup \{\infty\} \to \Gamma_{v_k \vee w_l} \cup \{\infty\}$ be the homomorphism of ordered monoids such that $v_k \vee w_l = g \circ (v_k \vee v_{i_l})$. We obtain

$$f_{w_l,v_k}(\beta_l) = f_{w_l,v_k}(f_{v_{i_l},w_l}(\alpha_{i_l})) = g(f_{v_{i_l},v_k}(\alpha_{i_l})) = g(f_{v_k,v_{i_l}}(\alpha_k)) = f_{v_k,w_l}(\alpha_k).$$

Claim 3: (β_k,β_l) is compatible for $1 \le k,l \le m$. We may assume that w_k, w_l are dependent. Then $v_{i_k} \vee v_{i_l} \le w_k \vee w_l$ by Proposition 2.13. Let $h : \Gamma_{v_{i_k} \vee v_{i_l}} \cup \{\infty\} \to \Gamma_{w_k \vee w_l} \cup \{\infty\}$ be the homomorphism of ordered monoids such that $w_k \vee w_l = h \circ (v_{i_k} \vee v_{i_l})$. We obtain

$$\begin{aligned} f_{w_k,w_l}(\beta_k) &= f_{w_k,w_l}(f_{v_{i_k},w_k}(\alpha_{i_k})) = h(f_{v_{i_k},v_{i_l}}(\alpha_{i_k})) \\ &= h(f_{v_{i_l},v_{i_k}}(\alpha_{i_l})) = f_{w_l,w_k}(f_{v_{i_l},w_l}(\alpha_{i_l})) \\ &= f_{w_l,w_k}(\beta_l). \end{aligned}$$

Since $(\alpha_1,\ldots,\alpha_n,\beta_1,\ldots,\beta_n)$ is compatible, there is by assumption some $x \in R$ such that $v_i(x) = \alpha_i$ for $1 \le i \le n$ and $w_j(x) = \beta_j$ for $1 \le j \le m$. Hence we are done. □

Theorem 5.11. *Let $v_1,\ldots,v_n$ be Manis valuations on R. The following are equivalent.*

(1) $v_1,\ldots,v_n$ have the inverse property.
(2) The approximation theorem in the neighbourhood of zero holds for $v_1,\ldots,v_n$.

Proof. (1) ⇒ (2): By Definitions 3 and 2 and Proposition 2 we can assume that $n = 2$. We write $v := v_1$ and $w := v_2$. By Proposition 3 we can assume that v and w are dependent. Let $(\alpha,\beta) \in \Gamma_v \times \Gamma_w$ be compatible. Then $f_{v,w}(\alpha) = f_{w,v}(\beta)$ in $\Gamma_{v \vee w}$. Hence there is some $y \in R$ with $(v \vee w)(y) = f_{v,w}(\alpha) = f_{w,v}(\beta)$. We get $v(y) - \alpha \in H_{v,w}$ and $w(y) - \beta \in H_{w,v}$. By Corollary 4 there is some $z \in R$ such that $v(z) = -v(y) + \alpha$ and $w(z) = -w(y) + \beta$. With $x := yz$ we get $v(x) = \alpha$ and $w(x) = \beta$.

(2) ⇒ (1): Let $x \in R$. For $1 \le i \le n$ we take $\alpha_i := -v_i(x)$ if $v_i(x) < \infty$ and $\alpha_i := 0$ if $v_i(x) = \infty$.

Claim: $(\alpha_1,\ldots,\alpha_n) \in \Gamma_{v_1} \times \ldots \times \Gamma_{v_n}$ is compatible.

Proof of the Claim: It clearly suffices to check the following case: Let $k \neq l$ such that $v_k(x) < \infty$ and $v_l(x) = \infty$. Then (α_k, α_l) is compatible. Since $x \in \operatorname{supp} v_l \setminus \operatorname{supp} v_k$ we get that v_k and v_l are independent. We have $\Gamma_{v,w} = \{0\}$ and are done.

By the assumption there is some $x' \in R$ such that $v_i(x') = \alpha_i$ for $1 \leq i \leq n$. By construction $v_i(x') = -v_i(x)$ for all $1 \leq i \leq n$ such that $v_i(x) \neq \infty$. □

Corollary 5.12. *Let $v_1, \ldots, v_n$ be Manis valuations on R. If $A_{v_1} \cap \ldots \cap A_{v_n}$ is Prüfer in R then the approximation theorem in the neighbourhood of zero holds for $v_1, \ldots, v_n$.*

Proof. By Theorem 4.18 and Corollary 4.9 the Manis valuations $v_1, \ldots, v_n$ have the inverse property. We get the claim by Theorem 11. □

Corollary 5.13. *Assume that the approximation theorem in the neighbourhood of zero holds for Manis valuations $v_1, \ldots, v_n$. Let $1 \leq k \leq n$. Then the approximation theorem in the neighbourhood of zero holds for $v_1, \ldots, v_k$.*

Proof. By Theorem 11 $v_1, \ldots, v_n$ have the inverse property. By Remark 3.1(c) we know that $v_1, \ldots, v_k$ have the inverse property. Again by Theorem 11 the approximation theorem in the neighbourhood of zero holds for $v_1, \ldots, v_k$. □

Corollary 5.14. *Let $v_1, \ldots, v_n$ be non-trivial Manis valuations on R such that the approximation theorem in the neighbourhood of zero holds for $v_1, \ldots, v_n$. Let $1 \leq k \leq n$. Then a compatible tuple $(\alpha_1, \ldots, \alpha_k) \in \Gamma_{v_1} \times \ldots \times \Gamma_{v_k}$ can be enlarged to a compatible tuple $(\alpha_1, \ldots, \alpha_n) \in \Gamma_{v_1} \times \ldots \times \Gamma_{v_n}$.*

Proof. Let $(\alpha_1, \ldots, \alpha_k) \in \Gamma_{v_1} \times \ldots \times \Gamma_{v_k}$ be compatible. By Corollary 13 the approximation theorem in the neighbourhood of zero holds for $v_1, \ldots, v_k$. Hence there is some $x \in R$ such that $v_i(x) = \alpha_i$ for all $1 \leq i \leq k$. By Theorem 11 $v_1, \ldots, v_n$ have the inverse property. Let $I := \{1 \leq i \leq n \mid v_i(x) < \infty\}$. Note that $I \supset \{1, \ldots, k\}$. By Corollary 3.16 there is some $y \in R$ such that $v_i(x) < v_i(y) < \infty$ for all $i \in I$ and $v_i(y) < \infty$ for all $i \notin I$. Let $z := x + y$. Then $v_i(z) = v_i(x) = \alpha_i$ for all $1 \leq i \leq k$ and $v_i(z) < \infty$ for all $k + 1 \leq i \leq n$. Let $\alpha_i := v_i(z)$ for $k + 1 \leq i \leq n$. Then $(\alpha_1, \ldots, \alpha_n) \in \Gamma_{v_1} \times \ldots \times \Gamma_{v_n}$ is compatible by Remark 5(a). □

In [Al-M] an example of two incomparable Manis valuations not fulfilling the approximation theorem in the neighbourhood of zero from [Ar] is formulated:

Example 5.15. Let v be a valuation on a field K with value group isomorphic to $\mathbb{Z} \oplus \mathbb{Z}$ equipped with the lexicographical order. Let $H := \{0\} \oplus \mathbb{Z}$ and let $w := v/H$. Then $\Gamma_w \cong \mathbb{Z}$. The valuations $\overline{v}$ and $\overline{w}$ defined by

$$\overline{v}(a_0 + a_1 X + \ldots + a_n X^n) := \min\{v(a_k) + k \cdot (1,0) \mid 1 \leq k \leq n\}$$

resp.

$$\overline{w}(a_0 + a_1 X + \ldots + a_n X^n) := \min\{w(a_k) + k \cdot 1 \mid 1 \leq k \leq n\}$$

are Manis valuations on $K[X]$ that do not satisfy the approximation theorem in the neighbourhood of zero.

We want to consider the case of families with finite avoidance. Some notations have to be introduced.

Definition 4. Let $(v_i \mid i \in I)$ be a family of Manis valuations on R. Let $i \in I$ and let $\alpha \in \Gamma_{v_i} \cup \{\infty\}$. If $\alpha \notin \{0, \infty\}$ let H_α denote the largest convex subgroup of Γ_{v_i} that does not contain α. We set

$$I(\alpha) := \begin{cases} \{i\} & \text{if } \alpha \in \{0,\infty\}, \\ \{j \in I \mid v_j \leq v_i/H_\alpha\} & \text{if } \alpha \notin \{0,\infty\}. \end{cases}$$

Remark 5.16. Let in the above situation $\alpha \notin \{0, \infty\}$. Let $w := v_i/H_\alpha$. Then w is a non-trivial Manis valuation on R with $v_i \leq w$ (cf. Scholium 1.7). In particular $i \in I(\alpha)$. (So $i \in I(\alpha)$ for all $i \in I$.) If $0 < \alpha < \infty$ and $\alpha = v_i(x)$ for some $x \in R$ then $\mathfrak{p}_w$ is the smallest v_i-convex prime ideal of A_{v_i} containing x (cf. Scholium 1.10).

Proposition 5.17. *Let $(v_i \mid i \in I)$ be a family of Manis valuations on R. Let $i \in I$ and $\alpha \in \Gamma_{v_i} \setminus \{0\}$. Then*

$$I(\alpha) = \{j \in I \mid f_{v_i,v_j}(\alpha) \neq 0\}.$$

Proof. Let $w := v_i/H_\alpha$. We have that w is non-trivial with $v_i \leq w$. Hence we obtain for $j \in I$ $v_j \leq w$ iff $v_i \vee v_j \leq w$ iff $H_{v_i,v_j} \subset H_\alpha$ iff $\alpha \notin H_{v_i,v_j}$ iff $f_{v_i,v_j}(\alpha) \neq 0$ in Γ_{v_i,v_j}. □

Remark 5.18. Let $(v_i \mid i \in I)$ be a family of Manis valuations on R having finite avoidance. Let $i \in I$ and let $\alpha \in \Gamma_{v_i} \cup \infty$. Then $I(\alpha)$ is finite.

Proof. If $\alpha \in \{0, \infty\}$ nothing is to show. If $\alpha \notin \{0, \infty\}$ then we are done by the definition and Remark 4.14. □

Definition 5. Let $(v_i \mid i \in I)$ be a family of Manis valuations on R and let $i_1, \ldots, i_n \in I$. Then $(\alpha_1, \ldots, \alpha_n) \in \Gamma_{v_{i_1}} \times \ldots \times \Gamma_{v_{i_n}}$ is called *$\{i_1, \ldots, i_n\}$-complete* if $\bigcup_{1 \leq k \leq n} I(\alpha_k) = \{i_1, \ldots, i_n\}$.

Remark 5.19. a) Note that $\bigcup_{1 \leq k \leq n} I(\alpha_k) \supset \{i_1, \ldots, i_n\}$ for any $i_1, \ldots, i_n \in I$ and $(\alpha_1, \ldots, \alpha_n) \in \Gamma_{v_{i_1}} \times \ldots \times \Gamma_{v_{i_n}}$ by Remark 16.

b) If $(v_i \mid i \in I)$ consists of pairwise independent Manis valuations then every $(\alpha_1, \ldots, \alpha_n) \in \Gamma_{v_{i_1}} \times \ldots \times \Gamma_{v_{i_n}}$ is $\{i_1, \ldots, i_n\}$-complete for all $i_1, \ldots, i_n \in I$ by Proposition 17.

Remark 5.20. Let $(v_i \mid i \in I)$ be a family of Manis valuations on R and let $i_1, \ldots, i_n \in I$. Let $(\alpha_1, \ldots, \alpha_n) \in \Gamma_{v_{i_1}} \times \ldots \times \Gamma_{v_{i_n}}$ be compatible and $\{i_1, \ldots, i_n\}$-complete. Let $j_1, \ldots, j_m \in I \setminus \{i_1, \ldots, i_n\}$. Then $(\alpha_1, \ldots, \alpha_n, 0, \ldots, 0) \in \Gamma_{v_{i_1}} \times \ldots \times \Gamma_{v_{i_n}} \times \Gamma_{v_{j_1}} \times \ldots \times \Gamma_{v_{j_m}}$ is compatible.

Proof. Let $1 \le k \le n$ and $1 \le l \le m$. If $\alpha_k = 0$ then clearly $f_{v_{i_k}, v_{j_l}}(\alpha_k) = 0$. If $\alpha_k \neq 0$ then $f_{v_{i_k}, v_{j_l}}(\alpha_k) = 0$ by Proposition 17 since $j_l \notin I(\alpha_k)$. □

Definition 6. Let $(v_i \mid i \in I)$ be a family of Manis valuations on R. We say that the *approximation theorem in the neighbourhood of zero* holds for the family if for each $i_1, \ldots, i_n \in I$ and each compatible and $\{i_1, \ldots, i_n\}$-complete tuple $(\alpha_1, \ldots, \alpha_n) \in \Gamma_{v_{i_1}} \times \ldots \times \Gamma_{v_{i_n}}$ there is some $x \in R$ with $v_{i_p}(x) = \alpha_p$ for $1 \le p \le n$ and $v_j(x) \ge 0$ for all $j \in I \setminus \{i_1, \ldots, i_n\}$.

Remark 5.21. If I is finite then Definition 6 coincides with Definition 3.

Proof. Let $I = \{1, \ldots, n\}$.

a) We show that the approximation theorem in the neighbourhood of zero in the sense of Definition 6 implies the one in the sense of Definition 3. To see this let $(\alpha_1, \ldots, \alpha_n) \in \Gamma_{v_1} \times \ldots \times \Gamma_{v_n}$ be compatible. By Remark 19(a) it is $\{1, \ldots, n\}$-complete and we are done.
b) We show that the approximation theorem in the neighbourhood of zero in the sense of Definition 3 implies the one in the sense of Definition 6. For this let $i_1 \ldots, i_k \in \{1, \ldots, n\}$. Without restriction we can assume that $i_1 = 1, \ldots, i_k = k$. Let $(\alpha_1, \ldots, \alpha_k) \in \Gamma_{v_1} \times \ldots \times \Gamma_{v_k}$ be compatible and $\{1, \ldots, k\}$-complete. Then $(\alpha_1, \ldots, \alpha_k, 0, \ldots, 0) \in \Gamma_{v_1} \times \ldots \times \Gamma_{v_n}$ is compatible by Remark 20. Hence there is some $x \in R$ such that $v_i(x) = \alpha_i$ for $1 \le i \le k$ and $v_i(x) = 0$ for $k + 1 \le i \le n$ and we are done. □

The approximation theorem in the neighbourhood of zero implies the finite avoidance inverse property.

Theorem 5.22. *Let $(v_i \mid i \in I)$ be a family of Manis valuations on R having finite avoidance. If the approximation theorem in the neighbourhood of zero holds for the family then it has the finite avoidance inverse property.*

Proof. By Theorem 4.21 it is enough to show that every v_i is A-essential. Let $i_0 \in I$. We set $\mathfrak{p} := \text{cent}_A(v_{i_0})$. We have to show that $A_{v_{i_0}} \subset A_{[\mathfrak{p}]}$. Let $x \in A_{v_{i_0}}$. Since the given family has finite avoidance we find $i_1, \ldots, i_n \in I$ such that $v_{i_k}(x) < 0$ for $1 \le k \le n$ and $v_j(x) \ge 0$ for all $j \in J := I \setminus \{i_1, \ldots, i_n\}$.

Case 1: $v_{i_0}(x) = 0$. Then the tuple $(v_{i_0}(x), v_{i_1}(x), \ldots, v_{i_n}(x)) \in \Gamma_{v_{i_0}} \times \Gamma_{v_{i_1}} \times \ldots \times \Gamma_{v_{i_n}}$ is compatible and $\{i_0, \ldots, i_n\}$-complete. The tuple $(-v_{i_0}(x), -v_{i_1}(x), \ldots, -v_{i_n}(x)) \in \Gamma_{v_{i_0}} \times \Gamma_{v_{i_1}} \times \ldots \times \Gamma_{v_{i_n}}$ is then also compatible and $\{i_0, \ldots, i_n\}$-complete. By assumption we find some $y \in \bigcap_{j \in J} A_{v_j}$ such that $v_{i_0}(y) = -v_{i_0}(x) = 0$ and $v_{i_k}(y) = -v_{i_k}(x) \ge 0$ for $1 \le k \le n$. Hence $y \in A \setminus \mathfrak{p}$ and $xy \in A$.

Case 2: $v_{i_0}(x) > 0$. Then $v_{i_0}(1 + x) = 0$, $v_{i_k}(1 + x) < 0$ for $1 \le k \le n$ and $v_j(1 + x) \ge 0$ for all $j \in I \setminus \{i_1, \ldots, i_n\}$. By Case 1 we have $1 + x \in A_{[\mathfrak{p}]}$ and therefore $x \in A_{[\mathfrak{p}]}$. □

We are not able to prove the other implication. We introduce the following notion.

Definition 7. Let $(v_i \mid i \in I)$ be a family of Manis valuations on R. We say that it has the *strong finite avoidance inverse property* if it has the finite avoidance inverse property and the following holds for all $i_1, \dots, i_n \in I$ and all R-overrings B of $\bigcap_{i \in I} A_{v_i}$: If $v_{i_1}|_B, \dots, v_{i_n}|_B$ are dependent then $v_{i_1}, \dots, v_{i_n}$ are dependent and $\bigvee v_{i_k}|B = (\bigvee v_{i_k})|_B$.

Remark 5.23. Let $(v_i \mid i \in I)$ be a family of Manis valuations on R such that $\bigcap_{i \in I} A_{v_i}$ is Prüfer in R. The the family has the strong finite avoidance inverse property.

Proof. This follows from Proposition 4.13. □

Theorem 5.24. *Let $(v_i \mid i \in I)$ be a family of Manis valuations on R having finite avoidance. If the family has the strong finite avoidance inverse property then the approximation theorem in the neighbourhood of zero holds for it.*

Proof. Let $i_1, \dots, i_n \in I$ and let $(\alpha_1, \dots, \alpha_n) \in \Gamma_{v_{i_1}} \times \dots \times \Gamma_{v_{i_n}}$ be compatible and $\{i_1, \dots, i_n\}$-complete. By Remark 4.20(b) the valuations $v_{i_1}, \dots, v_{i_n}$ have the inverse property. By Theorem 11 the approximation theorem in the neighbourhood of zero holds for $v_{i_1}, \dots, v_{i_n}$. Hence there is some $y \in R$ such that $v_{i_k}(y) = \alpha_k$ for $1 \leq k \leq n$. Since $(-\alpha_1, \dots, -\alpha_n)$ is also compatible there is some $y' \in R$ such that $v_{i_k}(y') = -\alpha_k$ for $1 \leq k \leq n$. Since the family has finite avoidance there is a finite subset J of I containing $i_1, \dots, i_n$ such that $v_j(y) \geq 0$ and $v_j(y') \geq 0$ for all $j \in I \setminus J$. Let $B := \bigcap_{j \in I \setminus J} A_{v_j}$. Then $y, y' \in B$. We write $J \setminus \{i_1, \dots, i_n\}$ as $\{i_{n+1}, \dots, i_m\}$ for some $m \geq n$. For $1 \leq k \leq m$ we set $w_k := v_{i_k}|_B$. The family $(v_i \mid i \in I)$ has the finite avoidance inverse property by assumption. Therefore the valuations $w_1, \dots, w_m$ are Manis on B and have the inverse property. By Theorem 11 the approximation theorem in the neighbourhood of zero holds for $w_1, \dots, w_m$. Since $y, y' \in B$ we have $\alpha_k \in \Gamma_{w_k}$ and $\alpha_k = w_k(y)$ for $1 \leq k \leq n$. We set $\alpha_k := 0 \in \Gamma_{w_k}$ for $n + 1 \leq k \leq m$.

Claim: The tuple $(\alpha_1, \dots, \alpha_m) \in \Gamma_{w_1} \times \dots \times \Gamma_{w_m}$ is compatible.

Proof of the Claim: Let $1 \leq k < l \leq m$. We show that (α_k, α_l) is compatible. We distinguish three cases.

Case 1: $1 \leq k, l \leq n$. Then $w_k(y) = \alpha_k$ and $w_l(y) = \alpha_l$, so (α_k, α_l) is compatible.

Case 2: $1 \leq k \leq n$ and $n + 1 \leq l \leq m$. Since $i_l \notin I(\alpha_k)$ we get $f_{v_{i_k}, v_{i_l}}(\alpha_k) = 0$ in $\Gamma_{v_{i_k} \vee v_{i_l}}$ by Proposition 17. By the strong finite avoidance inverse property and the definition of special restriction we obtain $f_{w_k, w_l}(\alpha_k) = 0$ in Γ_{w_k, w_l}. So (α_k, α_l) is compatible.

Case 3: $n + 1 \leq k < l \leq m$. This is obvious.

Since the approximation theorem in the neighbourhood of zero holds for $w_1, \dots, w_m$ there is some $x \in B$ such that $w_k(x) = \alpha_k$ for $1 \leq k \leq m$. Then $v_{i_k}(x) = \alpha_k$ for $1 \leq k \leq n$. It remains to show that $v_j(x) \geq 0$ for all $j \in I \setminus \{i_1, \dots, i_n\}$. If $j = i_k$ for some $n + 1 \leq k \leq m$ then $v_j(x) = w_k(x) = 0$. If $j \notin J$ then $v_j(x) \geq 0$ since $x \in B$. □

Corollary 5.25. *Let $(v_i \mid i \in I)$ be a family of Manis valuations on R having finite avoidance. If $\bigcap_{i \in i} A_{v_i}$ is Prüfer in R then the approximation theorem in the neighbourhood of zero holds for $(v_i \mid i \in I)$.*

Proof. By Remark 23 the family $(v_i, \mid i \in I)$ has the strong finite avoidance inverse property. We get the claim by Theorem 24. □

6 The General Approximation Theorem

We formulate the general approximation theorem both in the finite and in the infinite case. Note that our reasoning in the finite case also contains Gräter's general approximation theorem (cf. [Gr]). Assuming finite avoidance the general approximation theorem implies the approximation theorem in the neighbourhood of zero and is implied by the Prüfer condition if the valuations are additionally pairwise independent. The theory of distributive submodules (cf. Chap. II) is heavily used.

Definition 1. i) Let v, w be Manis valuations on R. Let $(\alpha, \beta) \in \Gamma_v \times \Gamma_w$ and $a, b \in R$. We call the tuple (α, β, a, b) *weakly compatible* if $(v \vee w)(a - b) \geq \min\{f_{v,w}(\alpha), f_{w,v}(\beta)\}$.
ii) Let $v_1, \ldots, v_n$ be Manis valuations on R. Let $(\alpha_1, \ldots, \alpha_n, a_1, \ldots, a_n) \in \Gamma_{v_1} \times \ldots \times \Gamma_{v_k} \times R^n$. The tuple $(\alpha_1, \ldots, \alpha_n, a_1, \ldots, a_n)$ is called *weakly compatible* if $(\alpha_i, \alpha_j, a_i, a_j)$ is weakly compatible for every $1 \leq i, j \leq n$.

Remark 6.1. Let $v_1, \ldots, v_n$ be Manis valuations on R.

a) Let $(\alpha_1, \ldots, \alpha_n, a_1, \ldots, a_n) \in \Gamma_{v_1} \times \ldots \times \Gamma_{v_n} \times R^n$. Let $x \in R$ such that $v_1(x - a_1) \geq \alpha_1, \ldots, v_n(x - a_n) \geq \alpha_n$. Then $(\alpha_1, \ldots, \alpha_n, a_1, \ldots, a_n)$ is weakly compatible.
b) For $1 \leq i \leq n$ let $\Delta_i := \bigcap_{j \neq i} H_{v_i, v_j}$. We set $w_i := v_i / \Delta_i$. Let $a_i \in A_{w_i}$ and $\varepsilon \in \Delta_i$. Then the tuple

$$(\varepsilon_1, \ldots, \varepsilon_n, a_1, \ldots, a_n) \in \Gamma_{v_1} \times \ldots \times \Gamma_{v_n} \times R^n$$

is weakly compatible.

Proof. Let $1 \leq i, j \leq n$.

a): We set $v := v_i, w := v_j, \alpha := \alpha_i, \beta := \alpha_j, a := a_i$ and $b := a_j$. We have

$$\begin{aligned}(v \vee w)(a - b) &= (v \vee w)(a - x + x - b) \geq \min\{(v \vee w)(a - x), (v \vee w)(b - x)\} \\ &= \min\{f_{v,w}(v(a - x)), f_{w,v}(w(b - x))\} \geq \min\{f_{v,w}(\alpha), f_{w,v}(\beta)\}.\end{aligned}$$

b): We may assume that $i \neq j$ and that v_i and v_j are dependent. We have $(v_i \vee v_j)(a_i - a_j) \geq (v_i \vee v_j)(a_i)$. Since $\Delta_i \subset H_{v_i, v_j}$ we see $w_i \leq v_i \vee v_j$. Since $a_i \in A_{w_i}$ we get $(v_i \vee v_j)(a_i) \geq 0$. Clearly $f_{v_i, v_j}(\varepsilon_i) = f_{v_j, v_i}(\varepsilon_j) = 0$ and we are done. □

Proposition 6.2. *Let $v_1, \ldots, v_n$ be Manis valuations on R. Then the following are equivalent.*

(1) For every $i \neq j$ v_i and v_j are independent.
(2) Every $(\alpha_1, \ldots, \alpha_n, a_1, \ldots, a_n) \in \Gamma_{v_1} \times \ldots \times \Gamma_{v_n} \times R^n$ is weakly compatible.

Proof. It clearly suffices to do the proof in the case $n = 2$. We set $v := v_1, w := v_2, \alpha := \alpha_1, \beta := \alpha_2, a := a_1$ and $b := b_1$.

(1) $\Rightarrow$ (2): This follows immediately from $\Gamma_{v,w} = \{0\}$ (cf. Definition 1 in Sect. 5).

(2) $\Rightarrow$ (1): Assume that v, w are dependent. We choose $x \in R \setminus A_{v \vee w}$. Then $(0, 0, x, 0) \in \Gamma_v \times \Gamma_w \times R^2$ is not compatible since $(v \vee w)(x) < 0$. □

Definition 2. Let $v_1, \ldots, v_n$ be Manis valuations on R. The *general approximation theorem* holds for $v_1, \ldots, v_n$ if for every weakly compatible $(\alpha_1, \ldots, \alpha_n, a_1, \ldots, a_n) \in \Gamma_{v_1} \times \ldots \times \Gamma_{v_k}{}^{\bullet} \times R^n$ there is some $x \in R$ such that $v_1(x - a_1) \geq \alpha_1, \ldots, v_n(x - a_n) \geq \alpha_n$.

Note that by Remark 1(b) our definition of a general approximation theorem contains the one of Gräter in [Gr].

Remark 6.3. Let $v_1, \ldots, v_n$ be Manis valuations on R and let $w_1, \ldots, w_m$ be trivial Manis valuations on R. The general approximation theorem holds for $v_1, \ldots, v_n$ iff it holds for $v_1, \ldots, v_n, w_1, \ldots, w_m$. □

It is clear by the preceding remark that the notion of general approximation theorem is vacuous in the case of trivial valuations. In the case of non-trivial valuations we get that the general approximation theorem implies the approximation theorem in the neighbourhood of zero.

Theorem 6.4. *Let $v_1, \ldots, v_n$ be non-trivial Manis valuations on R. If the general approximation theorem holds for $v_1, \ldots, v_n$ then also the approximation theorem in the neighbourhood of zero.*

Proof. We can clearly assume that the valuations $v_1, \ldots, v_n$ are pairwise non-isomorphic. Without restriction we write the valuations as $v_1, \ldots, v_k, w_1, \ldots, w_l$ such that the following properties hold.

i) v_i and v_j are incomparable for $i \neq j$.
ii) There is $\varphi : \{1, \ldots, l\} \to \{1, \ldots, k\}$ such that $v_{\varphi(j)} \leq w_j$ for all $1 \leq j \leq l$.

Let $(\alpha_1, \ldots, \alpha_k, \beta_1, \ldots, \beta_l) \in \Gamma_{v_1} \times \ldots \times \Gamma_{v_k} \times \Gamma_{w_1} \times \ldots \times \Gamma_{w_l}$ be compatible. For $1 \leq i \leq k$ let $x_i \in R$ such that $v_i(x_i) = \alpha_i$. For $1 \leq j \leq l$ let $y_j := x_{\varphi(j)}$. Then $w_j(y_j) = \beta_j$ for $1 \leq j \leq l$. Since v_i is non-trivial and $v_i \not\geq v_j$ for all $j \neq i$ we get that $H_{v_i, v_j} \neq \{0\}$ for all $j \neq i$. Since v_i is non-trivial and $v_i \not\geq w_j$ for all $1 \leq j \leq l$ we get that $H_{v_i, w_j} \neq \{0\}$ for all $1 \leq j \leq l$. Since these are convex subgroups of Γ_{v_i} we obtain that

$$\Delta_i := \bigcap_{j \neq i} H_{v_i, v_j} \cap \bigcap_{1 \leq j \leq l} H_{v_i, w_j} \neq \{0\}.$$

Let $\delta_i \in \Delta_i$ with $\delta_i > 0$ and let $\varepsilon_i := \alpha_i + \delta_i$.

Claim: The tuple $(\varepsilon_1, \dots, \varepsilon_k, \beta_1, \dots, \beta_l, x_1, \dots, x_k, y_1, \dots, y_l)$ is weakly compatible.

Proof of the Claim: We distinguish three cases.

Subclaim 1: $(\varepsilon_i, \varepsilon_j, x_i, x_j)$ is weakly compatible for $1 \le i, j \le k$. We have

$$\begin{aligned}(v_i \vee v_j)(x_i - x_j) &\ge \min\{(v_i \vee v_j)(x_i), (v_i \vee v_j)(x_j)\} \\ &= \min\{f_{v_i,v_j}(v_i(x_i)), f_{v_j,v_i}(v_j(x_j))\} \\ &= \min\{f_{v_i,v_j}(\alpha_i), f_{v_j,v_i}(\alpha_j)\} \\ &= \min\{f_{v_i,v_j}(\varepsilon_i), f_{v_j,v_i}(\varepsilon_j)\}. \quad .\end{aligned}$$

Subclaim 2: $(\varepsilon_i, \beta_j, x_i, y_j)$ is weakly compatible for $1 \le i \le k$ and $1 \le j \le l$. We may assume that v_i and w_j are dependent. Then $v_i \vee v_{\varphi(j)} \le v_i \vee w_j$ by Proposition 2.13. Let $g : \Gamma_{v_i \vee v_{\varphi(j)}} \cup \{\infty\} \to \Gamma_{v_i \vee w_j} \cup \{\infty\}$ be the homomorphism of ordered monoids such that $v_i \vee w_j = g \circ (v_i \vee v_{\varphi(j)})$. We have

$$\begin{aligned}(v_i \vee w_j)(x_i - y_j) &\ge \min\{(v_i \vee w_j)(x_i), (v_i \vee w_j)(y_j)\} \\ &= \min\{f_{v_i,w_j}(v_i(x_i)), f_{w_j,v_i}(w_j(y_j))\} \\ &= \min\{g(f_{v_i,v_{\varphi(j)}}(v_i(x_i))), g(f_{v_{\varphi(j)},v_i}(v_{\varphi(j)}(x_{\varphi(j)})))\} \\ &= g\left(\min\{f_{v_i,v_{\varphi(j)}}(v_i(x_i)), f_{v_{\varphi(j)},v_i}(v_{\varphi(j)}(x_{\varphi(j)}))\}\right) \\ &= g\left(\min\{f_{v_i,v_{\varphi(j)}}(\alpha_i), f_{v_{\varphi(j)},v_i}(\alpha_{\varphi(j)})\}\right) \\ &= g\left(\min\{f_{v_i,v_{\varphi(j)}}(\varepsilon_i), f_{v_{\varphi(j)},v_i}(\alpha_{\varphi(j)})\}\right) \\ &= \min\{g(f_{v_i,v_{\varphi(j)}}(\varepsilon_i)), g(f_{v_{\varphi(j)},v_i}(\alpha_{\varphi(j)}))\} \\ &= \min\{f_{v_i,w_j}(\varepsilon_i), f_{w_j,v_i}(\beta_j)\}.\end{aligned}$$

Subclaim 3: $(\beta_i, \beta_j, y_i, y_j)$ is weakly compatible for $1 \le i, j \le k$. We may assume that w_i and w_j are dependent. Then $v_{\varphi(i)} \vee v_{\varphi(j)} \le w_i \vee w_j$ by Proposition 2.13. Let $h : \Gamma_{v_{\varphi(i)} \vee v_{\varphi(j)}} \cup \{\infty\} \to \Gamma_{w_i \vee w_j} \cup \{\infty\}$ be the homomorphism of ordered monoids such that $v_{w_i \vee w_j} = h \circ (v_{\varphi(i)} \vee v_{\varphi(j)})$. We have

$$\begin{aligned}(w_i \vee w_j)(y_i - y_j) &\ge \min\{(w_i \vee w_j)(y_i), (w_i \vee w_j)(y_j)\} \\ &= \min\{f_{w_i,w_j}(w_i(y_i)), f_{w_j,w_i}(w_j(y_j))\} \\ &= \min\{h(f_{v_{\varphi(i)},v_{\varphi(j)}}(v_{\varphi(i)}(x_{\varphi(i)}))), h(f_{v_{\varphi(j)},v_{\varphi(i)}}(v_{\varphi(j)}(x_{\varphi(j)})))\} \\ &= h\left(\min\{f_{v_{\varphi(i)},v_{\varphi(j)}}(v_{\varphi(i)}(x_{\varphi(i)})), f_{v_{\varphi(j)},v_{\varphi(i)}}(v_{\varphi(j)}(x_{\varphi(j)}))\}\right) \\ &= h\left(\min\{f_{v_{\varphi(i)},v_{\varphi(j)}}(\alpha_{\varphi(i)}), f_{v_{\varphi(j)},v_{\varphi(i)}}(\alpha_{\varphi(j)})\}\right)\end{aligned}$$

$$= \min\{h(f_{v_{\varphi(i)},v_{\varphi(j)}}(\alpha_{\varphi(i)})), h(f_{v_{\varphi(j)},v_{\varphi(i)}}(\alpha_{\varphi(j)}))\}$$
$$= \min\{f_{w_i,w_j}(\beta_i), f_{w_j,w_i}(\beta_j)\}.$$

Since the general approximation theorem holds for $v_1, \ldots, v_n$ we find by the claim some $x \in R$ such that $v_i(x - x_i) \geq \varepsilon_i$ for $1 \leq i \leq k$ (and $w_j(x - y_j) \geq \beta_j$ for $1 \leq j \leq l$). Then

$$v_i(x) = \min\{v_i(x_i), v_i(x - x_i)\} = \alpha_i$$

for $1 \leq i \leq k$ and

$$w_j(x) = f_{v_{\varphi(j)},w_j}(v_{\varphi(j)}(x)) = f_{v_{\varphi(j)},w_j}(\alpha_j) = \beta_j$$

for $1 \leq j \leq l$. □

Corollary 6.5. *Let $v_1, \ldots, v_n$ be non-trivial Manis valuations on R. If the general approximation theorem holds for $v_1, \ldots, v_n$ then $v_1, \ldots, v_n$ have the inverse property.*

Proof. By Theorem 4 the approximation theorem in the neighbourhood of zero holds for $v_1, \ldots, v_n$. By Theorem 5.11 $v_1, \ldots, v_n$ have the inverse property. □

Next, we show that the general approximation theorem holds for $v_1, \ldots, v_n$ if $v_1, \ldots, v_n$ are pairwise independent and $A_{v_1} \cap \ldots \cap A_{v_n}$ is Prüfer in R. For this, we need some preparation.

Definition 3. Let A be a ring, M an A-module and $N_1, \ldots, N_k$ submodules of M. We say that the *Chinese Remainder Theorem (CRT)* holds for the submodules $N_1, \ldots, N_k$ of M, if for any elements $x_1, \ldots, x_k$ of M with $x_i \equiv x_j \mod (N_i + N_j)$ and for every $1 \leq i, j \leq k$ there exists some $x \in M$ with $x \equiv x_i \mod N_i$ for all $1 \leq i \leq k$.

Remark 6.6. Conversely, if $x \equiv x_i \mod N_i$ for some $x \in R$ then certainly $x_i \equiv x_j \mod (N_i + N_j)$ for every $1 \leq i, j \leq k$.

Lemma 6.7. *Let M be an A-module. Then CRT holds for any two submodules N_1 and N_2 of M.*

Proof. Let $x_1, x_2 \in M$ with $x_1 - x_2 \in N_1 + N_2$. Then there are $a_1 \in N_1, a_2 \in N_2$ such that $x_1 - x_2 = a_1 + a_2$. Then $x := x_1 - a_1 = x_2 + a_2$ does the job. □

Proposition 6.8. *Let M be an A-module and let $N_1, \ldots, N_k$ be distributive submodules of M (cf. [Vol. I, Definition 1 in II §5]). Then CRT holds for $N_1, \ldots, N_k$.*

Proof. We do induction on k.

$k = 1$: This is obvious.

$k-1 \to k$: Let $x_1, \ldots, x_k \in M$ with $x_i \equiv x_j \mod N_i + N_j$ for all $1 \leq i, j \leq k$. By the inductive hypothesis there is some $y \in M$ such that $y - x_i \in N_i$ for all

$1 \leq i \leq k-1$. It is enough to find some $x \in R$ with $x - y \in \bigcap_{i=1}^{k-1} N_i$ and $x - x_k \in N_k$. By Lemma 7 it suffices to show that $y - x_k \in \left(\bigcap_{i=1}^{k-1} N_i\right) + N_k$. Since N_k is distributive we get by applying [Vol. I, Proposition II.5.2(2)] iterated that

$$\left(\bigcap_{i=1}^{k-1} N_i\right) + N_k = \bigcap_{i=1}^{k-1} (N_i + N_k).$$

But for each $1 \leq i \leq k-1$ we have

$$y - x_k = y - x_i + x_i - x_k \in N_i + N_i + N_k = N_i + N_k.$$

□

Corollary 6.9. *Let A be Prüfer in R. Then CRT holds for finitely many R-regular A-submodules of R.*

Proof. By [Vol. I, Example II.5.1] every R-regular A-submodule of R is a distributive submodule of R. Now we can apply Proposition 8. □

Proposition 6.10. *Let $v_1, \ldots, v_n$ be Manis valuations on R such that $A_{v_1} \cap \ldots \cap A_{v_n}$ is Prüfer in R. Let $a_1, \ldots, a_n \in R$ and $\varepsilon_1 \in \Gamma_{v_1}, \ldots, \varepsilon_n \in \Gamma_{v_n}$. Then the following are equivalent.*

(1) There is some $x \in R$ such that $v_i(x - a_i) \geq \varepsilon_i$ for all $1 \leq i \leq n$.
(2) For each $1 \leq i, j \leq n$ there is some $x \in R$ with $v_i(x - a_i) \geq \varepsilon_i$ and $v_j(a - a_j) \geq \varepsilon_j$.

Proof. Let $A := A_{v_1} \cap \ldots \cap A_{v_n}$.

(1) ⇒ (2): This is trivial.

(2) ⇒ (1): Let $I_k := \{y \in R \mid v_k(y) \geq \varepsilon_k\}$ for $1 \leq k \leq n$. We have to show that there is some $x \in R$ such that $x \equiv a_k \mod I_k$ for $1 \leq k \leq n$. Note that $I_k = R$ if v_k is trivial. We have that I_k is a v_k-convex A_{v_k}-submodule of R properly containing $\operatorname{supp} v_k$. By [Vol. I, Theorem III.2.2] we get that I_k is an R-regular A_{v_k}-submodule and therefore an R-regular A-module for all $1 \leq k \leq n$. Hence CRT holds for $I_1, \ldots, I_n$ by Corollary 9. Hence it is enough to show that $a_k - a_l \in I_k + I_l$ for all $1 \leq k, l \leq n$. By Remark 6 we are done if we find some $x \in R$ with $x - a_k \in I_k$ and $x - a_l \in I_l$, i.e. $v_k(x - a_k) \geq \varepsilon_k$ and $v_l(x - a_l) \geq \varepsilon_l$. But this holds by (2). □

Theorem 6.11. *Let $v_1, \ldots, v_n$ be pairwise independent Manis valuations on R such that $A_{v_1} \cap \cdots \cap A_{v_n}$ is Prüfer in R. Then the general approximation theorem holds for $v_1, \ldots, v_n$.*

Proof. By Definition 1, Proposition 10 and [Vol. I, Corollary I.5.3] it is enough to show the case $n = 2$. So let $(\alpha_1, \alpha_2, a_1, a_2) \in \Gamma_{v_1} \times \Gamma_{v_2} \times R^2$ be weakly compatible. We set $I_k := \{y \in R \mid v_k(y) \geq \alpha_k\}$ for $1 \leq k \leq 2$. Then I_1, I_2 are A-modules. By Lemma 7 we have to show that $a_1 - a_2 \in I_1 + I_2$.

Case 1: $v_k(a_1 - a_2) \geq \alpha_k$ for $k = 1$ or $k = 2$. Then $a_1 - a_2 \in I_k \subset I_1 + I_2$ and we are done.

Case 2: $v_k(a_1 - a_2) < \alpha_k$ for $1 \leq k \leq 2$. Let $\gamma_k := \alpha_k - v_k(a_1 - a_2) > 0$. Since v_1 and v_2 are independent and since $A := A_{v_1} \cap A_{v_2}$ is Prüfer we find by Proposition 5.8 and Corollary 5.12 $x_1, x_2 \in R$ such that $v_1(x_1) = \gamma_1, v_2(x_1) = 0$ and $v_1(x_2) = 0, v_2(x_2) = \gamma_2$. We have $v_1(x_1(a_1 - a_2)) = \alpha_1$. So

$$x_1(a_1 - a_2) \in I_1 \subset (I_1 + I_2) \subset (I_1 + I_2)A_{v_2}.$$

By [Vol. I, Theorem III.2.2] I_2 is an R-regular A_{v_2}-module. Hence $(I_1 + I_2)A_{v_2}$ is an R-regular A_{v_2}-module. Again by [Vol. I, Theorem III.2.2] we see that $(I_1 + I_2)A_{v_2}$ is v_2-convex. Since $v_2(x_1) = 0$ we get $a_1 - a_2 \in (I_1 + I_2)A_{v_2}$. In the same way we obtain $a_1 - a_2 \in (I_1 + I_2)A_{v_1}$. Hence

$$a_1 - a_2 \in \bigcap_{k=1,2} (I_1 + I_2)A_{v_k} = I_1 + I_2$$

where the latter equality holds by [Vol. I, Theorem II.1.4(4)]. □

For the following, recall Definitions 4 and 5 from Sect. 5.

Definition 4. Let $(v_i \mid i \in I)$ be a family of Manis valuations on R and let $i_1, \ldots, i_n \in I$. Then $(\alpha_1, \ldots, \alpha_n, a_1, \ldots, a_n) \in \Gamma_{v_1} \times \ldots \times \Gamma_{v_n} \times R^n$ is called $\{i_1, \ldots, i_n\}$- *complete* if

$$\bigcup_{k=1}^{n} I(\alpha_k) \cup \bigcup_{k=1}^{n} I(v_k(a_k)) = \{i_1, \ldots, i_n\}.$$

Remark 6.12. Let $(v_i \mid i \in I)$ be a family of Manis valuations on R and let $i_1, \ldots, i_n \in I$. Let $(\alpha_1, \ldots, \alpha_n, a_1, \ldots, a_n) \in \Gamma_{v_{i_1}} \times \ldots \times \Gamma_{v_{i_n}} \times R^n$ be $\{i_1, \ldots, i_n\}$-complete. Then $(\alpha_1, \ldots, \alpha_n)$ is $\{i_1, \ldots, i_n\}$-complete.

Proof. This follows immediately from Remark 5.19(a). □

Remark 6.13. Let $(v_i \mid i \in I)$ be a family of Manis valuations on R and let $i_1, \ldots, i_n \in I$. Let $(\alpha_1, \ldots, \alpha_n, a_1, \ldots, a_n) \in \Gamma_{v_{i_1}} \times \ldots \times \Gamma_{v_{i_n}} \times R^n$ be weakly compatible and $\{i_1, \ldots, i_n\}$-complete. Let $j_1, \ldots, j_m \in I \setminus \{i_1, \ldots, i_n\}$. Then

$$(\alpha_1, \ldots, \alpha_n, 0, \ldots, 0, a_1, \ldots, a_n, 0, \ldots, 0) \in$$

$$\Gamma_{v_{i_1}} \times \ldots \times \Gamma_{v_{i_n}} \times \Gamma_{v_{j_1}} \times \ldots \times \Gamma_{v_{j_m}} \times R^{n+m}$$

is weakly compatible and $\{i_1, \ldots, i_n, j_1, \ldots, j_m\}$-complete.

Proof. Let $1 \leq p \leq n$ and $1 \leq q \leq m$.

a) We show that $f_{v_{i_p},v_{j_q}}(\alpha_p) = 0$. If $\alpha_p = 0$ this is clear. If $\alpha_p \neq 0$ then $f_{v_{i_p},v_{j_q}}(\alpha_p) = 0$ by Proposition 5.17 since $j_q \notin I(\alpha_p)$.
b) We show that $(v_{i_p} \vee v_{j_q})(a_p) \geq 0$. We have $(v_{i_p} \vee v_{j_q})(a_p) = f_{v_{i_p},v_{j_q}}(v_{i_p}(a_p))$. If $v_{i_p}(a_p) = 0$ or $v_{i_p}(a_p) = \infty$ nothing is to show. Therefore we assume the $v_{i_p}(a_p) \in \Gamma_{v_i} \setminus \{0\}$. Since $j_q \notin I(v_p(a_p))$ we get $f_{v_{i_p},v_{j_q}}(v_{i_p}(a_p)) = 0$ be Proposition 5.17.

By (a) and (b) we get that

$$(\alpha_1, \ldots, \alpha_n, 0, \ldots, 0, a_1, \ldots, a_n, 0, \ldots, 0) \in$$

$$\Gamma_{v_{i_1}} \times \ldots \times \Gamma_{v_{in}} \times \Gamma_{v_{j1}} \times \ldots \times \Gamma_{v_{jm}} \times R^{n+m}$$

is weakly compatible. It is clear by Definition 4 in Sect. 5 that it is $\{i_1, \ldots, i_n, j_1, \ldots, j_m\}$-complete. □

Definition 5. Let $(v_i \mid i \in I)$ be a family of Manis valuations on R. We say that the *general approximation theorem* holds for the family if for each $i_1, \ldots, i_n \in I$ and each weakly compatible and $\{i_1, \ldots, i_n\}$-complete tuple $(\alpha_1, \ldots, \alpha_n, a_1, \ldots, a_n) \in \Gamma_{v_{i_1}} \times \ldots \times \Gamma_{v_{in}} \times R^n$ there is some $x \in R$ with $v_{i_k}(x - a_i) \geq \alpha_k$ for $1 \leq k \leq n$ and $v_j(x) \geq 0$ for all $j \in I \setminus \{i_1, \ldots, i_n\}$.

Remark 6.14. If I is finite then Definition 5 coincides with Definition 2.

Proof. Let $I = \{1, \ldots, n\}$.

a) We show that the general approximation theorem in the sense of Definition 5 implies the one in the sense of Definition 2. To see this let $(\alpha_1, \ldots, \alpha_n, a_1, \ldots, a_n) \in \Gamma_{v_1} \times \ldots \times \Gamma_{v_n} \times R^n$ be weakly compatible. By Remark 5.19(a) it is $\{1, \ldots, n\}$-complete and we are done.
b) We show that the general approximation theorem in the sense of Definition 2 implies the one in the sense of Definition 5. For this let $i_1 \ldots, i_k \in \{1, \ldots, n\}$. Without restriction we can assume that $i_1 = 1, \ldots, i_k = k$. Let $(\alpha_1, \ldots, \alpha_k, a_1, \ldots, a_k) \in \Gamma_{v_1} \times \ldots \times \Gamma_{v_k} \times R^k$ be weakly compatible and $\{1, \ldots, k\}$-complete. Then

$$(\alpha_1, \ldots, \alpha_k, 0, \ldots, 0, a_1, \ldots, a_k, 0, \ldots, 0) \in$$

$$\Gamma_{v_1} \times \ldots \times \Gamma_{v_n} \times R^n$$

is weakly compatible by Remark 13. Hence there is some $x \in R$ such that $v_i(x - a_i) \geq \alpha_i$ for $1 \leq i \leq k$ and $v_i(x) \geq 0$ for $k + 1 \leq i \leq n$ and we are done. □

We can generalize Theorem 4 to families with finite avoidance.

Theorem 6.15. *Let $(v_i \mid i \in I)$ be a family of non-trivial Manis valuations on R with finite avoidance. If the general approximation theorem holds for $(v_i \mid i \in I)$ then also the approximation theorem in the neighbourhood of zero.*

Proof. We can clearly assume that the valuations $v_i, i \in I$, are pairwise non-isomorphic. We write $I = K \dot{\cup} L$ such that the following holds.

i) The valuations in the family $(v_k \mid k \in K)$ are pairwise incomparable.
ii) There is $\varphi : L \to K$ such that $v_{\varphi(l)} \leq v_l$ for all $l \in L$.

(Note that there are no infinite descending chains by Remark 4.14.) Let $i_1, \ldots, i_n \in K$ and $j_1, \ldots, j_m \in L$. Let

$$(\alpha_1, \ldots, \alpha_n, \beta_1, \ldots, \beta_m) \in \Gamma_{v_{i_1}} \times \ldots \times \Gamma_{v_{i_k}} \times \Gamma_{v_{j_1}} \times \ldots \times \Gamma_{v_{j_l}}$$

be compatible and $\{i_1, \ldots, i_n, j_1, \ldots, j_m\}$-complete. Applying Remark 5.20 we can assume that $\varphi(j_l) \in \{i_1, \ldots, i_n\}$ for all $1 \leq l \leq m$. For $1 \leq k \leq n$ let $x_k \in R$ such that $v_{i_k}(x_k) = \alpha_k$. For $1 \leq l \leq m$ let $y_l := x_{\varphi(j_l)}$. Then $v_{j_l}(y_l) = \beta_l$ for $1 \leq l \leq m$. Clearly the tuple

$$(\alpha_1, \ldots, \alpha_n, \beta_1, \ldots, \beta_m, x_1, \ldots, x_n, y_1, \ldots, y_m) \in$$

$$\Gamma_{v_{i_1}} \times \ldots \times \Gamma_{v_{i_n}} \times \Gamma_{v_{j_1}} \times \ldots \times \Gamma_{v_{j_m}} \times R^{n+m}$$

is $\{i_1, \ldots, i_n, j_1, \ldots, j_m\}$-complete.

Claim : Let $\Delta := \bigcap_{p \in I \setminus \{i_1\}} H_{i_1,p}$. Then $\Delta \neq \{0\}$.

Proof of the Claim: Since $v_{i_1} \neq v_p$ for all $p \neq i_1$ and since $i_1 \in K$ we have $H_{i_1,p} \neq \{0\}$ for all $p \neq i_1$. Assume that $\Delta = \{0\}$. Then we find a sequence $p_1, p_2, \ldots$ in $I \setminus \{i_1\}$ such that

$$\Gamma_{v_{i_1}} \supsetneqq H_{i_1,p_1} \supsetneqq H_{i_1,p_2} \supsetneqq H_{i_1,p_3} \supsetneqq \ldots.$$

So

$$v_{i_1} \vee v_{p_1} > v_{i_1} \vee v_{p_2} > v_{i_1} \vee v_{p_3} > \ldots.$$

Since $v_{i_1} \vee v_{p_1}$ is non-trivial we find some $x \in R$ such that $(v_{i_1} \vee v_{p_1})(x) < 0$. Then $(v_{i_1} \vee v_{p_n})(x) < 0$ for all $n \in \mathbb{N}$. We get $v_{p_n}(x) < 0$ for all $p \in \mathbb{N}$. This contradicts the condition that the family has finite avoidance.

For $1 \leq k \leq n$ we find by the claim $\delta_i \in \bigcap_{p \in I \setminus \{i_k\}} H_{i_k,p}$ with $\delta_i > 0$. Let $\varepsilon_k := \alpha_k + \delta_k$ for $1 \leq k \leq n$. We see as in the proof of Theorem 4 that the tuple $(\varepsilon_1, \ldots, \varepsilon_n, \beta_1, \ldots, \beta_m, x_1, \ldots, x_n, y_1, \ldots, y_m)$ is weakly compatible. By construction it is clearly $\{i_1, \ldots, i_n, j_1, \ldots, j_m\}$-complete.

Since the general approximation theorem holds for $(v_i \mid i \in I)$ we find some $x \in R$ such that $v_{i_k}(x - x_k) \geq \varepsilon_k$ for $1 \leq k \leq n$, $v_{j_l}(x - y_l) \geq \beta_l$ for $1 \leq l \leq m$ and $v_p(x) \geq 0$ for $p \in I \setminus \{i_1, \ldots, i_n, j_1, \ldots, j_m\}$. As in the proof of Theorem 4 we see that $v_{i_k}(x) = \alpha_k$ for $1 \leq k \leq n$ and $v_{j_l}(x) = \beta_l$ for $1 \leq l \leq m$ and are done. □

Theorem 6.16. *Let* $(v_i \mid i \in I)$ *be a family of pairwise independent Manis valuations on* R *having finite avoidance. If* $\bigcap_{i\in I} A_{v_i}$ *is Prüfer in* R *then the general approximation theorem holds for* $(v_i \mid i \in I)$.

Proof. We set $A := \bigcap_{i\in I} A_{v_i}$. Let $i_1,\ldots,i_n \in I$. Let $(\alpha_1,\ldots,\alpha_n,a_1,\ldots,a_n) \in \Gamma_{v_{i_1}} \times \ldots \times \Gamma_{v_{i_n}} \times R^n$. (The tuple is weakly compatible by Proposition 2 and $\{i_1,\ldots,i_n\}$-complete by Remark 5.19(b)) For $1 \le k \le n$ let $x_k \in R$ with $v_{i_k}(x_k) = \alpha_k$ and $y_k \in R$ with $v_{i_k}(y_k) = -\alpha_k$. Since $(v_i \mid i \in I)$ has finite avoidance there is a finite subset J of I containing $i_1,\ldots,i_n$ such that $v_j(x_k) \ge 0, v_j(y_k) \ge 0$ and $v_j(a_k) \ge 0$ for all $1 \le k \le n$ and all $j \in I\setminus J$. Let $B := \bigcap_{j\in I\setminus J} A_{v_j}$. Then $x_k, y_k, a_k \in B$ for all $1 \le k \le n$. We write $J\setminus\{i_1,\ldots,i_n\}$ as $\{i_{n+1},\ldots,i_m\}$ for some $m \ge n$. For $1 \le k \le m$ we set $w_k := v_{i_k}|_B$. By Corollary 4.10 $w_1,\ldots,w_m$ are Manis valuations. By Proposition 4.13 they are pairwise independent. By construction $\alpha_k \in \Gamma_{w_k}$ for $1 \le k \le n$. The tuple

$$(\alpha_1,\ldots,\alpha_n,0,\ldots,0,a_1,\ldots,a_n,0,\ldots,0) \in \Gamma_{w_1} \times \ldots \times \Gamma_{w_m} \times B^m.$$

is weakly compatible and $\{i_1,\ldots,i_m\}$-complete. We have

$$\bigcap_{1\le k\le m} A_{w_k} = \bigcap_{1\le k\le m} A_{v_{i_k}} \cap B = A.$$

Hence $\bigcap_{1\le k\le m} A_{w_k}$ is Prüfer in B by [Vol. I, Corollary I.5.3]. Applying Theorem 11 there is some $x \in B$ such that $w_k(x-a_k) \ge \alpha_k$ for $1 \le k \le n$ and $w_k(x) \ge 0$ for $n+1 \le k \le m$. Therefore $v_{i_k}(x-a_k) \ge \alpha_k$ for $1 \le k \le n$ and $v_j(x) \ge 0$ for $j \in I\setminus\{i_1,\ldots,i_n\}$. □

7 The Reinforced Approximation Theorem

We present the remarkable result of Gräter [Gr$_2$] on the reinforced approximation theorem. It implies the approximation theorem in the neighbourhood of zero. Assuming finite avoidance the reinforced approximation theorem for pairwise non-isomorphic PM-valuations is equivalent to the Prüfer condition. The concept of maximally dominant Manis valuations is introduced and used.

Definition 1. i) Let v, w be Manis valuation on R. Let $(\alpha,\beta,a,b) \in \Gamma_v \times \Gamma_w \times R^2$. We call the tuple (α,β,a,b) *compatible* if it is weakly compatible (cf. Sect. 6, Definition 1) and if the tuple $(\alpha,\beta) \in \Gamma_v \times \Gamma_w$ is compatible (cf. Sect. 5, Definition 2).

ii) Let $v_1,\ldots,v_n$ be Manis valuations on R. Let $(\alpha_1,\ldots,\alpha_n,a_1,\ldots,a_n) \in \Gamma_{v_1} \times \ldots \times \Gamma_{v_k} \times R^n$. The tuple $(\alpha_1,\ldots,\alpha_n,a_1,\ldots,a_n)$ is called *compatible* if $(\alpha_i,\alpha_j,a_i,a_j)$ is compatible for every $1 \le i, j \le n$.

Remark 7.1. Let $v_1, \ldots, v_n$ be Manis valuations on R. Let $(\alpha_1, \ldots, \alpha_n) \in \Gamma_{v_1} \times \ldots \times \Gamma_{v_n}$. The tuple $(\alpha_1, \ldots, \alpha_n)$ is compatible (in the sense of Definition 2 in Sect. 5) iff $(\alpha_1, \ldots, \alpha_n, 0, \ldots, 0) \in \Gamma_{v_1} \times \ldots \times \Gamma_{v_n} \times R^n$ is compatible (in the sense of Definition 1).

Remark 7.2. Let v, w be Manis valuations on R. A tuple $(\alpha, \beta, a, b) \in \Gamma_v \times \Gamma_w \times R^2$ is compatible iff $(\alpha, \beta) \in \Gamma_v \times \Gamma_w$ is compatible and the following holds:

$$v(a-b) < \alpha \Longrightarrow f_{v,w}(\alpha) = (v \vee w)(a-b) \text{ in } \Gamma_{v,w},$$

or

$$w(a-b) < \beta \Longrightarrow f_{w,v}(\beta) = (v \vee w)(a-b) \text{ in } \Gamma_{v,w}.$$

Remark 7.3. Let $v_1, \ldots, v_n$ be Manis valuations on R.

i) Let $x \in R \setminus \bigcup_{1 \le i \le n} \operatorname{supp} v_i$ and let $a_i \in R$ for $1 \le i \le n$. If $v_i(a_i) \ge v_i(x)$ for all $1 \le i \le n$ then $(v_1(x), \ldots, v_n(x), a_1, \ldots, a_n) \in \Gamma_{v_1} \times \ldots \times \Gamma_{v_n} \times R^n$ is compatible.
ii) Let $(\alpha_1, \ldots, \alpha_n) \in \Gamma_{v_1} \times \ldots \times \Gamma_{v_n}$ be compatible and let $x, a_1, \ldots, a_n \in R$ such that $v_i(x - a_i) \ge \alpha_i$ for all $1 \le i \le n$. Then $(\alpha_1, \ldots, \alpha_n, a_1, \ldots, a_n) \in \Gamma_{v_1} \times \ldots \times \Gamma_{v_n} \times R^n$ is compatible.

Proof. i): The tuple $(v_1(x), \ldots, v_n(x)) \in \Gamma_{v_1} \times \ldots \times \Gamma_{v_n}$ is compatible by Remark 5.5(a). We show that $(v_1(x), \ldots, v_n(x), a_1, \ldots, a_n)$ is weakly compatible. Let $1 \le i, j \le n$. We set $v := v_i, w := v_j, \alpha := v(x), \beta := w(x), a := a_i$, and $b := a_j$. Then

$$\begin{aligned}(v \vee w)(a-b)) &= (v \vee w)(a - x + x - b) \ge \min\{(v \vee w)(a-x), (v \vee w)(b-x)\} \\ &= \min\{f_{v,w}(v(a-x)), f_{w,v}(w(b-x))\} \\ &\ge \min\{f_{v,w}(v(x)), f_{v,w}(v(a)), f_{w,v}(w(x)), f_{w,v}(w(b))\} \\ &= f_{v,w}(\alpha)(= f_{w,v}(\beta)).\end{aligned}$$

ii): This follows from Remark 6.1(a). □

Definition 2. Let $v_1, \ldots, v_n$ be Manis valuations on R. The *reinforced approximation theorem* holds for $v_1, \ldots, v_n$ if for every compatible $(\alpha_1, \ldots, \alpha_n, a_1, \ldots, a_n) \in \Gamma_{v_1} \times \ldots \times \Gamma_{v_n} \times R^n$ there is some $x \in R$ such that $v_1(x - a_1) = \alpha_1, \ldots, v_n(x - a_n) = \alpha_n$.

Remark 7.4. (cf. [Al-M, p. 107]) Let $v_1, \ldots, v_n$ be Manis valuations and let

$$(\alpha_1, \ldots, \alpha_n, a_1, \ldots, a_n) \in \Gamma_{v_1} \times \ldots \times \Gamma_{v_n} \times R^n.$$

Assume that there is some $x \in R$ such that $v_i(x - a_i) = \alpha_i$ for all $1 \le i \le n$. Then the above tuple is not necessary compatible.

Proof. Let v_1, v_2 be dependent Manis valuations on R. Let $\alpha_1 \in \Gamma_{v_1}$ and $\alpha_2 \in \Gamma_{v_2}$ with $f_{v_1,v_2}(\alpha_1) < f_{v_2,v_1}(\alpha_2)$. Let $a_1, a_2 \in R$ with $v_1(a_1) = \alpha_1$ and $v_2(a_2) = \alpha_2$. Taking $x = 0$ we have $v_i(x - a_i) = \alpha_i$ for $1 \leq i \leq 2$. But (α_1, α_2) is not compatible. □

Remark 7.5. Let $v_1, \ldots, v_n, w_1, \ldots, w_m$ be Manis valuations on R such that for every $1 \leq j \leq m$ there is some $1 \leq i \leq n$ with $v_i \leq w_j$. If the reinforced approximation theorem holds for $v_1, \ldots, v_n, w_1, \ldots, w_m$ then also for $v_1, \ldots, v_n$.

Proof. For $1 \leq j \leq m$ we choose $1 \leq i_j \leq n$ such that $v_{i_j} \leq w_j$. Let

$$(\alpha_1, \ldots, \alpha_n, a_1, \ldots, a_n) \in \Gamma_{v_1} \times \ldots \times \Gamma_{v_n} \times R^n$$

be compatible. For $1 \leq j \leq m$ let $\beta_j := f_{v_{i_j},w_j}(\alpha_{i_j})$ and $b_j := a_{i_j}$. We see as in the proof of Remark 5.10, (2) ⇒ (1), that the tuple

$$(\alpha_1, \ldots, \alpha_n, \beta_1, \ldots, \beta_m, a_1, \ldots, a_n, b_1, \ldots, b_m) \in$$

$$\Gamma_{v_1} \times \ldots \times \Gamma_{v_n} \times \Gamma_{w_1} \times \ldots \times \Gamma_{w_m} \times R^{n+m}$$

is compatible. By assumption there is some $x \in R$ such that $v_i(x - a_i) = \alpha_i$ for all $1 \leq i \leq n$. □

Remark 7.6. Let $v_1, \ldots, v_n$ be non-trivial Manis valuations on R and let $w_1, \ldots, w_m$ be trivial Manis valuations on R. If the reinforced approximation theorem holds for $v_1, \ldots, v_n, w_1, \ldots, w_m$ then also for $v_1, \ldots, v_n$.

Proof. Let $(\alpha_1, \ldots, \alpha_n, a_1, \ldots, a_n) \in \Gamma_{v_1} \times \ldots \times \Gamma_{v_n} \times R^n$ be compatible. Then

$$(\alpha_1, \ldots, \alpha_n, 0, \ldots, 0, a_1, \ldots, a_n, 0, \ldots, 0) \in$$
$$\Gamma_{v_1} \times \ldots \times \Gamma_{v_n} \times \Gamma_{w_1} \times \ldots \times \Gamma_{w_m} \times R^{n+m}$$

is compatible since v_i, w_j are independent and w_j, w_l are independent for $1 \leq i \leq n$ and $1 \leq j \neq l \leq m$. By assumption there is some $x \in R$ such that $v_i(x - a_i) = \alpha_i$ for all $1 \leq i \leq n$. □

Proposition 7.7. *Let $v_1, \ldots, v_n$ be Manis valuations on R. If the reinforced approximation theorem holds for $v_1, \ldots, v_n$ then also the approximation theorem in the neighbourhood of zero.*

Proof. Let $(\alpha_1, \ldots, \alpha_n) \in \Gamma_{v_1} \times \ldots \times \Gamma_{v_n}$ be compatible. By Remark 1 the tuple $(\alpha_1, \ldots, \alpha_n, 0, \ldots, 0) \in \Gamma_{v_1} \times \ldots \times \Gamma_{v_n} \times R^n$ is compatible. Hence there is some $x \in R$ such that $v_i(x) = \alpha_i$ for all $1 \leq i \leq n$. □

Corollary 7.8. *Let $v_1, \ldots, v_n$ be Manis valuations on R. If the reinforced approximation theorem holds for $v_1, \ldots, v_n$ then $v_1, \ldots, v_n$ have the inverse property.*

Proof. By Proposition 7 the approximation theorem in the neighbourhood of zero holds for $v_1, \ldots, v_n$. By Theorem 5.11 $v_1, \ldots, v_n$ have the inverse property. □

Proposition 7.9. *Let $v_1, \ldots, v_n$ be Manis valuations on R such that $A_{v_1} \cap \cdots \cap A_{v_n}$ is Prüfer in R. Let $(\alpha_1, \ldots, \alpha_n, a_1, \ldots, a_n) \in \Gamma_{v_1} \times \ldots \times \Gamma_{v_n} \times R^n$ be compatible. Then there is some $x \in R$ such that $v_1(x - a_1) \geq \alpha_1, \ldots, v_n(x - a_n) \geq \alpha_n$.*

Proof. We can adapt the proof of Theorem 6.11. By Proposition 6.10 and [Vol. I, Corollary I.5.3] it is enough to show the case $n = 2$. So let $(\alpha_1, \alpha_2, a_1, a_2) \in \Gamma_{v_1} \times \Gamma_{v_2} \times R^2$ be compatible. We set $I_k := \{y \in R \mid v_k(y) \geq \alpha_i\}$ for $1 \leq k \leq 2$. By Lemma 6.7 we have to show that $a_1 - a_2 \in I_1 + I_2$.

Case 1: $v_k(a_1 - a_2) \geq \alpha_k$ for $k = 1$ or $k = 2$. Then $a_1 - a_2 \in I_k \subset I_1 + I_2$ and we are done.

Case 2: $v_k(a_1 - a_2) < \alpha_k$ for $1 \leq k \leq 2$. Then

$$(v_1 \vee v_2)(a_1 - a_2) = f_{v_1,v_2}(\alpha_1) = f_{v_2,v_1}(\alpha_2)$$

by Remark 2. Let $\gamma_k := \alpha_k - v_k(a_1 - a_2) > 0$. Then $f_{v_1,v_2}(\gamma_1) = f_{v_2,v_1}(\gamma_2) = 0$. So $(\gamma_1, 0)$ and $(0, \gamma_2)$ are compatible. Since $A := A_{v_1} \cap A_{v_2}$ is Prüfer we find by Corollary 5.12 $x_1, x_2 \in R$ such that $v_1(x_1) = \gamma_1, v_2(x_1) = 0$ and $v_1(x_2) = 0, v_2(x_2) = \gamma_2$. We have $v_1(x_1(a_1 - a_2)) = \alpha_1$. Now we can proceed as in the proof of Theorem 6.11. □

The main result is that the reinforced approximation theorem holds for pairwise non-isomorphic non-trivial Manis valuations $v_1, \ldots, v_n$ if $A_{v_1} \cap \ldots \cap A_{v_n}$ is Prüfer in R.

Theorem 7.10. *Let $v_1, \ldots, v_n$ be pairwise non-isomorphic non-trivial Manis valuations on R such that $A_{v_1} \cap \cdots \cap A_{v_n}$ is Prüfer in R. Then the reinforced approximation theorem holds for $v_1, \ldots, v_n$.*

Proof. Without restriction we write the valuations as $v_1, \ldots, v_k, w_1, \ldots, w_l$ such that the following properties hold.

i) v_i and v_j are incomparable for $i \neq j$.
ii) There is $\varphi : \{1, \ldots, l\} \to \{1, \ldots, k\}$ such that $v_{\varphi(j)} \leq w_j$ for all $1 \leq j \leq l$.

Let $(\alpha_1, \ldots, \alpha_k, \beta_1, \ldots, \beta_l, a_1, \ldots, a_k, b_1, \ldots, b_l) \in \Gamma_{v_1} \times \ldots \times \Gamma_{v_k} \times \Gamma_{w_1} \times \ldots \times \Gamma_{w_l} \times R^{k+l}$ be compatible. Since $(\alpha_1, \ldots, \alpha_k, \beta_1, \ldots, \beta_l)$ is compatible there is by Corollary 5.12 some $x' \in R$ such that $v_i(x') = \alpha_i$ for $1 \leq i \leq k$ and $w_j(x') = \beta_j$ for $1 \leq j \leq l$. As in the proof of Theorem 6.4 we find for $1 \leq i \leq k$ $\varepsilon_i > \alpha_i$ such that $(\varepsilon_1, \ldots, \varepsilon_k, \beta_1, \ldots, \beta_l, a_1, \ldots, a_k, b_1, \ldots, b_l)$ is compatible. By Proposition 9 there is some $x'' \in R$ such that $v_i(x'' - a_i) \geq \varepsilon_i$ for all $1 \leq i \leq k$ and $w_j(x'' - b_j) \geq \beta_j$ for all $1 \leq j \leq l$. Let $y := x' + x''$. Then $v_i(y - a_i) = \alpha_i$ for $1 \leq i \leq k$ and $w_j(y - b_j) \geq \beta_j$ for $1 \leq j \leq l$. For $1 \leq i \leq k$ we set

$$\Delta_i := \bigcap_{j \neq i} H_{v_i,v_j} \cap \bigcap_{1 \leq j \leq l} H_{v_i,w_j}.$$

Then Δ_i is a convex subgroup of Γ_{v_i} distinct from $\{0\}$. Hence we find some $\delta'_i \in \Delta_i$ such that $\delta'_i > 0$ and

$$\varepsilon'_i := \alpha_i + \delta'_i \notin \{v_i(y - b_1), \ldots, v_i(y - b_l)\}$$

for $1 \leq i \leq k$. Then $(\varepsilon'_1, \ldots, \varepsilon'_k, \beta_1, \ldots, \beta_l)$ is compatible. By Corollary 5.12 we find some $y' \in R$ such that $v_i(y') = \varepsilon'_i$ for $1 \leq i \leq k$ and $w_j(y') = \beta_j$ for $1 \leq j \leq l$. Let $x := y + y'$. Then $v_i(x - a_i) = \alpha_i$ for $1 \leq i \leq n$ and $w_j(x - b_j) \geq \beta_j$ for $1 \leq j \leq l$. We show that $w_j(x - b_j) = \beta_j$ for $1 \leq j \leq l$. To see this we fix $1 \leq j \leq l$. Since $\varepsilon'_{\varphi(j)} \neq v_{\varphi(j)}(y - b_j)$ we have

$$v_{\varphi(j)}(x - b_j) = v_{\varphi(j)}(y - b_j + y') = \min\{v_{\varphi(j)}(y - b_j), \varepsilon'_{\varphi(j)}\}.$$

We distinguish two cases.

Case 1: $v_{\varphi(j)}(x - b_j) = \varepsilon'_{\varphi(j)}$. Then

$$\begin{aligned} w_j(x - b_j) &= f_{v_{\varphi(j)}, w_j}(v_{\varphi(j)}(x - b_j)) = f_{v_{\varphi(j)}, w_j}(\varepsilon'_{\varphi(j)}) \\ &= f_{v_{\varphi(j)}, w_j}(v_{\varphi(j)}(y')) = w_j(y') = \beta_j. \end{aligned}$$

Case 2: $v_{\varphi(j)}(x - b_j) < \varepsilon'_{\varphi(j)}$ (and $v_{\varphi(j)}(x - b_j) = v_{\varphi(j)}(y - b_j)$). Then

$$\begin{aligned} w_j(x - b_j) &= f_{v_{\varphi(j)}, w_j}(v_{\varphi(j)}(x - b_j)) \leq f_{v_{\varphi(j)}, w_j}(\varepsilon'_{\varphi(j)}) \\ &= f_{v_{\varphi(j)}, w_j}(v_{\varphi(j)}(y')) = w_j(y') = \beta_j. \end{aligned}$$

But $w_j(x - b_j) \geq \beta_j$ by above. Hence equality holds. □

Corollary 7.11. *Let $v_1, \ldots, v_n$ be pairwise non-isomorphic Manis valuations on R. If $A_{v_1} \cap \ldots \cap A_{v_n}$ is Prüfer in R then the reinforced approximation theorem holds for $v_1, \ldots, v_n$.*

Proof. We can assume that $v_1, \ldots, v_k$ are trivial and $v_{k+1}, \ldots, v_n$ are non-trivial for some $0 \leq k \leq n$. We do induction on k.

$k = 0$: This is covered by Theorem 10.

$k - 1 \to k$: After switching the order we can assume that $\operatorname{supp} v_1$ is minimal in $\{\operatorname{supp} v_i \mid 1 \leq i \leq k\}$ with respect to inclusion. Let $(\alpha_1, \ldots, \alpha_n, a_1, \ldots, a_n) \in \Gamma_{v_1} \times \ldots \times \Gamma_{v_n} \times R^n$ be compatible. Note that $\alpha_i = 0$ for $1 \leq i \leq k$. The tuple $(\alpha_2, \ldots, \alpha_n, a_2, \ldots, a_n) \in \Gamma_{v_2} \times \ldots \times \Gamma_{v_n} \times R^{n-1}$ is also compatible. By the inductive hypothesis there is some $x' \in R$ with $v_i(x' - a_i) = \alpha_i$ for $2 \leq i \leq n$. If $v_1(x' - a_1) = 0$ we are done. Assume that $v_1(x' - a_1) = \infty$. We have $\operatorname{supp} v_1 \neq \operatorname{supp} v_i$ for all $2 \leq i \leq k$ by assumption. By Theorem 4.18 and Corollary 4.9 $v_1, \ldots, v_n$ have the inverse property. Therefore Claim 2 in the proof of Theorem 4.18 holds. Hence there is some $x'' \in R$ with

$$v_1(x'') = 0, v_2(x'') = \infty, \dots, v_k(x'') = \infty, v_{k+1}(x'') > \alpha_{k+1}, \dots, v_n(x'') > \alpha_n.$$

Let $x := x' + x''$. Then $v_i(x - a_i) = \alpha_i$ for all $1 \leq i \leq n$. □

Remark 7.12. It is in general necessary in Corollary 11 that the valuations are non-isomorphic.

Proof. Let R be the field $\mathbb{Z}/2\mathbb{Z}$. Let v be the trivial valuation on R (with $\operatorname{supp} v = \{0\}$). Then the reinforced approximation theorem does not hold for v, v for the following reason: The tuple $(0, 0, 1, 0) \in \Gamma_v \times \Gamma_v \times R^2$ is clearly compatible but there is no $x \in R$ such that $v(x-1) = 0$ and $v(x) = 0$ since there is no $x \in R$ with $x - 1 = 1$ and $x = 1$. □

The converse of Theorem 10 holds if the valuations are PM. We need the notion of maximal dominance.

Definition 3. A Manis valuation v on R is said to be *maximally dominant* if for all non-trivial Manis valuations w on R with $v \leq w$ the ideal $\mathfrak{p}_w$ is maximal in A_w.

Remark 7.13.

i) A trivial valuation is maximally dominant.
ii) Let v be a non-trivial Manis valuation on R that is maximally dominant. Then $\mathfrak{p}_v$ is a maximal ideal of A_v.

Proposition 7.14. *Let v be a non-trivial Manis valuation on R. Then the following are equivalent:*

(1) A_v is Prüfer in R (i.e. v is a PM-valuation).
(2) v is maximally dominant.
(3) $\mathfrak{p}_v$ is a maximal ideal of A_v and for any maximal ideal $\mathfrak{m}$ of A_v with $\mathfrak{m} \neq \mathfrak{p}_v$ there is no prime ideal $\mathfrak{p}$ of A_v such that $\operatorname{supp} v \subsetneqq \mathfrak{p} \subset \mathfrak{p}_v$ and $\mathfrak{p} \subset \mathfrak{m}$.

Proof. Let $A := A_v$.

(1) ⇒ (2): Let w be a non-trivial Manis valuation on R with $v \leq w$. Then $A_w \supset A_v$ by Proposition 1.8 and hence A_w is Prüfer by [Vol. I, Corollary I.5.3]. Therefore $\mathfrak{p}_w$ is a maximal ideal of A_w by [Vol. I, Corollary III.1.4] and [Vol. I, Proposition I.2.3].

(2) ⇒ (3): By Remark 13(ii) $\mathfrak{p}_v$ is a maximal ideal of A. Let $\mathfrak{m}$ be a maximal ideal of A with $\mathfrak{m} \neq \mathfrak{p}_v$. Suppose that there is some prime ideal $\mathfrak{p}$ of A such that $\operatorname{supp} v \subsetneqq \mathfrak{p} \subset \mathfrak{p}_v$ and $\mathfrak{p} \subset \mathfrak{m}$. By Scholium 1.10 there is a coarsening w of v such that $A_w = A_{[\mathfrak{p}]}$ and $\mathfrak{p}_w = \mathfrak{p}$ (namely $v^{\mathfrak{p}}$, cf. Definition 2 in Sect. 1). By Proposition 1.12 w is non-trivial. By assumption $\mathfrak{p}$ is a maximal ideal of $A_w =: B$. By [Vol. I, Proposition I.1.17] $v|_B$ is Manis.

Claim: $\operatorname{supp} v|_B = \mathfrak{p}$.

Proof of the Claim: We show that $\operatorname{supp} v|_B \subset \mathfrak{p}$. Let $x \in B \setminus \mathfrak{p}$. If $v(x) \leq 0$ then $v|_B(x) = v(x)$ and therefore $x \notin \operatorname{supp} v|_B$. So we assume that $v(x) > 0$. Then $x \in A \setminus \mathfrak{p}$. Since $\operatorname{supp} v \subset \mathfrak{p}$ and v is Manis there is some $y \in R$ with $v(y) = -v(x)$.

Hence $yx \in A$ and therefore $y \in A_{[\mathfrak{p}]} = B$. So $v|_B(x) = v(x) < \infty$. We show that $\mathfrak{p} \subset \operatorname{supp} v|_B$. Let $x \in \mathfrak{p}$. Then $v(x) > 0$. Assume that $x \notin \operatorname{supp} v|_B$. Then there is some $y \in B$ with $v(x) \leq -v(y)$. Since $B = A_{[\mathfrak{p}]}$ there is some $s \in A \setminus \mathfrak{p}$ with $ys \in A$. So $v(ys) \geq 0$ and therefore $v(s) \geq -v(y)$. We get $v(x) \leq v(s)$ and therefore $s \in \mathfrak{p}$ since $\mathfrak{p}$ is v-convex by [Vol. I, Proposition I.1.10], contradiction.

By the claim we obtain that $v|_B$ has maximal support (cf. [Vol. I, Definition 7 in I §1]). Clearly $A = A_{v|_B}$. By above $\mathfrak{p}_v$ and $\mathfrak{m}$ are prime ideals of A that contain $\operatorname{supp} v|_B$. By [Vol. I, Proposition I.1.11 v) $\Rightarrow$ vi)], the two ideals m and $\mathfrak{p}_v$ of A are $v|_B$ convex. Therefore they are comparable. This contradicts the fact that $\mathfrak{p}_v$ and $\mathfrak{m}$ are distinct maximal ideals of A.

(3) $\Rightarrow$ (1): Let $\mathfrak{m}$ be any maximal ideal of A. If $\mathfrak{m} = \mathfrak{p}_v$ then $A_{[\mathfrak{m}]} = A$ and $\mathfrak{m}_{[\mathfrak{m}]} = \mathfrak{m}$ by [Vol. I, Lemma III.1.0], so $(A_{[\mathfrak{m}]}, \mathfrak{m}_{[\mathfrak{m}]})$ is a Manis pair of R. Let now $\mathfrak{m} \neq \mathfrak{p}_v$. We show that $A_{[\mathfrak{m}]} = R$ and are done. Suppose that $A_{[\mathfrak{m}]} \neq R$ and let $x \in R \setminus A_{[\mathfrak{m}]}$. Then $(A : x) \subset \mathfrak{m}$. We have that $(A : x)$ is v-convex. Thus $\mathfrak{p} := \sqrt{(A : x)}$ is a v-convex prime ideal of A with $\mathfrak{p} \subset \mathfrak{m}$. We have $\mathfrak{p} \supsetneqq \operatorname{supp}(v)$ since v is Manis and $\mathfrak{p} \subset \mathfrak{p}_v$ since $(A : x) \subset \mathfrak{p}_v$ because of $x \notin A = A_{[\mathfrak{p}_v]}$. This contradicts (3). □

Note that (1) $\Rightarrow$ (3) in the previous proof would also follow from Theorem 4.16 with $A := A_v$ and Scholium 1.10 combined with Proposition 1.12.

Corollary 7.15. *Let v be a non-trivial discrete Manis valuation on R and $A := A_v$. Then A is Prüfer in R if and only if $\mathfrak{p}_v$ is a maximal ideal of A.*

Proof. Since v is discrete the set of all non-trivial Manis valuations coarser than v contains only v. We get the claim by Proposition 14. □

Theorem 7.16. *Let $v_1, \dots, v_n$ be PM-valuations on R. If the reinforced approximation theorem holds for $v_1, \dots, v_n$ then $A_{v_1} \cap \dots \cap A_{v_n}$ is Prüfer in R.*

Proof. By Proposition 1.8 and Remark 5 we can assume that v_i, v_j are incomparable for $i \neq j$. By Remark 6 we can assume that v_i is non-trivial for all $1 \leq i \leq n$. Let $A := A_{v_1} \cap \dots \cap A_{v_n}$. The approximation theorem in the neighbourhood of zero holds for $v_1, \dots, v_n$ by Proposition 7. By Theorems 5.11 and 4.18 we get that v_i is A-essential for every $1 \leq i \leq n$. Therefore we show condition (ii) of Theorem 4.17 to get the claim. Let $\mathfrak{m}$ be a maximal ideal of A with $\mathfrak{m} \neq \operatorname{cent}_A(v_i)$ for $1 \leq i \leq n$. Let w be a non-trivial Manis valuation with $v_i \leq w$ for some $1 \leq i \leq n$. We have to show that $\operatorname{cent}_A(w) \not\subset \mathfrak{m}$. Without restriction we assume that $i = 1$ and set $v := v_i$. Assume that $\operatorname{cent}_A(w) \subset \mathfrak{m}$. Let $B_0 := A[\mathfrak{p}_w] = A + \mathfrak{p}_w$ and $\mathfrak{q}_0 := \mathfrak{m} + \mathfrak{p}_w$.

Claim 1: $\mathfrak{q}_0$ is a prime ideal of B_0 with $\mathfrak{q}_0 \cap A = \mathfrak{m}$.

Proof of Claim 1: It is clear that $\mathfrak{q}_0$ is an ideal of B_0 with $\mathfrak{q}_0 \cap A = \mathfrak{m}$. We show that it is prime. Let $x = a + p, y = b + q \in B_0$ where $a, b \in A$ and $p, q \in \mathfrak{p}_w$ such that $xy \in \mathfrak{q}_0$. Let $p' := aq + bp + pq \in \mathfrak{p}_w$. Then $xy = ab + p'$. Since $xy \in \mathfrak{q}_0$ there is some $m \in \mathfrak{m}$ and some $q' \in \mathfrak{p}_w$ such that $ab + p' = m + q'$. Hence $ab - m \in \operatorname{cent}_A(w) \subset \mathfrak{m}$ and therefore $ab \in \mathfrak{m}$. Since $\mathfrak{m}$ is prime in A we get $a \in \mathfrak{m}$ or $b \in \mathfrak{m}$. This gives that $x \in \mathfrak{q}_0$ or $y \in \mathfrak{q}_0$.

By [LM, Theorem 10.6] there is a Manis pair $(B, \mathfrak{q})$ of A_w with $B_0 \subset B$ and $\mathfrak{q} \cap B_0 = \mathfrak{q}_0$. Hence $\mathfrak{q} \supset \mathfrak{p}_w$ and $\mathfrak{q} \cap A = \mathfrak{m}$.

Claim 2: $(B, \mathfrak{q})$ is a Manis pair of R.

Proof of Claim 2: By [Vol. I, Theorem I.2.4] we have to find for $x \in R \setminus B$ some $y \in \mathfrak{q}$ such that $xy \in B \setminus \mathfrak{q}$. If $x \in A_w$ we are done since $(B, \mathfrak{q})$ is a Manis pair in A_w. So we assume that $x \in R \setminus A_w$. Then there is some $p \in \mathfrak{p}_w$ with $xp \in A_w \setminus \mathfrak{p}_w$. By assumption v is PM. Hence it is maximally dominant by Proposition 14, (1) $\Rightarrow$ (2). So $\mathfrak{p}_w$ is a maximal ideal of A_w. Hence $xpA_w + \mathfrak{p}_w = A_w$ and we find $a \in A_w$ and $p' \in \mathfrak{p}_w$ with $xpa + p' = 1$. Since $\mathfrak{p}_w \subset \mathfrak{q}$ we get that $xpa \notin \mathfrak{q}$. Let $q := pa \in \mathfrak{p}_w$. Then $xq \in A_w \setminus \mathfrak{q}$. If $xq \in B$ we take $y := q$. If $xq \in A_w \setminus B$ there is some $q' \in \mathfrak{q}$ such that $xqq' \in B \setminus \mathfrak{q}$. Then we take $y := qq'$.

Let u be the Manis valuation on R corresponding to $(B, \mathfrak{q})$. Since w and u are non-trivial we get $u \leq w$ by Proposition 1.8.

Claim 3: $v_i \not\leq u$ for $1 \leq i \leq n$.

Proof of Claim 3: Assume that there is some $1 \leq i \leq n$ such that $v_i \leq u$. Then $\mathfrak{p}_u \subset \mathfrak{p}_{v_i}$. We have $\mathfrak{p}_u = \mathfrak{q}$ and obtain $\mathrm{cent}_A(u) = \mathfrak{q} \cap A = \mathfrak{m}$. Hence $\mathfrak{m} \subset \mathrm{cent}_A(v_i)$. This contradicts the fact that $\mathfrak{m}$ is a maximal ideal of A distinct from $\mathrm{cent}_A(v_i)$.

We may assume that there is some $1 \leq l \leq n$ such that

$$\{v_1, \ldots, v_l\} = \{v_i \mid v_i \leq w\}.$$

In the proof of Claim 2 we have seen that $\mathfrak{p}_w$ is a maximal ideal of A_w. Hence $K := A_w/\mathfrak{p}_w$ is a field. For $1 \leq i \leq l$ we define

$$\overline{v_i} : K \to \Gamma_{v_i} \cup \{\infty\}, \overline{v_i}(x + \mathfrak{p}_w) = \begin{cases} v_i(x) & x \in A_w \setminus \mathfrak{p}_w, \\ & \text{if} \\ \infty & x \in \mathfrak{p}_w. \end{cases}$$

and we set

$$\overline{u} : K \to \Gamma_u \cup \{\infty\}, \overline{u}(x + \mathfrak{p}_w) = \begin{cases} u(x) & x \in A_w \setminus \mathfrak{p}_w, \\ & \text{if} \\ \infty & x \in \mathfrak{p}_w. \end{cases}$$

Since $\mathfrak{p}_w$ is v_i-convex for every $1 \leq i \leq l$ and u-convex these are well defined valuations on k. We have $\mathfrak{p}_{\overline{v_i}} = \mathrm{cent}_{A_w}(v_i)/\mathfrak{p}_w = \mathfrak{p}_{v_i}/\mathfrak{p}_w$ for $1 \leq i \leq l$ and $\mathfrak{p}_{\overline{u}} = \mathrm{cent}_{A_w}(u)/\mathfrak{p}_w = \mathfrak{p}_u/\mathfrak{p}_w$. As in Claim 3 we see that $\overline{v_i} \not\leq \overline{u}$ for $1 \leq i \leq l$.

Claim 4: There is some $y \in K$ with $\overline{u}(y) < 0$ and $\overline{v_i}(y) = 0$ for all $1 \leq i \leq l$.

Proof of Claim 4: Since $\overline{v_i} \not\leq \overline{u}$ for all $1 \leq i \leq l$ we have that $\bigcap_{1 \leq i \leq l} H_{\overline{u}, \overline{v_i}} \neq \{0\}$. Let $\alpha \in \bigcap_{1 \leq i \leq l} H_{\overline{u}, \overline{v_i}}$ with $\alpha < 0$. Then $(\alpha, 0, \ldots, 0) \in \Gamma_{\overline{u}} \times \Gamma_{\overline{v_1}} \times \ldots \times \Gamma_{\overline{v_l}}$ is compatible. By Remark 3.1(d) valuations on a field have the inverse property.

Hence we find by Theorem 5.11 some $y \in K$ with $\overline{u}(y) = \alpha < 0$ and $\overline{v_i}(y) = 0$ for $1 \leq i \leq l$.

From Claim 4 we obtain some $x \in A_w$ with $u(x) < 0$ and $v_i(x) = 0$ for $1 \leq i \leq l$. We fix this x.

Claim 5: Let $J := \{(i, j) \in \{1, \ldots, l\} \times \{l+1, \ldots, n\} \mid v_i, v_j \text{ dependent}\}$. Then there is some $z \in \mathfrak{p}_w$ with $z \in A_{v_i \vee v_j} \setminus \mathfrak{p}_{v_i \vee v_j}$ for all $(i, j) \in J$.

Proof of Claim 5: Assume that the assertion does not hold. Then

$$\mathrm{cent}_A(w) \subset \bigcup_{(i,j)\in J} \mathrm{cent}_A(v_i \vee v_j).$$

So there is a pair $(i, j) \in J$ with $\mathrm{cent}_A(w) \subset \mathrm{cent}_A(v_i \vee v_j)$. Then $A_{[\mathrm{cent}_A(w)]} \supset A_{[\mathrm{cent}_A(v_i \vee v_j)]}$. Since v_i is A-essential we get that w is also A-essential by Proposition 4.5. By the same argument we see that $v_i \vee v_j$ is A-essential. So $A_w \supset A_{v_i \vee v_j}$. We have seen at the beginning of the proof that the approximation theorem in the neighbourhood of zero holds for $v_1, \ldots, v_n$. Hence by Theorem 5.11 and Proposition 3.21 w, v_i, v_j have the inverse property. By Proposition 3.22 we obtain that $v_i \vee v_j \leq w$. Therefore $v_j \leq w$, contradiction.

By Claim 5 we find some $z \in \mathfrak{p}_w$ with $z \in A_{v_i \vee v_j} \setminus \mathfrak{p}_{v_i \vee v_j}$ for all $(i, j) \in J$. Let $\alpha_i := v_i(z)$ for $1 \leq i \leq l$. Then $\alpha_i > 0$ since $z \in \mathfrak{p}_w \subset \mathfrak{p}_{v_i}$ for $1 \leq i \leq l$. Since $z \in A_{v_i \vee v_j} \setminus \mathfrak{p}_{v_i \vee v_j}$ for all $(i, j) \in J$ the tuple

$$(\alpha_1, \ldots, \alpha_l, 0, \ldots, 0, x, \ldots, x, 0, \ldots, 0) \in \Gamma_{v_1} \times \ldots \times \Gamma_{v_n} \times R^n$$

is compatible. Since the reinforced approximation theorem holds for $v_1, \ldots, v_n$ there is some $a \in R$ with $v_i(a - x) = \alpha_i$ for $1 \leq i \leq l$ and $v_j(a) = 0$ for $l + 1 \leq j \leq n$. For $1 \leq i \leq l$ we get $v_i(a) = v_i(a - x + x) = 0$ since $v_i(x) = 0$ and $\alpha_i > 0$. So $a \in A$. We show that $u(a - x) > 0$. We have $v \leq w$. Therefore

$$w(a - x) = f_{v,w}(v(a - x)) = f_{v,w}(\alpha_1) = f_{v,w}(v(z)) = w(z) > 0$$

since $z \in \mathfrak{p}_w$. Since $u \leq w$ we get $u(a - x) > 0$. So $u(a) = u(x + a - x) < 0$. Hence we have found some $a \in A$ with $u(a) < 0$. But $A \subset A_u$ by construction, contradiction. □

Corollary 7.17. *Let $v_1, \ldots, v_n$ be pairwise non-isomorphic non-trivial PM-valuations on R satisfying the reinforced approximation theorem. Let $w_1, \ldots, w_m$ be pairwise non-isomorphic trivial Manis valuations on R. Then the reinforced approximation theorem holds for $v_1, \ldots, v_n, w_1, \ldots, w_m$.*

Proof. By Theorem 16 $\bigcap_{1\leq i\leq n} A_{v_i}$ is Prüfer in R. Since $\bigcap_{1\leq i\leq n} A_{v_i} \cap \bigcap_{1\leq j\leq m} A_{w_j} = \bigcap_{1\leq i\leq n} A_{v_i}$ we get the claim by Corollary 11. □

Corollary 7.18. *Let $v_1, \ldots, v_n$ be pairwise independent PM-valuations on R. If the reinforced approximation theorem holds for $v_1, \ldots, v_n$ then also the general approximation theorem.*

Proof. By Theorem 16 $A_{v_1} \cap \ldots \cap A_{v_n}$ is Prüfer in R. By Theorem 6.11 the general approximation theorem holds for $v_1, \ldots, v_n$. □

Definition 4. Let $(v_i \mid i \in I)$ be a family of Manis valuations on R. We say that the *reinforced approximation theorem* holds for the family if for each $i_1, \ldots, i_n \in I$ and each compatible and $\{i_1, \ldots, i_n\}$-complete tuple $(\alpha_1, \ldots, \alpha_n, a_1, \ldots, a_n) \in \Gamma_{v_{i_1}} \times \ldots \times \Gamma_{v_{i_n}} \times R^n$ there is some $x \in R$ with $v_{i_k}(x - a_k) = \alpha_k$ for $1 \leq k \leq n$ and $v_j(x) \geq 0$ for all $j \in I \setminus \{i_1, \ldots, i_n\}$.

Remark 7.19. Let $(v_i \mid i \in I)$ be a family of Manis valuations on R and let $i_1, \ldots, i_n \in I$. Let $(\alpha_1, \ldots, \alpha_n, a_1, \ldots, a_n) \in \Gamma_{v_{i_1}} \times \ldots \times \Gamma_{v_{i_n}} \times R^n$ be compatible and $\{i_1, \ldots, i_n\}$-complete. Let $j_1, \ldots, j_m \in I \setminus \{i_1, \ldots, i_n\}$. Then

$$(\alpha_1, \ldots, \alpha_n, 0, \ldots, 0, a_1, \ldots, a_n, 0, \ldots, 0) \in$$

$$\Gamma_{v_{i_1}} \times \ldots \times \Gamma_{v_{in}} \times \Gamma_{v_{j_1}} \times \ldots \times \Gamma_{v_{jm}} \times R^{n+m}$$

is compatible and $\{i_1, \ldots, i_n, j_1, \ldots, j_m\}$-complete.

Proof. Let $1 \leq k \leq n$ and $1 \leq l \leq m$. If $\alpha_k = 0$ then clearly $f_{v_{i_k}, v_{j_l}}(\alpha_k) = 0$. If $\alpha_k \neq 0$ then also $f_{v_{i_k}, v_{j_l}}(\alpha_l) = 0$ by Proposition 5.17 since $j_q \notin I(\alpha_i)$. By Proposition 5.17 we get also $(v_{i_k} \vee v_{j_l})(a_k) = 0$ in $\Gamma_{v_{i_k}, v_{j_l}}$ since $j_q \notin I(v_{i_k}(a_k))$. So the above tuple is compatible. It is clear by Definition 4 in Sect. 5 that the tuple is $\{i_1, \ldots, i_n, j_1, \ldots, j_m\}$-complete. □

Remark 7.20. If I is finite then Definition 4 coincides with Definition 2.

Proof. Let $I = \{1, \ldots, n\}$.

a) We show that the reinforced approximation theorem in the sense of Definition 4 implies the one in the sense of Definition 2. To see this let $(\alpha_1, \ldots, \alpha_n, a_1, \ldots, a_n) \in \Gamma_{v_1} \times \ldots \times \Gamma_{v_n} \times R^n$ be compatible. By Remark 5.19(a) it is $\{1, \ldots, n\}$-complete and we are done.
b) We show that the reinforced approximation theorem in the sense of Definition 2 implies the approximation theorem in the sense of Definition 4. For this let $i_1 \ldots, i_k \in \{1, \ldots, n\}$. Without restriction we can assume that $i_1 = 1, \ldots, i_k = k$. Let $(\alpha_1, \ldots, \alpha_k, a_1, \ldots, a_k) \in \Gamma_{v_1} \times \ldots \times \Gamma_{v_k} \times R^k$ be compatible and $\{1, \ldots, k\}$-complete. Then

$$(\alpha_1, \ldots, \alpha_k, 0, \ldots, 0, a_1, \ldots, a_k, 0, \ldots, 0) \in \Gamma_{v_1} \times \ldots \times \Gamma_{v_n} \times R^n$$

is compatible by Remark 19. Hence there is some $x \in R$ such that $v_i(x - a_i) = \alpha_i$ for $1 \leq i \leq k$ and $v_i(x) = 0$ for $k + 1 \leq i \leq n$ and we are done. □

Proposition 7.21. *Let $(v_i \mid i \in I)$ be a family of Manis valuations on R. If the reinforced approximation holds then the approximation theorem in the neighbourhood of zero.*

Proof. Let $i_1, \ldots, i_n \in I$. Let $(\alpha_1, \ldots, \alpha_n) \in \Gamma_{v_{i_1}} \times \ldots \times \Gamma_{v_{i_n}}$ be compatible and $\{i_1, \ldots, i_n\}$-complete. Then

$$(\alpha_1, \ldots, \alpha_n, 0, \ldots, 0) \in \Gamma_{v_{i_1}} \times \ldots \times \Gamma_{v_{i_n}} \times R^n$$

is compatible and $\{i_1, \ldots, i_n\}$-complete. Since the reinforced approximation theorem holds there is some $x \in R$ such that $v_{i_k}(x) = \alpha_k$ for $1 \leq k \leq n$ and $v_j(x) \geq 0$ for $j \in I \setminus \{i_1, \ldots, i_n\}$. □

We extend the above relationship between the reinforced approximation theorem and Prüfer rings to families having finite avoidance.

Theorem 7.22. *Let $(v_i \mid i \in I)$ be a family of pairwise non-isomorphic Manis valuations on R having finite avoidance. If $\bigcap_{i \in I} A_{v_i}$ is Prüfer in R then the reinforced approximation theorem holds for the family.*

Proof. We set $A := \bigcap_{i \in I} A_{v_i}$. Let $i_1, \ldots, i_n \in I$ and let $(\alpha_1, \ldots, \alpha_n, a_1, \ldots, a_n) \in \Gamma_{v_{i_1}} \times \ldots \times \Gamma_{v_{i_n}} \times R^n$ be compatible and $\{i_1, \ldots, i_n\}$-complete. Since A is Prüfer also $\bigcap_{k=1}^{n} A_{v_{i_k}}$ is Prüfer in R by [Vol. I, Corollary I.5.3]. By Corollary 5.12 the approximation theorem in the neighbourhood of zero holds for $v_{i_1}, \ldots, v_{i_n}$. Hence there are $y, y' \in R$ such that $v_{i_k}(y) = \alpha_i$ and $v_{i_k}(y') = -\alpha_i$ for $1 \leq k \leq n$. By Corollary 11 the reinforced approximation theorem holds for $v_{i_1}, \ldots, v_{i_n}$. Hence there is some $z \in R$ such that $v_{i_k}(z - a_i) = \alpha_i$ for $1 \leq k \leq n$. Since the family has finite avoidance there is a finite subset J of I containing $i_1, \ldots, i_n$ such that $v_i(y) \geq 0, v_i(y') \geq 0, v_i(z) \geq 0$ and $v_i(a_k) \geq 0$ for all $i \in I \setminus J$ and all $1 \leq k \leq n$. Let $B := \bigcap_{i \in I \setminus J} A_{v_i}$. Then $y, y', z, a_1, \ldots, a_n \in B$. We write $J \setminus \{i_1, \ldots, i_n\}$ as $\{i_{n+1}, \ldots, i_m\}$ for some $m \geq n$. For $1 \leq k \leq m$ we set $w_k := v_{i_k}|_B$.

Claim 1: The valuations $w_1, \ldots, w_k$ are pairwise non-isomorphic and Manis.

Proof of Claim 1: By Corollary 4.10 the valuations are Manis. Assume that there are $r \neq s$ such that $w_r \cong w_s$. Then $\mathfrak{p}_{w_r} = \mathfrak{p}_{w_s}$. Since $\mathfrak{p}_{w_r} = \mathfrak{p}_{v_{i_r}} \cap B$ and $A \subset B$ we obtain $\mathrm{cent}_A(v_{i_r}) = \mathrm{cent}_A(w_r)$. Similarly $\mathrm{cent}_A(v_{i_s}) = \mathrm{cent}_A(w_s)$. Hence $\mathrm{cent}_A(v_{i_r}) = \mathrm{cent}_A(v_{i_s})$. By Corollary 1.17 and Proposition 1.2 we get that v_{i_r} and v_{i_s} are isomorphic, contradiction.

Clearly $\bigcap_{1 \leq k \leq m} A_{w_k} = A$. By [Vol. I, Corollary I.5.3] A is Prüfer in B. Hence the reinforced approximation theorem holds for $w_1, \ldots, w_m$ by Corollary 11 and Claim 1. Since $y, y' \in B$ we have $\alpha_i \in \Gamma_{w_k}$ for $1 \leq k \leq n$. We set $\alpha_k := 0 \in \Gamma_{w_k}$ and $a_k := 0 \in B$ for $n + 1 \leq k \leq m$.

Claim 2: The tuple $(\alpha_1, \ldots, \alpha_m, a_1, \ldots, a_m) \in \Gamma_{w_1} \times \ldots \times \Gamma_{w_m} \times B^m$ is compatible.

Proof of Claim 2: Let $1 \leq k < l \leq m$. We show that $(\alpha_k, \alpha_l, a_k, a_l)$ is compatible. We distinguish three cases.

Case 1: $1 \le k, l \le n$. Then $w_k(y) = \alpha_k$ and $w_l(y) = \alpha_l$, so (α_k, α_l) is compatible. We have $w_k(z - a_k) = \alpha_k$ and $w_l(z - a_l) = \alpha_l$, so $(\alpha_k, \alpha_l, a_k, a_l)$ is compatible by Remark 3(ii).

Case 2: $1 \le k \le n$ and $n + 1 \le l \le m$. Since $i_l \notin I(\alpha_k)$ we get $f_{v_{i_k}, v_{i_l}}(\alpha_k) = 0$ in $\Gamma_{v_{i_k}, v_{i_l}}$ by Proposition 5.17. Hence $f_{w_k, w_l}(\alpha_i) = 0$ in Γ_{w_k, w_l} by Proposition 4.13. So (α_k, α_l) is compatible. Since $i_l \notin I(v_{i_k}(a_i))$ we get $f_{v_{i_k}, v_{i_l}}(v_{i_k}(a_k)) = 0$ in $\Gamma_{v_{i_k} \vee v_{i_l}}$ by Proposition 5.17. By Proposition 4.13 we obtain $f_{w_k, w_l}(w_k(a_k)) = 0$ in Γ_{w_k, w_l}. Since $f_{w_k, w_l}(\alpha_k) = 0$ in Γ_{w_k, w_l} as just seen we get the claim.

Case 3: $n + 1 \le k < l \le m$. This is obvious.

Since the reinforced approximation theorem holds for $w_1, \ldots, w_m$ there is some $x \in B$ such that $w_k(x - a_k) = \alpha_k$ for $1 \le k \le m$. Then $v_{i_k}(x - a_k) = \alpha_k$ for $1 \le k \le n$. It remains to show that $v_i(x) \ge 0$ for all $i \in I \setminus \{i_1, \ldots, i_n\}$. If $i = i_k$ for some $n + 1 \le k \le m$ then $v_i(x) = w_k(x) = 0$. If $i \notin J$ then $v_i(x) \ge 0$ since $x \in B$. □

Theorem 7.23. *Let $(v_i \mid i \in I)$ be a family of PM-valuations on R with finite avoidance. If the reinforced approximation theorem holds for the family then $\bigcap_{i \in I} A_{v_i}$ is Prüfer in R.*

Proof. We can clearly assume that v_i is non-trivial for all $i \in I$ (cf. Remark 6). Let $A := \bigcap_{i \in I} A_{v_i}$. The approximation theorem in the neighbourhood of zero holds for $(v_i \mid i \in I)$ by Proposition 21. By Theorems 5.22 and 4.21 every v_i is A-essential. Therefore we show condition (ii) of Theorem 4.17 to get the claim. Let $\mathfrak{m}$ be a maximal ideal of A with $\mathfrak{m} \ne \mathrm{cent}_A(v_i)$ for all $i \in I$. Let w be a non-trivial Manis valuation with $v_{i_1} \le w$ for some $i_1 \in I$. We have to show that $\mathrm{cent}_A(w) \not\subset \mathfrak{m}$. We set $v := v_{i_1}$. By assumption v is PM. Hence it is maximally dominant by Proposition 14, (1) $\Rightarrow$ (2). So $\mathfrak{p}_w$ is a maximal ideal of A_w.

Assume that $\mathrm{cent}_A(w) \subset \mathfrak{m}$. Let $B_0 := A[\mathfrak{p}_w] = A + \mathfrak{p}_w$ and $\mathfrak{q}_0 := \mathfrak{m} + \mathfrak{p}_w$. As in the proof of Theorem 16 (Claim 1 and Claim 2) we find a Manis pair $(B, \mathfrak{q})$ of R such that $B_0 \subset B$ and $\mathfrak{q} \cap B_0 = \mathfrak{q}_0$. Then $\mathfrak{q} \supset \mathfrak{p}_w$ and $\mathfrak{q} \cap A = \mathfrak{m}$. Let u be the Manis valuation on R corresponding to $(B, \mathfrak{q})$. Since w and u are non-trivial we get $u \le w$ by Proposition 1.8. Again as in the proof of Theorem 16 (Claim 3) we have $v_i \not\le u$ for all $i \in I$. By Remark 4.14 the set $J := \{i \in I \mid v_i \le w\}$ is finite. Let $J = \{i_1, \ldots, i_n\}$. By above $\mathfrak{p}_w$ is a maximal ideal of A_w. Hence $K := A_w/\mathfrak{p}_w$ is a field. For $1 \le k \le n$ we define

$$\overline{v_{i_k}} : K \to \Gamma_{v_i} \cup \{\infty\}, \overline{v_{i_k}}(x + \mathfrak{p}_w) = \begin{cases} v_{i_k}(x) & \text{if } x \in A_w \setminus \mathfrak{p}_w, \\ \infty & \text{if } x \in \mathfrak{p}_w. \end{cases}$$

and we set

$$\overline{u} : K \to \Gamma_u \cup \{\infty\}, \overline{u}(x + \mathfrak{p}_w) = \begin{cases} u(x) & \text{if } x \in A_w \setminus \mathfrak{p}_w, \\ \infty & \text{if } x \in \mathfrak{p}_w. \end{cases}$$

Since $\mathfrak{p}_w$ is v_{i_k}-convex for every $1 \le k \le n$ and u-convex these are well defined valuations on K. We have $\mathfrak{p}_{\overline{v_{i_k}}} = \mathrm{cent}_{A_w}(v_{i_k})/\mathfrak{p}_w = \mathfrak{p}_{v_{i_k}}/\mathfrak{p}_w$ for $1 \le k \le n$ and $\mathfrak{p}_{\overline{u}} = \mathrm{cent}_{A_w}(u)/\mathfrak{p}_w = \mathfrak{p}_u/\mathfrak{p}_w$. As above we get that $\overline{v_{i_k}} \not\le \overline{u}$ for $1 \le k \le n$. As in the proof of Theorem 16 (Claim 4) we find some $y \in K$ with $\overline{u}(y) < 0$ and $\overline{v_{i_k}}(y) = 0$ for all $1 \le k \le n$. Hence we find some $x \in A_w$ with $u(x) < 0$ and $v_{i_k}(x) = 0$ for $1 \le k \le n$. We fix this x.

Claim A: There is some $z \in \mathfrak{p}_w \setminus \operatorname{supp} w$ such that $J = I(v(z))$.

Proof of Claim A: Note that given $z \in \mathfrak{p}_w \setminus \operatorname{supp} w$ we have $J \subset I(v(z))$. To see this let $z \in \mathfrak{p}_w \setminus \operatorname{supp} w$. Then $0 < w(z) < \infty$. We have $v \vee v_{i_k} \le w$ for all $1 \le k \le n$. Hence $0 < (v \vee v_{i_k})(z) < \infty$ for all $1 \le k \le n$. Therefore $f_{v,v_{i_k}}(v(z)) \ne 0$ for all $1 \le k \le n$. By Remark 5.17 we obtain $J \subset I(v(z))$.

By the above we have to find some $z \in \mathfrak{p}_w \setminus \operatorname{supp} w$ such that $I(v(z)) \subset J$. Let H be the convex subgroup of Γ_v such that $w = v/H$ (cf. Remarks 1.13(b)). Assume that there is no convex subgroup $\tilde{H}$ of Γ_v with $H \subsetneqq \tilde{H} \subsetneqq \Gamma_v$. Let then $z \in \mathfrak{p}_w \setminus \operatorname{supp} w$ be arbitrary. Since $0 < w(z) < \infty$ we have $v(z) \notin H$. By the assumption we get $H_{v(z)} = H$ (cf. Definition 4 in Sect. 5). Therefore $v/H_{v(z)} = v/H = w$. By Definition 4 in Sect. 5 we obtain

$$I(v(z)) = \{i \in I \mid v_i \le v/H_{v(z)}\} = \{i \in I \mid v_i \le w\} = J$$

and are done. So we assume that there is a convex subgroup $\tilde{H}$ of Γ_v with $H \subsetneqq \tilde{H} \subsetneqq \Gamma_v$. Let $\tilde{w} := v/\tilde{H}$. Then $\tilde{w}$ is non-trivial and $w \le \tilde{w}$. By Remark 4.14 the set

$$\tilde{J} := \{i \in I \mid v_i \le \tilde{w}\}$$

is finite and contains clearly J. Let $\tilde{J} \setminus J := \{i_{n+1}, \ldots, i_m\}$. Arguing similarly to above we obtain some $z \in A_{\tilde{w}}$ such that $w(z) > 0$ and $v_{i_{n+1}}(z) = \ldots = v_{i_m}(z) = 0$. Since $v_{i_k} \le \tilde{w}$ for all $n+1 \le k \le m$ we obtain $z \notin \mathfrak{p}_{\tilde{w}}$. So $z \in \mathfrak{p}_w \setminus \mathfrak{p}_{\tilde{w}}$. Note that necessarily $\tilde{w}(z) = 0$ since $\tilde{w}(z) \le 0$ and $w \le \tilde{w}$. Since $\operatorname{supp} w = \operatorname{supp} \tilde{w}$ we have $z \in \mathfrak{p}_w \setminus \operatorname{supp} w$. Since $w(z) > 0$ and $\tilde{w}(z) = 0$ we have $H \subset H_{v(z)} \subsetneqq \tilde{H}$. So $w \le v/H_{v(z)} < \tilde{w}$ and $v/H_{v(z)}(z) > 0$. Let $i \in I(v(z))$. Then $v_i \le v/H_{v(z)}$. We get $v_i(z) > 0$. Moreover, $v_i \le \tilde{w}$. So $i \in \tilde{J}$. But $v_{i_{n+1}}(z) = \ldots = v_{i_m}(z) = 0$. So $i \in J$ and Claim A is proven

We choose z as in Claim A. Let $\alpha_k := v_{i_k}(z)$ for $1 \le k \le n$. Since $z \in \mathfrak{p}_w \setminus \operatorname{supp} w$ we have $0 < w(z) < \infty$. Since $v_{i_k} \le w$ for all $1 \le k \le n$ we get $0 < \alpha_k < \infty$ for all $1 \le k \le n$.

Claim B: The tuple $(\alpha_1, \ldots, \alpha_n, x, \ldots, x) \in \Gamma_{v_{i_1}} \times \ldots \times \Gamma_{v_{i_n}} \times R^n$ is compatible and J-complete.

Proof of Claim B. The tuple is clearly compatible. By Claim A $I(\alpha_1) = J$. Since $v_{i_k}(x) = 0$ for all $1 \le k \le n$ it remains to show that $I(\alpha_k) \subset J$ for all $2 \le k \le n$. Fix $2 \le k \le n$. Let $j \in I(\alpha_k)$. Then $v_j \le v_{i_k}/H_{\alpha_{i_k}}$. By Claim A $i_k \in I(\alpha_1)$. Therefore $v_{i_k} \le v_{i_1}/H_{\alpha_1}$. We conclude that $v_j \le v_{i_1}/H_{\alpha_1}$ and so $j \in I(\alpha_1) = J$.

Applying the reinforced approximation theorem to the tuple of Claim B we can finish the proof as the proof of Theorem 16. □

Corollary 7.24. *Let* $(v_i \mid i \in I)$ *be a family of pairwise independent PM-valuations on* R *with finite avoidance. If the reinforced approximation theorem holds for* $(v_i \mid i \in I)$ *then also the general approximation theorem.*

Proof. By Theorem 23 $\bigcap_{i \in I} A_{v_i}$ is Prüfer in R. By Theorem 6.16 the general approximation theorem holds for $(v_i \mid i \in I)$. □

Chapter 3
Kronecker Extensions and Star Operations

Summary. The all over idea of the present chapter is to associate to any ring extension $A \subset R$ a commuting square of ring extensions

$$(\dagger) \qquad \begin{array}{ccc} R & \xrightarrow{j} & T \\ \uparrow & & \uparrow \\ A & \longrightarrow & B \end{array}$$

such that B is Prüfer in T and there exists a process $v \mapsto v^*$ which associates with v in a suitable family $\mathfrak{M}$ of valuations of R over A a special valuation v of T over B such that $v^* \circ j = v$. ("Over A" means that $A_v \supset A$.)

Since $B \subset T$ is Prüfer, the set $\mathfrak{M}^* = \{v^* \mid v \in \mathfrak{M}\}$ will consist of PM-valuations and thus will be amenable to the methods of previous chapters. Then one may hope to obtain results about A-modules in R from well known facts about B-modules in T, in particular exploiting facts from the multiplicative ideal theory of Prüfer extensions in Chap. II.

The extensions $B \subset T$ occurring in the square ($\dagger$) will be the "Kronecker extensions’" quoted in the title of the chapter. Actually they will be *Bezout extensions*, a special class of Prüfer extensions which has been exhibited in [Vol. I, Chap. II §10].

The definition of Kronecker extensions relies on the notion of content $c_A(f)$ of a polynomial $f \in R[X]$ over A in Sect. 1 and on the all important Dedekind–Mertens formula (Theorem 1.1 below), which generalizes the Gaußian content formula over principal ideal domains. Given a multiplicative filter $\mathcal{G}$ of ideals of R, we define in Sect. 2 the ring $R(X, \mathcal{G})$ as the localization $S_{\mathcal{G}}^{-1} R[X]$ with respect to the multiplicative set $S_{\mathcal{G}} \subset R[X]$ consisting of all $f \in R[X]$ with $c_R(f) \in \mathcal{G}$, and then introduce the Kronecker subrings B of $R(X, \mathcal{G})$ following a splendid idea of Halter-Koch (there for R a domain, cf. [Ha-K], and [Ha-K$_1$, Exercises 20.4–20.9]).

M. Knebusch and T. Kaiser, *Manis Valuations and Prüfer Extensions II*,
Lecture Notes in Mathematics 2103, DOI 10.1007/978-3-319-03212-2_3,

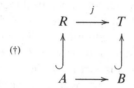

$R(X,\mathcal{G})$ contains a unique minimal Kronecker subring $R(X,\mathcal{G})_{\text{kr}}$. It is Prüfer in $R(X,\mathcal{G})$, even Bezout.

Given a subring A of R we obtain the commuting square (†) from above by taking $T := R(X,\mathcal{G})$, the natural homomorphism $j : R \to R(X,\mathcal{G})$, and $B := AR(X,\mathcal{G})_{\text{kr}}$, the subring of T generated by A and $R(X,\mathcal{G})_{\text{kr}}$. It has all the desired properties listed above.

For every valuation v on R with $\operatorname{supp} v \notin \mathcal{G}$ there exists a unique valuation v^* on $R(X,\mathcal{G})$ with $v^*(X) = 0$ and $v^* \circ j = v$, called the *Gauß extension of* v. At the end of Sect. 2 we identify the valuations v on R over A with $A_{v^*} \supset AR(X,\mathcal{G})_{\text{kr}} = B$ as "*$\mathcal{G}$-special valuations*" over A, and then have an isomorphism $v \mapsto v^*$ of the poset consisting of these valuations v to the restricted PM-spectrum of $S(R(X,\mathcal{G}/B)$.

In Sect. 3 we define star operations $I \mapsto I^*$ on the set $J(A, R)$ of A-submodules of R. Every such operation $*$ gives us a Kronecker subring $A(X,\mathcal{G},*)$ of $R(X,\mathcal{G})$, defined by use of the contents $c_A(f)$ of polynomials $f \in R[X]$. We then prove that these subrings are all the Kronecker subrings of $R(X,\mathcal{G})$ containing A. Indeed, if B is a Kronecker subring of $R(X,\mathcal{G})$, we have a canonical star operation $*$ at hands with $A(X,\mathcal{G},*) = B$, namely $I^* := j^{-1}(IB)$ (cf. Theorem 3.11).

In Sects. 4–7 the basics of a multiplicative ideal theory related to star operations are displayed. In the classical theory, where A is a domain and R is its field of fractions, star operations are defined on the set $\operatorname{Fract}(A, R)$ of "fractional ideals" of A. These are the A-submodules I of R with $I \neq \{0\}$ and $I \subset d^{-1}A$ for some $d \in A \setminus \{0\}$. Our main problem is that for a ring extension $A \subset R$ the ideals Ad with $d \in A \setminus \{0\}$ in general are not invertible. We define R-fractional ideals of A only in Sect. 4, Definition 5 as those $I \in J(A, R)$ for which there exists an R-invertible ideal $\mathfrak{a}$ of A with $\mathfrak{a} \subset I \subset \mathfrak{a}^{-1}$, while already in Sect. 3 we introduce star operations on the whole semiring $J(A, R)$ of A-submodules of R. On the subset $J_*(A, R)$ consisting of all $I \in J(A, R)$ with $I = I^*$ (called "star-A-modules" in R) we introduce a *star product* $I \circ J := (IJ)^*$ and similarly a star sum which turns $J_*(A, R)$ into a semiring. We then have the group $\tilde{D}_*(A, R)$ of units of $J_*(A, R)$ at hands. Aiming at a "star-divisor theory", as known from the classical setting, this group seems to be too big. But

$$D_*(A, R) := \tilde{D}_*(A, R) \cap \operatorname{Fract}(A, R)$$

and/or a slightly smaller group $D_*^f(A, R)$ (cf. Sect. 4, Definition 6) seem to be right counterparts of the classical star divisor groups. For the expert in classical divisor theory it should now be plausible, that a meaningful star-divisor theory can be established in arbitrary ring extensions. But the book stops here.

We further develop and discuss counterparts to various topics in the classical theory of star operations, in particular the cancellation property "e.a.b" (= endlich arithmetisch brauchbar, cf. [Gi, §32]) in Sect. 5 and star operations of finite type in Sect. 6.

In Sect. 7 we define a Kronecker subring $A(X,\mathcal{G},*)$ of $R(X,\mathcal{G})$ also for a *partial star operation* $*$. This is a star operation, suitably defined, on the subset $\Phi(\mathcal{H}/A)$ of $J(A, R)$ consisting of all $I \in J(A, R)$ with $IR \in \mathcal{H}$ for $\mathcal{H}$ a multiplicative filter of ideals R with $\mathscr{H} \supset \mathscr{G}$. We then have a diagram (†) as above with $B := A(X, \mathscr{G}, *)$. We dwell on the following intriguing point: The diagram (†) gives us a star operation α on the whole $J(A, R)$, namely the canonical star associated to $B \subset R(X, \mathscr{G})$. We call α the *Kronecker companion* of the partial star $*$. We analyse α in terms of $*$ (cf. Theorem 7.6) and state some of the properties of α.

Given a ring extension $A \subset R$ and a multiplicative filter $\mathscr{H}$ of ideals of R, we define in Sect. 9 *"semistar operations"*[1] and *"weak star operations"* on $\Phi(\mathscr{H}/A)$, in the following called "semistars" and "weak stars", which both generalize star operations on $\Phi(\mathscr{H}/A)$, weak stars being more general than semistars. If $\mathscr{H} \supset \mathscr{G}$ then a weak star $*$ on $\Phi(\mathscr{H}/A)$ is still good enough to give us a Kronecker subring $A(X, \mathscr{G}, *)$ of $R(X, \mathscr{G})$, and thus provides us with a Kronecker companion α which is a star (!) on $J(A, R)$. Both semistars and weak stars form a much more flexible variety than stars. For example, the composite $\alpha \circ \beta$ of two semistars α, β on $\Phi(\mathscr{H}/A)$ is again a semistar, ditto for weak stars.

Given a problem on A-modules in R it may be a piece of art to design a star operation fitting to the problem. Here semistars and weak stars can be helpful. In Sect. 10 we expound a construction which, starting from a family $(\alpha_\lambda \mid \lambda \in \Lambda)$ of weak stars on $\Phi(\mathscr{H}/A)$, produces a star γ on $\Phi(\mathscr{H}/A)$, which is minimal with the property $I^\gamma \supset I^{\alpha_\lambda}$ for all $I \in \Phi(\mathscr{H}/A)$ and $\lambda \in \Lambda$.

Notations. In this chapter a lot of multiplicative ideal theory—well beyond [Vol. I, Chap. II]—will come into play. We will need the following notations. Let $A \subset R$ be a ring extension (as always, the rings are commutative with 1). We set

$J(A, R) :=$ the set of all A-submodules of R.
$J^f(A, R) :=$ the set of finitely generated A-submodules of R.
$J(A) := J(A, A) =$ the set of ideals of A. We use the notation $I \lhd A$ for I being an ideal of A.
$J^f(A) := J^f(A, A) =$ the set of finitely generated ideals of A.
$\Phi(R/A) := \{I \in J(A, R) | IR = R\}$, the set of R-regular A-submodules of R.
$\Phi^f(R/A) := \{I \in J^f(A, R) | IR = R\} = \Phi(R/A) \cap J^f(A, R)$.
$\mathscr{F}(R/A) := \Phi(R/A) \cap J(A) = \{I \lhd A \mid IR = R\}$, the set of R-regular ideals of A.
$\mathscr{F}^f(R/A) := \Phi(R/A) \cap J^f(A)$, the set of finitely generated R-regular ideals of A.

$J(A, R)$ is a commutative monoid (= commutative semigroup with 1) under the multiplication $(I, J) \mapsto IJ$, where IJ is the set of finite sums $\sum_{i=1}^n a_i b_i, a_i \in I, b_i \in J$. All the sets introduced so far are submonoids of $J(A, R)$. The notations $J(A)$ and $\mathscr{F}(R/A)$ occurred already in [Vol. I, Chap. II §3] as well as the following ones [Vol. I, Chap. II §6 & §4].

[1] "Semistar operation" here has another meaning than in the classical literature, cf. Comments, p. 139.

$D(A, R) :=$ set of R-invertible A-submodules of $R = \{I \in J(A, R) \mid \exists J \in J(A, R)$ with $IJ = A\}$.
$\mathrm{Inv}(A, R) = \{I \triangleleft A \mid I \in D(A, R)\}$.

Recall from [Vol. I, Chap. II] that the submonoid $D(A, R)$ of $J(A, R)$ is a group contained in $\Phi^f(R/A)$, and that $\mathrm{Inv}(A, R)$ is a submonoid of $\mathscr{F}(R/A)$. Recall also from [Vol. I, Chap. II] that in the case that A is Prüfer in R we have $\Phi^f(A, R) = D(A, R)$.

1 Kronecker Subrings of $R(X)$, Their Use for Special Valuations

We embark for a technical definition of Kronecker extensions in a special case. After that we will work on the idea described in the Summary (Diagram †) in the case that there $\mathfrak{M}$ is the set of special valuations of R over A. In later sections we will advance to more general sets $\mathfrak{M}$ of valuations and more general Kronecker extensions.

We start with the notion of "content" of a polynomial in one variable X. Let $A \subset R$ be any ring extension.

Definition 1 ([Gi, §28]). The *content over* A (or: A-content) of a polynomial $f = a_0 + a_1X + \ldots + a_dX^d \in R[X]$ is the A-submodule of R generated by the coefficients of f. We denote it by $c_A(f)$. Thus

$$c_A(f) = \sum_{j=0}^{d} Aa_j.$$

In the same way we can define the content of a polynomial in several variables [Gi, §28], but we will not need this here.

Notice that every $I \in J^f(A, R)$ occurs as the content of a polynomial $f(X) \in R[X]$ over A, of course not uniquely determined by I.

While the notion of content goes back to Gauß, it was Kronecker, who had the insight, how to gain a good theory of "divisors" ($\approx$ a multiplicative ideal theory in our language) in rings of numbers, by representing an ideal by a polynomial (in several variables) via the content and then working with polynomials and rational functions [Kr]. We refer to Chap. II of Hermann Weyl's book [W] for an accessible account of Kronecker's theory.

The following well known theorem will be crucial for our working with contents.

Theorem 1.1 (Dedekind–Mertens [Me]). *Let $f, g \in R[X]$ and $n \geq deg(f)$. Then*

$$(*) \qquad c_A(fg)c_A(g)^n = c_A(f)c_A(g)^{n+1}.$$

We call this equation the *"Dedekind–Mertens formula"*. It generalizes the Gauß formula

$$(**) \qquad c_A(fg) = c_A(f)c_A(g),$$

which is valid for A a principal ideal domain and R its field of quotients. For a proof of $(*)$ we refer to [Gi, §28].

Definition 2. We call a polynomial $f \in R[X]$ *R-unimodular*, if $c_R(f) = R$. We denote the set of these polynomials by S_R.

Due to the Dedekind–Mertens formula (with $A = R$), S_R is a multiplicative subset of $R[X]$. It is saturated in $R[X]$, since clearly $c_R(fg) \subset c_R(f)$ for $f, g \in R[X]$. In other words, for any $f, g \in R[X]$ we have $fg \in S_R$ iff $f \in S_R$ and $g \in S_R$.

Definition 3. Let $R(X)$ denote the localization of the ring $R[X]$ with respect to S_R,

$$R(X) := S_R^{-1}R[X].$$

As is well known, a polynomial $\sum_0^d a_j X^j \in R[X]$ is a zero divisor of the ring $R[X]$ iff there exists some $b \in R$ with $b \neq 0$, $ba_j = 0$ for $j = 0, \ldots, d$ (e.g. [N, p. 17]). Thus S_R contains only non-zero divisors of $R[X]$, hence $R[X] \subset R(X) \subset \mathrm{Quot}(R[X])$.

Remark 1.2. If A is Prüfer in R, the Gauß formula $(**)$ holds for any $f \in R[X]$ and $g \in S_R$. This follows from the Dedekind–Mertens formula $(*)$ since the A-module $c_A(g)$, being R-regular and finitely generated, now is R-invertible (cf. [Vol. I, Theorem II.1.13]). Of course, the Gauß formula holds as well for $f \in S_R$, $g \in R[X]$.

We now define "Kronecker subrings" of $R(X)$. The definition is an adaption to the present setting of the definition of a "Kronecker function ring" by Franz Halter-Koch [Ha-K, Definition 2.1] (Halter-Koch deals with the case that R is a field).

Definition 4. Let R be any ring.

a) We call a subring B of $R(X)$ *Kronecker in* $R(X)$ (or a Kronecker subring of $R(X)$), if the following two axioms hold.

   (1) $X \in B$.
   (2) For every R-unimodular $f = a_0 + a_1X + \cdots + a_dX^d \in R[X]$ the elements a_j/f of $R(X)(j = 0, \ldots, d)$ are contained in B.

b) If A is a subring of R, we call B a Kronecker subring of $R(X)$ *over* A if in addition $A \subset B$.

Comment. In the literature a lot can be found about "Kronecker function rings" in the presence of so called "star operations". We refer to [Gi, §40] and [Huc, §20–22].

Halter-Koch [Ha-K] had the insight to define Kronecker function rings in an axiomatic way without using star operations. We will follow this track even more radically than he does, introducing star operations only much later in Sect. 3 of this chapter.

Remarks 1.3.

i) It is clear from Definition 1.4 that $R(X)$ contains a unique smallest Kronecker subring $R(X)_{\mathrm{kr}}$. It is generated over the prime ring $\mathbb{Z} \cdot 1_R$ by X and all elements a_i/f with $f = a_0 + a_1X + \cdots + a_dX^d \in S_R$ and $i = 0, \ldots, d$. Given a subring A of R the Kronecker subrings of $R(X)$ over A are precisely the overrings of the ring $A \cdot R(X)_{\mathrm{kr}}$ in R.
ii) Let again A be a subring of R. Let $\sum_A$ denote the multiplicative subset of $A[X]$ consisting of the monic polynomials (i.e. highest coefficient = 1) in $A[X]$. Every Kronecker subring B of $R(X)$ over A contains the localization $A\langle X\rangle := \sum_A^{-1} A[X]$. In particular, every $f \in \sum_A$ is a unit of B and $A[X, X^{-1}] \subset A\langle X\rangle \subset B$.

Up to the end of this section we fix a ring extension $A \subset R$. It is evident that for any overring B of $A[X]$ in R and any $f \in R[X]$ we have $fB \subset c_A(f)B$. It turns out that for B Kronecker in R we have equality here for every R-unimodular f, and that this characterizes Kronecker subrings over A.

Proposition 1.4. *Let B be a subring of $R(X)$ containing $A[X]$. The following are equivalent.*

(1) *B is Kronecker in $R(X)$ over A.*
(2) *$c_A(f)B = fB$ for every $f \in S_R$.*

Proof. (1) ⇒ (2): We have

$$fB \subset \sum_0^d a_iX^iB \subset \sum_0^d a_iB = c_A(f)B = \sum_0^d \frac{a_i}{f} fB \subset fB,$$

the second inclusion being valid since $X \in B$ and the last one since $a_i/f \in B$. Thus $fB = c_A(f)B$.

(2)⇒ (1): Let $f = \sum_0^d a_iX^i \in S_R$ be given. Then $a_iB \subset c_A(f)B = fB$. Since f is a unit of $R(X)$ we conclude that $(a_i/f)B \subset B$, hence $a_i/f \in B$ for every $i \in \{0, \ldots, d\}$. □

Theorem 1.5. *If B is a Kronecker subring in $R(X)$ then B is Bezout in $R(X)$.*

Proof. Due to characterization of Bezout extensions in [Vol. I, Theorem II.10.2] we are done if we verify for a given $z \in R(X)$ that

$$(*) \qquad B + Bz = By$$

with some $y \in R(X)$. We apply Proposition 4 with A the prime ring $\mathbb{Z} \cdot 1_R$. Let $z = f/g$ with $f \in R[X], g \in S_R$. Choose some $n > \deg f$. Then $h := f + X^n g$ is R-unimodular since g is R-unimodular. We have

$$hB = (f + X^n g)B \subset fB + gB \subset c_A(f)B + c_A(g)B = c_A(h)B = hB.$$

Thus $hB = fB + gB$, and $(*)$ holds with $y = h/g$. □

We now have established a square (†) as indicated in the beginning of the section, taking $T = R(X)$ and $B = A \cdot R(X)_{\mathrm{kr}}$, with inclusion maps as horizontal arrows.

We go on to study valuations. Let $v : R \to \Gamma \cup \infty$ be a valuation over A. We define for any $I \in J^f(A, R)$ a value $v(I) \in \Gamma \cup \infty$ as follows. We choose a set of generators $a_1, \dots, a_r$ of the A-module I, $I = Aa_1 + \cdots + Aa_r$, and put

$$v(I) := \min(v(a_1), \dots, v(a_r)). \tag{1}$$

Clearly

$$v(I) = \inf\{v(x) | x \in I\}. \tag{2}$$

In particular $v(I)$ does not depend on the choice of generators of I. It is also evident that

$$v(I + J) = \min(v(I), v(J)), \quad v(IJ) = v(I) + v(J). \tag{3}$$

Starting from v we now define a valuation $v' : R[X] \to \Gamma \cup \infty$ on $R[X]$ by

$$v'(f) := v(c_A(f)). \tag{4}$$

We have to verify that v' is indeed a valuation. For $a \in R$ we have $v'(a) = v(a)$. In particular $v(0) = \infty$ and $v(1) = 0$. If $f, g \in R[X]$ are given, we have $c_A(f + g) \subset c_A(f) + c_A(g)$. We conclude that

$$\begin{aligned} v'(f + g) &= v(c_A(f + g)) \geq v(c_A(f) + c_A(g)) \\ &= \min\{v(c_A(f)), v(c_A(g))\} = \min\{v'(f), v'(g)\}. \end{aligned}$$

From the Dedekind–Mertens formula and (3) we obtain that for some $n \in \mathbb{N}$

$$v(c_A(fg)) + nv(c_A(g)) = v(c_A(f)) + (n + 1)v(c_A(g)),$$

i.e.

$$v'(fg) + nv'(g) = v'(f) + (n + 1)v'(g).$$

If $v'(g) \neq \infty$ it follows that $v'(fg) = v'(f) + v'(g)$. We always have $c_A(fg) \subset c_A(f)c_A(g)$, hence $v'(fg) \geq v'(f) + v'(g)$. Thus $v'(fg) = v'(f) + v'(g)$ also holds if $v'(g) = \infty$. We have proved that v' is a valuation on $R[X]$ and $v'|R = v$. □

Remarks.

a) Our definition of v' does not change if we replace A by another subring of R. We could have taken $A = \mathbb{Z} \cdot 1_R$.
b) It is possible to define v' directly by the formula

$$v'(\sum_0^d a_j \cdot X^j) = \min_{0 \leq j \leq d} v(a_j)$$

and to verify directly that this is a valuation on R, cf. [Bo, Chap. VI, Lemma 10.1] (Bourbaki assumes that R is a field, but this assumption is not used there). We mentioned this already in [Vol. I, Chap. III §3 (p. 197)].

If f is R-unimodular, then $c_A(f)R = R$, hence $v'(f) = v(c_A(f)) \neq \infty$. Thus v' extends to a valuation $v^* : R(X) \to \Gamma \cup \infty$ in a unique way by the formula

$$v^*(\frac{f}{g}) = v'(f) - v'(g) \tag{5}$$

for $f \in R[X], g \in S_R$.

Definition 5. We call v' the *Gauß extension of v to $R[X]$* and v^* the *Gauß extension of v to $R(X)$*.

Notice that $A_{v'} = A_v[X]$, $\mathfrak{p}_{v'} = \mathfrak{p}_v[X]$, supp v' = (supp v)$[X]$, $\Gamma_{v^*} = \Gamma_{v'} = \Gamma_v$. If v is special, then clearly v' is special, and this implies that v^* is special (cf. [Vol. I, Chap. I §1 p. 13]).

If a_j is a coefficient of a polynomial $f \in S_R$ then $a_jA \subset c_A(f)$ hence $v(a_j) \geq v'(f)$, hence $v^*(a_j/f) \geq 0$. Thus A_{v^*} contains the smallest Kronecker subring $B := A \cdot R(X)_{\text{kr}}$ of $R(X)$ over A. Since $R(X)$ is Bezout over B, it follows for v special that v^* is a BM (= Bezout–Manis) valuation of $R(X)$ over A. This had already been proved in [Vol. I, Chap. III] in a somewhat different way ([Vol. I, Proposition III.3.17]).[2]

Moreover the following holds.

Theorem 1.6. *The special valuations v of R over A correspond bijectively with the BM-valuations w of $R(X)$ over $A \cdot R(X)_{\text{kr}}$ via $w = v^*, v = w|R$.*

---

[2] In [Vol. I, Chap. III §3] we only aimed at giving some interesting examples of BM-valuations, and thus have been more brief than now.

We refrain from proving this now. Later a theorem will be proved (Theorem 2.9) that contains Theorem 6 as a special case.

2 $\mathscr{G}$-Special Valuations and the Ring $R(X,\mathscr{G})$

As before R may be any ring. We start out to define families $\mathfrak{M}$ of valuations on R, for which the idea indicated at the beginning of Sect. 1 will work. These will be the sets of "$\mathscr{G}$-special valuations over A" for a "multiplicative filter" $\mathscr{G}$ on R. We begin with three easy basic definitions.

Definition 1. A *multiplicative filter* $\mathscr{G}$ (of ideals) *on* R is a non empty subset $\mathscr{G}$ of $J(R)$ (cf. notations above) with the following properties.

(1) If $I \in \mathscr{G}$, $J \in J(R)$ and $I \subset J$, then $J \in \mathscr{G}$.
(2) If $I \in \mathscr{G}$, $J \in \mathscr{G}$, then $IJ \in \mathscr{G}$ (hence also $I \cap J \in \mathscr{G}$).

Clearly $\{R\}$ is the smallest and $J(R)$ is the biggest such filter.

Definition 2. We say that a multiplicative filter $\mathscr{G}$ on R is of *finite type*, if for every $I \in \mathscr{G}$ there exists a finitely generated ideal $I_0 \in \mathscr{G}$ with $I_0 \subset I$.

The following is obvious.

Remark 2.1. Any multiplicative filter $\mathscr{G}$ on R contains a unique biggest multiplicative filter of finite type, namely the set $\mathscr{G}_f$ consisting of all ideals I of R which contain some finitely generated ideal I_0 with $I_0 \in \mathscr{G}$. In other terms, $\mathscr{G}_f$ is the filter in $J(R)$ generated by $\mathscr{G} \cap J^f(R)$.

Often it will be only a notational convenience, that we work with filters in $J(R)$ instead of $J^f(R)$, and that we allow multiplicative filters which are not of finite type.

In the following a multiplicative filter $\mathscr{G}$ on R will be fixed for the whole section.

Definition 3. Given a subring A of R we call an A-submodule I of R *$\mathscr{G}$-regular* if $IR \in \mathscr{G}$. We denote the set of all these modules I by $\Phi(\mathscr{G}/A)$. We also introduce the set

$$\Phi^f(\mathscr{G}/A) = \Phi(\mathscr{G}/A) \cap J^f(A,R) = \{I \in J^f(A,R) \mid IR \in \mathscr{G}\}.$$

$\Phi(\mathscr{G}/A)$ is a multiplicative filter on $J(A,R)$ in the obvious sense. In the case $\mathscr{G} = \{R\}$ we are back to the notion of an R-regular A-module from [Vol. I, Chap. II]. Then, in the notation from the beginning of the chapter, $\Phi(\mathscr{G}/A) = \Phi(R/A), \Phi^f(\mathscr{G}/A) = \Phi^f(R/A)$.

Definition 4. A valuation $v : R \to \Gamma \cup \infty$ is called *$\mathscr{G}$-regular* if $\operatorname{supp} v \notin \mathscr{G}_f$.

Notice that every valuation v on R is $\{R\}$-regular, but none is $J(R)$-regular.

Lemma 2.2. *Given a $\mathcal{G}$-regular valuation v on R and a subring A of R with $A \subset A_v$, every $I \in \Phi^f(\mathcal{G}/A)$ has a finite value $v(I) \in \Gamma_v$.*[3]

Proof. $\operatorname{supp} v$ does not contain the set I, since otherwise we would have $IR \subset \operatorname{supp} v$, hence $\operatorname{supp} v \in \mathcal{G}_f$. □

We now are prepared to introduce "$\mathcal{G}$-special" valuations.

Definition 5. Given a $\mathcal{G}$-regular valuation v on R we define a subgroup $H_{v,\mathcal{G}}$ of Γ_v as follows. We choose a subring A of A_v. Then $H_{v,\mathcal{G}}$ denotes the smallest convex subgroup of Γ_v containing the values $v(I)$ with I running through $\Phi^f(\mathcal{G}/A)$.

Notice that the differences $v(I) - v(J)$ with $I, J \in \Phi^f(\mathcal{G}/A)$ form a subgroup of Γ_v. Thus $H_{v,\mathcal{G}}$ is the convex hull of this subgroup in Γ_v. We write

$$H_{v,\mathcal{G}} = \operatorname{conv}_{\Gamma_v}\{v(I) - v(J) \mid I, J \in \Phi^f(\mathcal{G}/A)\}.$$

This group $H_{v,\mathcal{G}}$ does not depend on the choice of the ring $A \subset A_v$ (we could take $A = A_v$ or $A = \mathbb{Z} \cdot 1_R$). Indeed, $H_{v,\mathcal{G}}$ is the smallest convex subgroup of Γ_v containing the elements $\min_{x \in S} v(x)$, where S runs trough the finite subsets $\{x_1, \dots, x_r\}$ of R with $\sum_1^r Rx_i \in \mathcal{G}$. Notice also that $H_{v,\mathcal{G}} = H_{v,\mathcal{G}_f}$.

Definition 6. We call a valuation v on R *$\mathcal{G}$-special* if v is $\mathcal{G}$-regular and $H_{v,\mathcal{G}} = \Gamma_v$.

It turns out that in the case $\mathcal{G} = \{R\}$ the $\mathcal{G}$-special valuations are the same as the special valuations defined in [Vol. I, Chap. I §1]. This follows from

Proposition 2.3. *Let v be any valuation on R. Then $H_{v,\{R\}} = c_v(\Gamma_v)$.*

Proof. We assume without loss of generality that $\Gamma = \Gamma_v$. We choose same ring $A \subset A_v$. The group $c_v(\Gamma)$ is the smallest convex subgroup of Γ containing the negative values $v(x) < 0$ with $x \in R$. Thus we are done if we verify that every such value is contained in $H_{v,\{R\}}$ and every $v(I)$, $I \in \Phi^f(R/A)$ is contained in $c_v(\Gamma)$.

a) Let $x \in R$ with $v(x) < 0$. Then $A + Ax \in \Phi^f(R/A)$ and $v(A + Ax) = v(x)$.
b) Let $I \in \Phi^f(R/A)$ be given. If $v(I) \le 0$, then of course $v(I) \in c_v(\Gamma)$. Assume now that $v(I) > 0$. We write $I = Aa_1 + \cdots + Aa_r$ with elements $a_j \in R$. Since $v(I) > 0$, every $v(a_j) > 0$. There exist elements $x_1, \dots, x_r \in R$ with

$$1 = x_1 a_1 + \cdots + x_r a_r.$$

It follows that

$$0 = v(1) \ge \min_i v(x_i a_i),$$

[3] $v(I)$ had been defined in Sect. 1.

hence $0 \geq v(x_k a_k)$ for some $k \in \{1, \dots, r\}$. We have $v(x_k) \leq -v(a_k) < 0$, hence $v(x_k) \in c_v(\Gamma)$, and then $v(a_k) \in c_v(\Gamma)$, due to the convexity of $c_v(\Gamma)$. Since

$$0 < v(I) \leq v(a_k),$$

we conclude that $v(I) \in c_v(\Gamma)$. □

We determine the group $H_{v,\mathscr{G}}$ in another case.

Example 2.4. Let $\mathfrak{q}$ be a prime ideal of R. It is plain that

$$\mathscr{G}_\mathfrak{q} := \{I \in J(R) \mid I \not\subset \mathfrak{q}\}$$

is a multiplicative filter of finite type on R. If v is a valuation on R then v is $\mathscr{G}_\mathfrak{q}$-regular iff $\operatorname{supp} v \subset \mathfrak{q}$.

Assume now that the prime ideal $\mathfrak{q}$ ist v-convex (N.B. Since $\operatorname{supp} v$ is the smallest v-convex ideal of R, this implies that v is $\mathscr{G}_\mathfrak{q}$-regular). If I is a finitely generated ideal of R not contained in $\mathfrak{q}$, then $v(I) = v(a)$ for some $a \in I$, and $a \notin \mathfrak{q}$ due to the convexity of $\mathfrak{q}$. Thus $H_{v,\mathscr{G}_\mathfrak{q}}$ is the smallest convex subgroup of Γ_v containing the values $v(a)$ with $a \notin \mathfrak{q}$, i.e. the set $v(R \backslash \mathfrak{q})$.

The valuation v is $\mathscr{G}_\mathfrak{q}$-special iff this group is the whole value group Γ_v. Clearly this happens iff $\mathfrak{q} = \operatorname{supp} v$. □

Thus every valuation on R is $\mathscr{G}$-special for a suitable multiplicative filter $\mathscr{G}$.

On the other hand, if the filter $\mathscr{G}$ on R is fixed, we can turn any $\mathscr{G}$-regular valuation $v : R \to \Gamma \cup \infty$ into a $\mathscr{G}$-special valuation by taking a suitable primary specialization.[4]

Proposition 2.5. *Let $\mathscr{G}$ be a multiplicative filter on a ring R and $v : R \to \Gamma \cup \infty$ a $\mathscr{G}$-regular valuation on R. Assume w.l.o.g. that $\Gamma = \Gamma_v$. The following is true.*

a) $H_{v,\mathscr{G}} \supset c_v(\Gamma)$, hence the primary specialization $w := v|H_{v,\mathscr{G}}$ exists.
b) $H_{v,\mathscr{G}}$ is the smallest convex subgroup U of Γ containing $c_v(\Gamma)$ such that $v|U$ is $\mathscr{G}$-regular.
c) $v|H_{v,\mathscr{G}}$ is $\mathscr{G}$-special.

Proof. a): $H_{v,\mathscr{G}} \supset H_{v,\{R\}}$, since $\mathscr{G} \supset \{R\}$, and $H_{v,\{R\}} = c_v(\Gamma)$ by Proposition 3.
b): We choose some subring $A \subset A_v$ of R. Let $w := v|U$ for some convex subgroup U of Γ containing $c_v(\Gamma)$. The valuation w is $\mathscr{G}$-regular if there does not exist any $I \in \Phi^f(\mathscr{G}/A)$ with $w(I) = \infty$, which means that $v(I) > U$ (i.e. $v(I) > \gamma$ for every $\gamma \in U$). If $U \supset H_{v,\mathscr{G}}$ this certainly holds since $v(I) \in H_{v,\mathscr{G}}$ for every $I \in \Phi^f(\mathscr{G}/A)$. Thus then $v|U$ is $\mathscr{G}$-regular. Assume now that $U \subsetneq H_{v,\mathscr{G}}$. Then there exist $K, L \in \Phi^f(\mathscr{G}/A)$, such that $v(K) - v(L) \notin U$, hence $v(K) \notin U$ or $v(L) \notin U$. Thus we have some $I \in \Phi^f(\mathscr{G}/A)$ with $v(I) \notin U$. Since U

[4] We call any valuation equivalent to $v|H$ for some convex subgroup H of Γ with $H \supset c_v(\Gamma)$ a *primary specialization* of v (cf. [HK] for this terminology).

contains all values $v(x), x \in R$, which are negative, it follows that $v(I) > U$, hence $w(I) = \infty$. The valuation $v|U$ is not $\mathscr{G}$-regular.

c): Let $H := H_{v,\mathscr{G}}$ and $w := v|H$. The valuation $w : R \to H \cup \infty$ is $\mathscr{G}$-regular, as proved, and $A_w = A_v \supset A$. For any $I \in \Phi^f(\mathscr{G}/A)$, we have $w(I) = v(I)$. This implies that $H_{w,\mathscr{G}} = H_{v,\mathscr{G}} = H$. Thus w is $\mathscr{G}$-special. □

In short, Proposition 2.5 tells us, that any $\mathscr{G}$-regular valuation has a unique primary specialization (up to equivalence) which is $\mathscr{G}$-special.

We return to the program sketched at the beginning of Sect. 1. Given a multiplicative filter $\mathscr{G}$ on R we want to realize the program taking there for $\mathfrak{M}$ the set of all $\mathscr{G}$-special valuations on R. We will proceed on the same line, as in Sect. 1, where we settled the case $\mathscr{G} = \{R\}$.

We first define the "correct analogue" of the ring $R(X)$ there. We will denote this ring by $R(X, \mathscr{G})$. Let $S_{\mathscr{G}}$ denote the set of polynomials $f \in R[X]$ with $c_R(f) \in \mathscr{G}$. It follows from the Dedekind–Mertens formula (Theorem 1.1) that $S_{\mathscr{G}}$ is a saturated multiplicative subset of $R[X]$. Indeed, for $f, g \in S_{\mathscr{G}}$ with some $n \in \mathbb{N}$

$$c_R(fg) \supset c_R(fg)c_R(g)^n = c_R(f)c_R(g)^{n+1} \in \mathscr{G},$$

hence $fg \in S_{\mathscr{G}}$. On the other hand, if $f, g \in R[X]$ and $fg \in S_{\mathscr{G}}$ then $c_R(f) \supset c_R(fg) \in \mathscr{G}$, hence $f \in S_{\mathscr{G}}$. We define

$$R(X, \mathscr{G}) := S_{\mathscr{G}}^{-1} R[X].$$

This is the ring of fractions f/g with $f \in R[X]$, $g \in S_{\mathscr{G}}$. If every ideal $I \in \mathscr{G}$ is *dense* in R (i.e. $x \in R, Ix = 0$ implies $I = 0$), the set $S_{\mathscr{G}}$ consists of non-zero divisors of $R[X]$, and we may regard $R(X, \mathscr{G})$ as an overring of $R[X]$ in Quot $R[X]$. In general we have a localization map $j'_{\mathscr{G}} : R[X] \to R(X, \mathscr{G})$ and its restriction $j_{\mathscr{G}} : R \to R(X, \mathscr{G})$, by which we regard $R(X, \mathscr{G})$ as an algebra over R or over $R[X]$. The map $j'_{\mathscr{G}}$ extends to a homomorphism $R(X) \to R(X, \mathscr{G})$, since $S_R \subset S_{\mathscr{G}}$. If $\mathscr{G}$ consists of dense ideals then we view R and $R(X)$ as subrings of $R(X, \mathscr{G})$, $R \subset R(X) \subset R(X, \mathscr{G})$.

It is now possible to define "Kronecker subrings" of $R(X, \mathscr{G})$ and to verify assertions completely analogous to Proposition 1.4 and Theorem 1.5 and other matters in exactly the same way as in Sect. 1 in the case $\mathscr{G} = \{R\}$, the only difference being, that we now have to use homomorphisms $j_{\mathscr{G}}, j'_{\mathscr{G}}$ instead of inclusion maps. We may safely leave all proofs to the reader.

Definition 7.

a) We call a subring B of $R(X, \mathscr{G})$ *Kronecker in* $R(X, \mathscr{G})$ (or: a *Kronecker subring of* $R(X, \mathscr{G})$), if the following two axioms hold.

(1) $X/1 \in B$,[5]
(2) For every polynomial $f \in S_{\mathscr{G}}$, $f = a_0 + a_1X + \cdots + a_dX^d$, the elements $a_j/f (j = 0,\dots,d)$ are contained in B.

b) Given a subring A of R we say that B is *Kronecker* in $R(X,\mathscr{G})$ over A, if in addition $j_{\mathscr{G}}(A) \subset B$.

Remark 2.6. Above we introduced the biggest multiplicative filter of finite type $\mathscr{G}_f$ of $\mathscr{G}$. The set $S_{\mathscr{G}}$ does not change if we replace $\mathscr{G}$ by $\mathscr{G}_f$. Thus $R(X,\mathscr{G}) = R(X,\mathscr{G}_f)$. Also the notion of a Kronecker subring of $R(X,\mathscr{G})$ does not change if we replace $\mathscr{G}$ by $\mathscr{G}_f$.

Proposition 2.7. *Let A be a subring of R and B be a subring of $R(X,\mathscr{G})$ containing $j'_{\mathscr{G}}(A[X]) = j_{\mathscr{G}}(A)[X]$. The following are equivalent.*

(1) B is Kronecker in $R(X,\mathscr{G})$ over A.
(2) $c_A(f)B = fB$ for every $f \in S_{\mathscr{G}}$. □

Here we have followed the usual notation in B as an $R[X]$-algebra via $j'_{\mathscr{G}}$: $c_A(f)B$ means $j_{\mathscr{G}}(c_A(f))B$, and fB means $j'_{\mathscr{G}}(f)B$, etc.

Theorem 2.8. *Every Kronecker subring of $R(X,\mathscr{G})$ is Bezout in $R(X,\mathscr{G})$.* □

$R(X,\mathscr{G})$ contains a unique smallest Kronecker subring, which we denote by $R(X,\mathscr{G})_{\mathrm{kr}}$. This subring is generated in $R(X,\mathscr{G})$ by $X/1$ and the elements a_j/f with $f = \sum_{j=0}^{d} a_jX^j \in S_{\mathscr{G}}$ and $0 \le j \le d$. If A is a subring of R then, of course, $A \cdot R(X,\mathscr{G})_{\mathrm{kr}} := j_{\mathscr{G}}(A) \cdot R(X,\mathscr{G})_{\mathrm{kr}}$ is the smallest Kronecker subring of $R(X,\mathscr{G})$ over A. We then have a commuting square

$$(\ddagger) \qquad \begin{array}{ccc} R & \xrightarrow{j_{\mathscr{G}}} & R(X,\mathscr{G}) \\ \uparrow & & \uparrow \\ A & \longrightarrow & B \end{array}$$

with $B := A \cdot R(X,\mathscr{G})_{\mathrm{kr}}$, and the lower horizontal arrow a restriction of $j_{\mathscr{G}}$. We want to prove that, based on this diagram, the $\mathscr{G}$-special valuations of R over A correspond uniquely with the BM-valuations of $R(X,\mathscr{G})$ over B in a natural way.

Let $v : R \to \Gamma \cup \infty$ be a $\mathscr{G}$-regular valuation. The associated Gauß extension $v' : R[X] \to \Gamma \cup \infty$ (cf. Sect. 1) has value $v'(g) \neq \infty$ for every $g \in S_{\mathscr{G}}$, due to the $\mathscr{G}$-regularity of v. Thus v' extends further to a valuation

$$v^* : R(X,\mathscr{G}) \to \Gamma \cup \infty,$$

[5] $X/1$ denotes the image of X in $R(X,\mathscr{G})$ under the map $j'_{\mathscr{G}}$.

which obeys the formula

$$v^*(\frac{f}{g}) = v(c_R(f)) - v(c_R(g))$$

for $f \in R[X], g \in S_{\mathscr{G}}$. We call v^* the *Gauß extension* of v to $R(X,\mathscr{G})$. Clearly $v^* \circ j_{\mathscr{G}} = v$. The word "extension" is meant in this sense. If w is any valuation on $R(X,\mathscr{G})$ we usually denote its "restriction" $w \circ j_{\mathscr{G}}$ to R by $w|R$, and its restriction $w \circ j'_{\mathscr{G}}$ to $R[X]$ by $w|R[X]$.

We now determine the characteristic subgroup $c_{v^*}(\Gamma)$ of v^*. We need two lemmas.

Lemma 2.9. *Assume that $B \subset T$ is a Bezout extension and $w : T \to \Gamma \cup \infty$ is a valuation of T over B. Then $w(T^*)$ is the group generated by the set $\{w(x) \mid x \in T, w(x) \leq 0\}$ in Γ.*[6] *Thus $c_w(\Gamma_w)$ is the convex hull of $w(T^*)$ in Γ_w.*

Proof. Let $U := w(T^*)$. It is trivial that for any $x \in T^*$ either $w(x) \leq 0$ or $w(x^{-1}) \leq 0$. Thus U is contained in the group generated by the negative part of the set $w(T)$. On the other hand, if $x \in T$ and $w(x) < 0$ then $B + Bx = By$ for some $y \in T$, and $w(x) = w(y)$. We have $T = T + Tx = Ty$, hence $y \in T^*$. This proves that both groups are equal. □

Lemma 2.10. *Let T be any ring and S be a saturated multiplicative subset of T. The group $(S^{-1}T)^*$ of units of the localization $S^{-1}T$ is the set of fractions s_1/s_2 with $s_1, s_2 \in S$. More precisely, if $a \in T, s \in S$ is given with $a/s \in (S^{-1}T)^*$ then $a \in S$.*

Proof. Easy. □

Proposition 2.11. *Let v be a $\mathscr{G}$-regular valuation on R. Then v^* has the characteristic group $c_{v^*}(\Gamma_v) = H_{v,\mathscr{G}}(\Gamma)$ (cf. Definition 5 above).*

Proof. We assume without loss of generality that $\Gamma_v = \Gamma$. We choose a subring A of R with $A \subset A_v$. Let $T := R(X,\mathscr{G})$. By Lemma 9 we know that $c_{v^*}(\Gamma)$ is the convex hull of the group $U := v^*(T^*)$ in Γ. By Lemma 10 we conclude that U consists of the elements $v^*(f) - v^*(g)$ with $f, g \in S_{\mathscr{G}}$. These are the elements $v(I) - v(J)$ with $I, J \in \Phi^f(\mathscr{G}/A)$. Thus $H_{v,\mathscr{G}}$ coincides with $c_{v^*}(\Gamma)$. □

Corollary 2.12. *A $\mathscr{G}$-regular valuation v on R is $\mathscr{G}$-special iff v^* is special.*

Proof. This is a direct consequence of Proposition 11, since v is $\mathscr{G}$-special iff $H_{v,\mathscr{G}} = \Gamma_v$, and v^* is special iff $c_{v^*}(\Gamma_v) = \Gamma_v$. □

But more is true: If v is $\mathscr{G}$-special then v^* is *BM* (= Bezout–Manis) due to the following general proposition.

[6] Recall that T^* denotes the group of units of T.

Proposition 2.13. *If v is a $\mathscr{G}$-regular valuation on R then $A_{v^*} \supset R(X,\mathscr{G})_{\mathrm{kr}}$.*

Proof. We have $v^*(X) = 0$. Let $A := A_v$. If $f = \sum_j a_j X^j \in S_{\mathscr{G}}$ then, for any j,

$$v^*(\frac{a_j}{f}) = v(a_j) - v^*(f) = v(a_j) - v(c_A(f)) \geq 0.$$

This implies the claim due to our description of $R(X,\mathscr{G})_{\mathrm{kr}}$ above. □

Definition 8. We call a valuation w on $R(X,\mathscr{G})$ *Kronecker*, if the ring A_w is Kronecker in $R(X,\mathscr{G})$.

In this terminology Proposition 12 says that the Gauß extension v^* of any $\mathscr{G}$-regular valuation v on R is Kronecker. We look at Kronecker valuations in general.

Proposition 2.14. *Let w be any valuation on $R(X,\mathscr{G})$ and $v := w|R$. The following are equivalent.*

i) w is Kronecker.
ii) v is $\mathscr{G}$-regular and $v^*(z) \leq w(z)$ for every $z \in R(X,\mathscr{G})$.
iii) v is $\mathscr{G}$-regular and $v^*(u) = w(u)$ for every unit u of $R(X,\mathscr{G})$.

Proof. i) $\Rightarrow$ ii): Let $B := A_w$ and $A := A_v = j_{\mathscr{G}}^{-1}(B)$. Suppose that v is not $\mathscr{G}$-regular, i.e. $\mathrm{supp}(v) \in \mathscr{G}_f$. We choose a finitely generated A-submodule I of R with $RI \in \mathscr{G}$ and $RI \subset \mathrm{supp}(v)$. We then choose a polynomial $g \in R[X]$ with $c_A(g) = I$. Now $c_R(g) = RI \in \mathscr{G}$, hence $g \in S_{\mathscr{G}}$. It follows that g is a unit of $R(X,\mathscr{G})$. This implies that $w(g) \neq \infty$. But $gB = c_A(g)B = IB \subset \mathrm{supp}(w)$, a contradiction. We have proved that v is $\mathscr{G}$-regular. We have $w(X) = 0$, since X and $1/X$ are elements of $B = A_w$. Let $f = \sum_i a_i X^i \in R[X]$ be given. Then

$$v^*(f) = \min_i v(a_i) = \min_i w(a_i X^i) \leq w(f).$$

In the case that $f \in S_{\mathscr{G}}$ the elements a_i/f lie in B, hence $w(a_i/f) \geq 0$, i.e. $v(a_i) \geq w(f)$. This implies $v^*(f) \geq w(f)$. We conclude that $v^*(f) = w(f)$ for every $f \in S_{\mathscr{G}}$, and then, that $v^*(z) \leq w(z)$ for $z \in R(X,\mathscr{G})$, since every such z is a fraction f/g with $f \in R[X], g \in S_{\mathscr{G}}$.

ii) $\Rightarrow$ iii): If $u \in R(X,\mathscr{G})^*$ we have $v^*(u) \leq w(u) \neq \infty$ and $v^*(u^{-1}) \leq w(u^{-1})$, hence $v^*(u) = w(u)$.

iii) $\Rightarrow$ i): Since X is a unit in $R(X,\mathscr{G})$ and $v^*(X) = 0$, we have $w(X) = 0$ by (iii), hence $X \in A_w^*$. If $f = \sum_j a_j X^j \in R[X]$, it follows that, for any j,

$$w(a_j) = w(a_j X^j) \geq \min_k w(a_k X^k) = \min_k v(a_k) = v^*(f).$$

If in addition $f \in S_{\mathscr{G}}$, then f is a unit in $R(X,\mathscr{G})$, hence $w(f) = v^*(f) \neq \infty$, hence $w(a_j/f) \geq 0$ for every j. This proves that A_w is Kronecker in $R(X,\mathscr{G})$ (recall Definition 7). □

We now are ready to prove that every special valuation w of $R(X,\mathcal{G})$ over $R(X,\mathcal{G})_{\text{kr}}$ is the Gauß extension v^* of some $\mathcal{G}$-special valuation v on R. We arrive at the main result of this section.

Theorem 2.15. *Let R be any ring and $\mathcal{G}$ a multiplicative filter on R. The $\mathcal{G}$-special valuations v on R correspond uniquely with the special valuations (hence BM-valuations) w of $R(X,\mathcal{G})$ over $R(X,\mathcal{G})_{\text{kr}}$ via $w = v^*, v = w|R$.*

Proof. If v is a $\mathcal{G}$-special valuation on R then we know by Corollary 12, that v^* is special, and by Proposition 13, that $A_{v^*} \supset R(X,\mathcal{G})_{\text{kr}}$ (i.e. v^* is Kronecker). Of course, $v^*|R = v$.

Let now a special Kronecker valuation w on $R(X,\mathcal{G})$ be given. The direction (i)$\Rightarrow$(ii) in Proposition 14 tells us that v is $\mathcal{G}$-regular and $v^*(z) \leq w(z)$ for every $z \in R(X,\mathcal{G})$. This implies $\mathfrak{p}_w \subset \mathfrak{p}_{v^*}$. Suppose there exists some $z \in R(X,\mathcal{G})$ with $v^*(z) < w(z)$. Since v^* is a Manis valuation, there exists some $z' \in R(X,\mathcal{G})$ with $v^*(zz') = 0$. Then

$$w(zz') = w(z) + w(z') \geq w(z) + v^*(z') > v^*(z) + v^*(z') = 0,$$

hence $zz' \in \mathfrak{p}_w$. This contradicts the fact that $zz' \notin \mathfrak{p}_{v^*}$, while $\mathfrak{p}_w \subset \mathfrak{p}_{v^*}$. Thus $v^* = w$. □

Remark 2.16. Theorem 15 is a very strong statement, since here we do not identify equivalent valuations. Nevertheless the formulation of the theorem is sloppy. Let us call a valuation $w : T \to \Gamma \cup \infty$ on a ring T *epimorphic*, if the group Γ is generated by the set of finite values of w, i.e. $\Gamma = \Gamma_w$. (Any valuation $w : T \to \Gamma \cup \infty$ has a unique "associated epimorphic valuation", which we obtain simply replacing Γ by Γ_w.) In Theorem 15 we should only admit epimorphic valuations. The $\mathcal{G}$-special epimorphic valuations on R correspond uniquely with the special epimorphic valuations on $R(X,\mathcal{G})$ over $R(X,\mathcal{G})_{\text{kr}}$.

Corollary 2.17. *Let $w : R(X,\mathcal{G}) \to \Gamma \cup \infty$ be a Kronecker valuation on $R(X,\mathcal{G})$ and $v := w|R$. Assume that v is $\mathcal{G}$-special (equivalently: v is $\mathcal{G}$-regular and v^* is special). Let w' denote the maximal primary specialization of w (i.e. $w' = w|c_w(\Gamma)$). Then $v^* = w'$ (more precisely: the epimorphic valuations associated to v^* and w' are equal).*

Proof. We have $v^*(z) \leq w(z)$ for every $z \in R(X,\mathcal{G})$ (Proposition 13). Trivially $w(z) \leq w'(z)$ for every $z \in R(X,\mathcal{G})$. Thus $v^*(z) \leq w'(z)$ for every $z \in R(X,\mathcal{G})$, and w' is again Kronecker since $A_{w'} = A_w$. Theorem 15 gives the claim. □

3 Star Operations: Definitions and Some Examples

We come to the second main topic of this chapter, the "star operations". Originally suitable star operations have been chiefly used to create a reasonable multiplicative ideal theory in the case of rings which are no longer Dedekind or Prüfer domains, in particular to define "groups of divisors", cf. [Bo, Chap. VII], [vdW, §131]. We will turn to this important theme only later, using star operations in the present section mainly to build Kronecker extensions, as is done in the case of integral domains e.g. in [Gi, Chap. V].

Definition 1. A *star operation on* $J(A, R)$ is a map $* : J(A, R) \to J(A, R)$ with the following four properties. For any $I, J \in J(A, R)$

(St1) $I \subset I^*$,
(St2) $I \subset J \Rightarrow I^* \subset J^*$,
(St3) $(I^*)^* = I^*$,
(St4) $IJ^* \subset (IJ)^*$.

We call the star operation *strict*, if in addition

(St5) $A^* = A$.

Comments.

1. There exists a very extended literature on star operations in the case that A is an integral domain and R its field of quotients (cf. [Gi, Chap. V] for the literature up to 1970), and a less extended literature in the case that A is a ring and R its total ring of quotients Quot(A) (cf. [Huc, §20]). There the axioms look somewhat different, having a stress on principal fractional deals.

   It does not seem to make enough sense to work with principal ideals or principal A-modules in the case of an arbitrary ring extension $A \subset R$.
2. In the literature our axiom St5 is incorporated in the definition of a star operation while the pendant to operations obeying axioms St1–St4 is usually called a semi-star operation. For our purposes the axiom St5 is much less important than the others (cf. Remark 4.3 below). Consequently we call the operations with St1–St4 "star operations" while we reserve the word "semi-star operation" for a somewhat weaker notion which will play an auxiliarly role later on (Sects. 9, 10).

Remark 3.1. In our definition of a star operation the axiom St4 may be replaced by the following one which at first glance looks weaker:

(St4′) $aI^* \subset (aI)^*$

for any $a \in R, I \in J(A, R)$. Indeed, suppose (St4′) holds. Then we have for $K, I \in J(A, R)$

$$KI^* = \bigcup_{a \in K} aI^* \subset \bigcup_{a \in K} (aI)^* \subset KI^*,$$

since by (St2) we have $(aI)^* \subset KI^*$ for any $a \in K$.

If several star operations come in to play we will use letters like $\alpha, \beta, \gamma \dots$ for them instead of just a star.

We give first examples of star operations.

Example 3.2. Assume that $A \subset R$ is a ring extension and $\mathscr{F}$ is a multiplicative filter on A of finite type. We define a map $* : J(A, R) \to J(A, R)$ by

$$I^* := I^R_{[\mathscr{F}]} := \{x \in R \mid (I : x) \in \mathscr{F}\}$$

for any $I \in J(A, R)$, calling $I^R_{[\mathscr{F}]}$ the "*$\mathscr{F}$-hull*" of I in R.

We verify that this is a star operation. Axioms St1 and St2 are obvious. Turning to St3, let $x \in (I^*)^*$ be given. Then $K := (I^* : x) \in \mathscr{F}$. We choose elements $a_1, \dots, a_r$ of K such that $K_0 := a_1A + \cdots + a_rA \in \mathscr{F}$. We have $a_ix \in I^*$, hence $L_i := (I : a_ix) \in \mathscr{F}$ and $a_ixL_i \subset I$. Let $L := L_1 \cap \cdots \cap L_r \in \mathscr{F}$. Then $a_ixL \subset I$ for $i = 1, \dots, r$. We conclude that $K_0Lx \subset I$. Since $K_0L \in \mathscr{F}$, this proves that $x \in I^*$. St3 is verified.

In order to verify St4, let $I, J \in J(A, R)$ be given. For $a \in I, x \in J^*$ we have $(IJ : ax) \supset (J : x) \in \mathscr{F}$. Thus $(IJ : ax) \in \mathscr{F}$, i.e. $ax \in (IJ)^*$. This proves that $IJ^* \subset (IJ)^*$.

Thus $I \mapsto I^*$ is indeed a star operation. We have $A^* = A^R_{[\mathscr{F}]}$. It may well happen that $A^* \supsetneqq A$. This ring A^* is known in the literature as the *$\mathscr{F}$-transform* of A in R (in the case $R = \operatorname{Quot}(A)$).

Example 3.3. Let $A \subset R$ and $B \subset T$ be ring extensions and let $\varphi : R \to T$ be a ring homomorphism with $\varphi(A) \subset B$. In this situation we call φ a morphism from the pair (R, A) to (T, B) and write $\varphi : (R, A) \to (T, B)$. Given such a morphism φ we define a map $* : J(A, R) \to J(A, R)$ by

$$I^* := \varphi^{-1}(\varphi(I)B)$$

for $I \in J(A, R)$. It can be verified in a straight forward way that this is a star operation. We call it the *star operation induced by the morphism*

$$\varphi : (R, A) \to (T, B),$$

or, *by the diagram*

$$\begin{array}{ccc} R & \xrightarrow{\varphi} & T \\ \uparrow & & \uparrow \\ A & \longrightarrow & B, \end{array}$$

the lower horizontal arrow being a restriction of φ. We have $A^* = \varphi^{-1}(B)$. Again it may well happen that $A^* \supsetneqq A$.

Later on we will often write IB for $\varphi(I)B$ if $I \in J(A, R)$, and $R \cap K$ for $\varphi^{-1}(K)$ if $K \in J(B, T)$.

Subexample 3.4. Let R and B be subrings of a ring T and A a subring of $R \cap B$. We define a star operation $*$ on $J(A, R)$ by $I^* := R \cap (IB)$. This star operation is strict iff $A = R \cap B$.

Subexample 3.5. Let A and B be subrings of a ring R with $A \subset B \subset R$. Then $I^* := IB$ defines a very trivial but reasonable star operation $J(A, R)$ (notice that $J(B, R) \subset J(A, R)$). This is the special case $R = T$ of Example 4.

We introduce still another star operation. Let $A \subset R$ be any ring extension. For $I \in J(A, R)$ we put

$$I^{-1} := [A :_R I] = \{x \in R \mid Ix \subset A\}.$$

Clearly $I^{-1} \in J(A, R)$ and $II^{-1} \subset A$. We then define[7] $I^\delta := (I^{-1})^{-1}$ (think of "δ" as "double inverse").

Proposition 3.6.

a) The map

$$\delta : J(A, R) \to J(A, R), I \mapsto I^\delta,$$

is a strict star operation.

b) For any $I \in J(A, R)$, $I^{-1} = (I^{-1})^\delta = (I^\delta)^{-1}$.

c) If $*$ is any strict star operation on $J(A, R)$ then $I^* \subset I^\delta$ for every $I \in J(A, R)$.

Proof. Since $II^{-1} \subset A$ we have $I \subset I^\delta$. Clearly $I \subset J$ implies $I^{-1} \supset J^{-1}$ and then $I^\delta \subset J^\delta$. Further $A^{-1} = A$, hence $A^\delta = A$. The axioms St1, St2, St5 are verified.

We prove St4′. Let $a \in R, I \in J(A, R), x \in I^\delta, y \in (aI)^{-1}$ be given. Then $xI^{-1} \subset A$ and $ay \in I^{-1}$, hence $axy \in A$, hence $ax \in ((aI)^{-1})^{-1}$. This proves that $aI^\delta \subset ((aI)^{-1})^{-1} = (aI)^\delta$.

Before proving St3 we verify assertion b) of the proposition. From $I^\delta \supset I$ we obtain $(I^\delta)^{-1} \subset I^{-1}$. But also $(I^\delta)^{-1} = ((I^{-1})^{-1})^{-1} = (I^{-1})^\delta \supset I^{-1}$. Thus $(I^\delta)^{-1} = (I^{-1})^\delta = I^{-1}$. From this we obtain $(I^\delta)^\delta = I^\delta$, i.e. St3. We now know that δ is a strict star operation.

Let finally $I \mapsto I^*$ be any strict star operation. Then $I^{-1}I^* \subset (I^{-1}I)^* \subset A^* = A$ by St4, St2, St5. Thus $I^* \subset (I^{-1})^{-1} = I^\delta$. □

We call δ the *double-inverse operation*. Historically this operation has been the prototype of all other star operations, cf. [vdW, §131], [Bo, Chap. VII], [Gi, §32, §34]. As far as we know it has been invented (for domains) by Emil Artin.

Let us take a brief look at the set of all star operations on $J(A, R)$.

[7] In most of the literature I^δ is denoted by I^v or I_v (the "v-operation"). We refuse to use the letter v here, since all too often in our book v denotes a valuation.

Definition 2. We denote the set of all star operations on $J(A,R)$ by $\mathrm{Star}(R/A)$ and the subset of strict star operations by $\mathrm{Star}_0(R/A)$. We introduce a partial ordering on $\mathrm{Star}(R/A)$ as follows: If $\alpha, \beta \in \mathrm{Star}(R/A)$ are given, we decree that $\alpha \leq \beta$ iff $I^\alpha \subset I^\beta$ for every $I \in J(A,R)$. We then say that the operation β is *coarser* than α, or, that α is *finer* than β.

Notice that $\mathrm{Star}(R/A)$ has a unique finest element, the "identity" star operation id: $I \mapsto I$, and a unique coarsest element, the "trivial" operation triv: $I \mapsto R$ for every $I \in J(A,R)$. By Proposition 6, $\mathrm{Star}_0(R/A)$ has a unique coarsest element, the operation δ, and, of course, id as the finest element.

Proposition 3.7. *If* $(\alpha_\lambda \mid \lambda \in \Lambda)$ *is a family in* $\mathrm{Star}(A,R)$ *then*

$$\alpha : J(A,R) \to J(A,R), I^\alpha := \bigcap_{\lambda \in \Lambda} I^{\alpha_\lambda},$$

is again a star operation. If some α_λ *is strict, then* α *is strict.*

Proof. A straight forward verification. □

Remark 3.8. It follows from Proposition 7 that $\mathrm{Star}(R/A)$ is a complete lattice. The infimum of a family (α_λ) is the star operation α described above and will consequently be denoted by $\bigwedge_\lambda \alpha_\lambda$. The supremum of the family (α_λ) is the infimum of the set of all $\gamma \in \mathrm{Star}(R/A)$ with $\gamma \geq \alpha_\lambda$ for every λ. $\mathrm{Star}_0(R/A)$ is a lower set in this lattice: If $\alpha \in \mathrm{Star}_0(R/A)$ and $\beta \leq \alpha$, then also $\beta \in \mathrm{Star}_0(R/A)$. Thus $\mathrm{Star}_0(R/A) = \{\alpha \in \mathrm{Star}(R/A) \mid \alpha \leq \delta\}$.

We turn to an application of star operations. Let a star operation $* : J(A,R) \to J(A,R)$ and also a multiplicative filter $\mathcal{G}$ on R be given. We define a subset $A(X, \mathcal{G}, *)$ of the ring $R(X, \mathcal{G})$ (introduced in Sect. 2), which will be shown to be a Kronecker subring of $R(X, \mathcal{G})$.

Definition 3. $A(X, \mathcal{G}, *)$ is the set of all quotients f/g with $f \in R[X], g \in S_{\mathcal{G}}$ (cf. Sect. 2), such that there exists some $H \in \Phi^f(\mathcal{G}/A)$ (i.e. $H \in J^f(A,R), HR \in \mathcal{G}$) with

$$(*) \qquad c_A(f)H \subset (c_A(g)H)^*.$$

Remark. Due to the axioms St1–St3 for a star operation this condition is equivalent to

$$(c_A(f)H)^* \subset (c_A(g)H)^*.$$

It is important to understand that condition $(*)$ in this definition is a property of the element $\xi = f/g \in R(X, \mathcal{G})$ not depending on the chosen presentation of ξ as a quotient of polynomials f, g.

Lemma 3.9. *Let $f, f_1 \in R[X], g, g_1 \in S_{\mathcal{G}}$, and $f/g = f_1/g_1$ in $R(X, \mathcal{G})$. Assume there exists some $H \in \Phi^f(\mathcal{G}/A)$ with $c_A(f)H \subset (c_A(g)H)^*$. Then there exists some $H_1 \in \Phi^f(\mathcal{G}/A)$ with $c_A(f_1)H_1 \subset (c_A(g_1)H_1)^*$.*

Proof. We have $fg_1h = f_1gh$ with some $h \in S_{\mathcal{G}}$. Due to the Dedekind–Mertens formula (Theorem 1.1) there exists some $U \in \Phi^f(\mathcal{G}/A)$ with

$$c_A(fg_1h)U = c_A(f)c_A(g_1)c_A(h)U,$$
$$c_A(f_1gh)U = c_A(f_1)c_A(g)c_A(h)U.$$

Using $c_A(f)H \subset (c_A(g)H)^*$ and St4 we obtain

$$\begin{aligned} c_A(f_1)c_A(g)c_A(h)UH &= c_A(f_1gh)UH = c_A(fg_1h)UH \\ &= c_A(f)c_A(g_1)c_A(h)UH \\ &\subset c_A(g_1)c_A(h)U(c_A(g)H)^* \\ &\subset (c_A(g_1)c_A(h)Uc_A(g)H)^*. \end{aligned}$$

With $H_1 := c_A(g)c_A(h)UH \in \Phi^f(\mathcal{G}/A)$ we have

$$c_A(f_1)H_1 \subset (c_A(g_1)H_1)^*.$$

□

Theorem 3.10. *The set $A(X, \mathcal{G}, *)$ is a Kronecker subring of $R(X, \mathcal{G})$ over A.*

Proof. a) Of course, the ring $j_{\mathcal{G}}(A) = \{a/1 \mid a \in A\}$ is a subset of $A(X, \mathcal{G}, *)$.
b) Let $\xi, \eta \in A(X, \mathcal{G}, *)$ be given. We choose $f, f_1 \in R[X], g, g_1 \in S_{\mathcal{G}}$ with $\xi = f/g, \eta = f_1/g_1$. We verify that $\xi - \eta$ and $\xi\eta$ are elements of $A(X, \mathcal{G}, *)$. There exist A-modules $H, H_1 \in \Phi^f(\mathcal{G}/A)$ with

$$c_A(f)H \subset (c_A(g)H)^*, \; c_A(f_1)H_1 \subset (c_A(g_1)H_1)^*. \tag{6}$$

By the Dedekind–Mertens formula (Theorem 1.1) there exist $n, n_1 \in \mathbb{N}$ such that

$$c_A(fg_1)c_A(g_1)^n = c_A(f)c_A(g_1)c_A(g_1)^n,$$
$$c_A(f_1g)c_A(g)^{n_1} = c_A(f_1)c_A(g)c_A(g)^{n_1}.$$

Thus, with $U := c_A(g_1)^n c_A(g)^{n_1}$,

$$c_A(fg_1)U = c_A(f)c_A(g_1)U, \; c_A(f_1g)U = c_A(f_1)c_A(g)U. \tag{7}$$

It follows from (6), (7) that

$$\begin{aligned} c_A(fg_1 - f_1g)UHH_1 &\subset [c_A(fg_1) + c_A(f_1g)]UHH_1 \\ &= c_A(f)c_A(g_1)UHH_1 + c_A(f_1)c_A(g)UHH_1 \\ &\subset (c_A(g_1)UH_1c_A(g)H)^* + (c_A(g)UHc_A(g_1)H_1)^* \\ &= (c_A(g)c_A(g_1)UHH_1)^*. \end{aligned}$$

Again by the Dedekind–Mertens formula, there exists some $V \in \Phi^f(\mathcal{G}/A)$ such that

$$c_A(gg_1)V = c_A(g)c_A(g_1)V. \tag{8}$$

We conclude that

$$c_A(fg_1 - f_1g)UVHH_1 \subset (c_A(gg_1)UVHH_1)^*,$$

and this proves that

$$\xi - \eta = \frac{fg_1 - f_1g}{gg_1} \in A(X, \mathcal{G}, *).$$

By a similar, in fact less massy computation one verifies that there is some $W \in \Phi^f(\mathcal{G}/A)$ such that

$$c_A(ff_1)W \subset (c_A(gg_1)W)^*.$$

Thus $\xi\eta \in A(X, \mathcal{G}, *)$. We now know that $A(X, \mathcal{G}, *)$ is a subring of $R(X, \mathcal{G})$ containing $j_{\mathcal{G}}(A)$.

c) Clearly $X/1 \in A(X, \mathcal{G}, *)$. If $f = \sum_i a_i X^i \in S_{\mathcal{G}}$ then $a_i/f \in A(X, \mathcal{G}, *)$ for every i, since $a_i A \subset c_A(f) \subset c_A(f)^*$. Thus $A(X, \mathcal{G}, *)$ is Kronecker in $R(X, \mathcal{G})$. □

Comment. In most of the literature it is assumed that the star operation is "e.a.b." (= "endlich arithmetisch brauchbar") in order to define "Kronecker function rings", which are instances of our rings $A(X, \mathcal{G}, *)$, cf. [Gi, p. 394ff]. "e.a.b" is a cancelation property which allows us to omit the factor "H" in the definition of $A(X, \mathcal{G}, *)$. We will give a precise definition of e.a.b only later in Sect. 5. The idea to define Kronecker function rings without assuming e.a.b seems to be recent, cf. [H-K], [FL].

Can we obtain every Kronecker subring of $R(X, \mathcal{G})$ over A by a star operation? This is indeed true.

Theorem 3.11. *Let $A \subset R$ be any ring extension, $\mathcal{G}$ a multiplicative filter on R, and B a Kronecker subring of $R(X,\mathcal{G})$ over A. We choose the star operation $I \mapsto I^*$ associated with the natural commuting square*

$$\begin{array}{ccc} R & \xrightarrow{j_{\mathcal{G}}} & R(X,\mathcal{G}) \\ \uparrow & & \uparrow \\ A & \longrightarrow & B, \end{array}$$

i.e. $I^* = R \cap (IB) := j_{\mathcal{G}}^{-1}(IB)$, (cf. Example 3). Then $B = A(X,\mathcal{G},*)$.

Proof. a) We first prove that $A(X,\mathcal{G},*)$ is contained in B. Let $f \in R[X], g \in S_{\mathcal{G}}$ be given with $c_A(f)H \subset (c_A(g)H)^*$ for some $H \in \Phi^f(\mathcal{G}/A)$. We have to verify that $f/g \in B$.

We choose some $h \in R[X]$ with $c_A(h) = H$ (notice that this is possible). Then $h \in S_{\mathcal{G}}$ and, with $\varphi := j_{\mathcal{G}}$,

$$c_A(f)c_A(h) \subset \varphi^{-1}(c_A(g)c_A(h)B),$$

in other terms,

$$\varphi(c_A(f)c_A(h)) \subset c_A(g)c_A(h)B.$$

Multiplying by B and taking into account that $c_A(u)B = uB$ for $u \in S_{\mathcal{G}}$ (Proposition 2.6), we obtain

$$c_A(f)c_A(h)B \subset c_A(g)c_A(h)B = ghB.$$

Since $X/1 \in B$ and $c_A(fh) \subset c_A(f)c_A(h)$ we have $fhB \subset c_A(f)c_A(h)B$. Thus $fhB \subset ghB$. Since $g/1$ and $h/1$ are units of B we may conclude that $fh/ghB \subset B$, i.e. $f/g = fh/gh \in B$.

b) We now verify the opposite inclusion $B \subset A(X,\mathcal{G},*)$. Let $f \in R[X], g \in S_{\mathcal{G}}$ be given with $f/g \in B$. We claim that $c_A(f)B \subset c_A(g)B$. From this it will follow by definition of the present star operation that $c_A(f)^* \subset c_A(g)^*$, and this will certainly imply that $f/g \in A(X,\mathcal{G},*)$. Our claim is equivalent to

$$(c_A(f) + c_A(g))B = c_A(g)B.$$

Let $d := \deg(g)$ and $h = g + X^{d+1}f$. Then $c_A(h) = c_A(f) + c_A(g)$ and $h \in S_{\mathcal{G}}$. Our claim translates to $c_A(h)B = c_A(g)B$. By Proposition 2.6 this means that $hb = gB$. Thus we have to verify that $h/g \in B$ and $g/h \in B$. Clearly

$$\frac{h}{g} = 1 + X^{d+1}\frac{f}{g} \in B.$$

Let $g = \sum_0^d a_i X^i$. Then

$$\frac{g}{h} = \sum_{i=0}^{d} \frac{a_i}{g + X^{d+1} f} X^i.$$

Each coefficient $a_i/(g + X^{d+1} f)$ is an element of $R(X, \mathcal{G})_{\mathrm{kr}}$. Thus indeed $g/h \in B$. The theorem is proved. □

Corollary 3.12. *Choosing the identity star operation* id : $I \to I$ *on* $J(A, R)$ *we have* $A(X, \mathcal{G}, id) = AR(X, \mathcal{G})_{\mathrm{kr}}$.

Proof. We have $\mathrm{id} \le \alpha$ for every $\alpha \in \mathrm{Star}(R/A)$, hence $A(X, \mathcal{G}, \mathrm{id}) \subset A(X, \mathcal{G}, \alpha)$. Theorems 10 and 11 tell us that $A(X, \mathcal{G}, \alpha)$ runs through all Kronecker subrings of $R(X, \mathcal{G})$ over A if α runs through $\mathrm{Star}(R/A)$. Thus $A(X, \mathcal{G}, \mathrm{id})$ is the smallest Kronecker subring of $R(X, \mathcal{G})$ over A. □

4 The Star Product and the Star Sum; Star Modules and Fractional Ideals

In this section we do first steps into a multiplicative ideal theory in connection with star operations, explicating how central notions of the classical "star-multiplicative ideal theory" (cf. e.g. [Gi]) can be extended to our general setting. In later sections we will not continue this study, developing and using star operations for other purposes.

In the following $A \subset R$ is a ring extension, and $* : J(A, R) \to J(A, R)$ is a star operation. We draw some consequences from the axioms St1–St4 (Sect. 3, Definition 1).

Proposition 4.1.

a) *If* $I, J \in J(A, R)$ *then* $(IJ)^* = (I^* J)^* = (IJ^*)^* = (I^* J^*)^*$.
b) *If in addition* I *is* R*-invertible then* $IJ^* = (IJ)^*$.

Proof. a) By St4 we have $IJ^* \subset (IJ)^*$. Using St2 and St3 we obtain $(IJ^*)^* \subset (IJ)^{**} = (IJ)^*$. On the other hand $IJ \subset IJ^*$, hence $(IJ)^* \subset (IJ^*)^*$. This proves that $(IJ)^* = (IJ^*)^*$ and, due to symmetry in I, J, also that $(IJ)^* = (I^* J)^*$. Replacing J by J^* we obtain $(IJ^*)^* = (I^* J^*)^*$.
b) By St4 we have $IJ^* \subset (IJ)^*$. Multiplying with I^{-1} and applying St4 again we obtain $J^* \subset I^{-1}(IJ)^* \subset J^*$, hence $J^* = I^{-1}(IJ)^*$, which gives the claim. □

Proposition 4.2.

a) *If* C *is an overring of* A *in* R *then* C^* *is again a subring of* R. *If* $I \in J(C, R)$ *then* $I^* \in J(C^*, R) \subset J(C, R)$ *and our star operations gives by restriction a star operation* $I \mapsto I^*$ *on* $J(C, R)$ *and a strict star operation on* $J(C^*, R)$.

b) In particular A^* is an overring of A in R and our star operation restricts to a strict star operation on $J(A^*, R)$.

c) For every $I \in J(A, R)$ we have $I^* = (IA^*)^*$.

Proof. a): We have $C \in J(A, R)$ and $C^*C^* \subset (CC)^* = C^*$ by Proposition 1(a). Thus C^* is a subring of R. It contains C. If $I \in J(C, R)$ then, again by Proposition 1, $C^*I^* \subset (CI)^* = I^*$. Thus I^* is a C^*-submodule of R. Thus our star operation restricts to a map $I \mapsto I^*$ from $J(C, R)$ to $J(C^*, R) \subset J(C, R)$. It is trivial that the map $I \mapsto I^*, J(C, R) \to J(C, R)$, obeys again the axioms St1–St4. We may replace C by C^*, and then obtain a star operation $I \mapsto I^*$ on $J(C^*, R)$. It is strict since $(C^*)^* = C^*$.

b): The special case $C = A$ of (a).

c): By Proposition 1 we have $I^* = IA^* = (IA^*)^*$. □

Remark 4.3. By part (b) and (c) of the proposition the map $* : J(A, R) \to J(A, R)$ factors as follows

$$* : J(A, R) \xrightarrow{\alpha} J(A^*, R) \xrightarrow{\beta} J(A^*, R) \xhookrightarrow{\gamma} J(A, R)$$

with $\alpha(I) := IA^*$, β a strict star operation, which maps $K \in J(A^*, R)$ to K^*, and γ the inclusion map from $J(A^*, R)$ to $J(A, R)$. This reveals that any star operation is closely related to a strict star operation. We call β the *strict star operation associated* to $* : J(A, R) \to J(A, R)$.

Definition 1.

a) We call an A-module $I \in J(A, R)$ a *star module*, more precisely an *A-star module* (in the case $I \subset A$: a *star ideal*) with respect to the given star operation, if $I^* = I$. The set of all star modules will be denoted by $J_*(A, R)$.

b) If $I, J \in J(A, R)$ are given we define the *star product* $I \circ J$ by

$$I \circ J = (IJ)^* = (I^*J^*)^*.$$

If necessary, we write more precisely $I \circ_* J$ to indicate the dependence on the chosen star operation. (If α is another star operation we have to distinguish between $I \circ_* J$ and $I \circ_\alpha J$.)

c) We say that $I \in J(A, R)$ is *star invertible*, if there exists some $K \in J(A, R)$ with $I \circ K = A^*$, and then call K a *star inverse* of I.

This is perhaps the right place to indicate the—as it seems—dominant motivation of people since a long time to be interested in star operations. In [Vol. I, Chap. II] we defined the group $D(A, R)$ of R-invertible A-submodules of R. This abelian group, together with its partial ordering by the inclusion relation, seems to be a primordial object of multiplicative ideal theory (in the classical setting, where A is an integral domain and R its field of quotients[8]).

If $A \subset R$ is a Prüfer extension we know that $D(A, R)$ is lattice ordered group (cf. [Vol. I, Remarks II.6.1]) and things are fine. But in more general cases there may not exist enough invertible ideals of A to make the group $D(A, R)$ an interesting object.

One way to remedy this, is to introduce an equivalence relation $\sim$ on $J(A, R)$ compatible with multiplication and to look for a large relevant subgroup of the monoid of equivalence classes $J(A, R)/\sim$. Our given star operation provides such an equivalence relation.

Definition 2. We call two A-modules I, J in R *star-equivalent*, and write $I \sim J$, or more precisely $I \sim_* J$, if $I^* = J^*$.

By Proposition 1 it is evident that the star product induces on $J(A, R)/\sim$ a product which makes $J(A, R)/\sim$ an abelian monoid. Now every star-equivalence class contains a unique star module. Thus we may identify $J(A, R)/\sim$ with the monoid $J_*(A, R)$, which is the set of all A-star-submodules of R, equipped with the star product. The set of equivalence classes of star invertible A-submodules of R is a subgroup of the monoid $J(A, R)/\sim$, and this can be identified with a subgroup of $J_*(A, R)$ which we will denote by $\tilde{D}_*(A, R)$.[9]

We usually will work with star modules instead of star-equivalence classes of modules. It is only a matter of taste. Altogether we fix the following notations.

Definition 3.

a) $J_*(A, R)$ denotes the set of all star A-submodules of R.
b) $\tilde{D}_*(A, R)$ denotes the set of all *star invertible* star modules in R. If $I \in \tilde{D}_*(A, R)$, then the unique star module J with $I \circ J = A^*$ is called the *star inverse* of I.
c) We call a star module I *star finite*, if there exists a finitely generated A-submodule I_0 of I with $I_0^* = I$. We then also say more briefly that I is a *finite star module*. We denote the set of all finite A-star submodules of R by $J_*^f(A, R)$.
d) $\tilde{D}_*^f(A, R)$ denotes the set of all $I \in J_*^f(A, R)$ such that the star inverse I is again star finite.

If the given star operation is denoted, say, by α, then we write $J_\alpha(A, R)$, $\tilde{D}_\alpha(A, R)$ etc.

Let us assume for a moment, without much loss of generality, that our star operator is strict, i.e. $A^* = A$. Then we know by Proposition 1(b) that every R-invertible

[8]In most of the literature an additive notation is chosen and the order relation is reversed: $0 = A$ is the neutral element; $I \leq J$ iff $I \supset J$.

[9]Below (Definition 6) we will introduce a subgroup of $\tilde{D}_*(A, R)$ conisting of star invertible "fractional ideals", for which we reserve the notation $D_*(A, R)$.

A-submodule I of R is a star module with star inverse I^{-1}. {Take there $J = I^{-1}$}. Thus $D(A, R)$ is a subgroup of $\tilde{D}_*(A, R)$.

In the general case we have $\tilde{D}_*(A, R) = \tilde{D}_*(A^*, R)$. We conclude that $D(A^*, R)$ is a subgroup of $\tilde{D}_*(A, R)$. It is even a subgroup of $\tilde{D}_*^f(A, R)$, as is easily seen.

If $I = I^*$ is a star invertible A-submodule of R we want to give a description of its star inverse. More generally it will be useful to study $[I : J]^*$ and $[I^* : J]$ for any A-submodules I, J of R.

Proposition 4.4. *Let $I, J \in J(A, R)$. Then*

$$[I : J]^* \subset [I^* : J] = [I^* : J]^* = [I^* : J^*].$$

Proof. a) Using St4 and St2 we obtain

$$[I : J]^* J \subset ([I : J]J)^* \subset I^*.$$

Thus $[I : J]^* \subset [I^* : J]$.

b) Replacing I by I^* this gives us $[I^* : J]^* \subset [I^* : J]$. Since the reverse inclusion is trivial, we conclude that $[I^* : J]^* = [I^* : J]$.

c) Since $J \subset J^*$ we have $[I^* : J] \supset [I^* : J^*]$. On the other hand, if $x \in [I^* : J]$ then $xJ \subset I^*$, hence $xJ^* \subset (xJ)^* \subset I^*$. Thus $x \in [I^* : J^*]$. This proves that $[I^* : J] = [I^* : J^*]$. □

Proposition 4.5. *Assume that I is star invertible.*

a) Then $[A^* : I]$ is the star inverse of I^*, i.e. the unique star module which is star inverse to I.

b) For any $J \in J(A, R)$ we have $J \circ [A^* : I] = [J^* : I]$.

Proof. a) Let K denote a star module with $I \circ K = A^*$. We have $I[A^* : I] \subset A^*$, hence $I \circ [A^* : I] \subset A^*$. Star multiplication by K gives $[A^* : I] \subset K$. On the other hand $IK \subset I \circ K = A^*$. Thus $K \subset [A^* : I]$, and we conclude that $K = [A^* : I]$.

b) We have $[J^* : I]I \subset J^*$, hence $[J^* : I] \circ I \subset J^*$. Star multiplication by $[A^* \circ I]$ gives $[J^* : I] \subset J^* \circ [A^* : I] = J \circ [A^* : I]$. On the other hand, $J^*[A^* : I]I \subset J^*A^* = J^*$, hence $J^*[A^* : I] \subset [J^* : I]$. Now $[J^* : I]$ is a star module by Proposition 4. Thus $J^* \circ [A^* : I] \subset [J^* : I]$. It follows that $J \circ [A^* : I] = J^* \circ [A^* : I] = [J^* : I]$. □

We add some general facts about intersections and sums of star modules.

Proposition 4.6. *If $(I_\lambda \mid \lambda \in \Lambda)$ is a family in $J(A, R)$ then*

$$\bigcap_\lambda I_\lambda^* = (\bigcap_\lambda I_\lambda^*)^*.$$

Thus any intersection of star modules in R is again a star module.

Proof. Let $\mu \in \Lambda$ be given. Then we conclude from $\bigcap_\lambda I_\lambda^* \subset I_\mu^*$ that $(\bigcap_\lambda I_\lambda^*)^* \subset I_\mu^*$. Thus $(\bigcap_\lambda I_\lambda^*)^* \subset \bigcap_\lambda I_\lambda^*$. The reverse inclusion is trivial. □

Proposition 4.7. *Let $(I_\lambda \mid \lambda \in \Lambda)$ be any family in $J(A, R)$.*

a) Then $(\sum\limits_\lambda I_\lambda)^* = (\sum\limits_\lambda I_\lambda^*)^*$.

b) If K is another A-submodule of R then

$$(\sum_\lambda I_\lambda) \circ K = (\sum_\lambda I_\lambda \circ K)^*.$$

Proof. a): For any $\mu \in \Lambda$ the module I_μ^* is contained in $(\sum_\lambda I_\lambda)^*$. Thus $\sum_\mu I_\mu^* \subset (\sum_\lambda I_\lambda)^*$, and we conclude that $(\sum_\lambda I_\lambda^*)^* \subset (\sum_\lambda I_\lambda)^*$. The reverse inclusion is trivial.

b): Using (a) we obtain

$$(\sum_\lambda I_\lambda) \circ K = [(\sum_\lambda I_\lambda)K]^* = (\sum_\lambda I_\lambda K)^* = [\sum_\lambda (I_\lambda K)^*]^* = (\sum_\lambda I_\lambda \circ K)^*.$$

□

Definition 4. The *star sum* of a family $(I_\lambda \mid \lambda \in \Lambda)$ in $J(A, R)$ is the star ideal $(\sum_\lambda I_\lambda)^*$. We denote it by $\sum_{*\lambda} I_\lambda$. In the case of finitely many elements $I_1 \ldots, I_r$ we also write $I_1 +_* I_2 +_* \cdots +_* I_r$.

It is now evident that the set $J_*(A, R)$ of all A-star modules in R is a complete lattice under the inclusion relation. The infimum of a family $(I_\lambda | \lambda \in \Lambda)$ of star ideals is $\bigcap_\lambda I_\lambda$ and the supremum is $\sum_{*\lambda} I_\lambda$. Moreover we have the law

$$(\sum\nolimits_{*\lambda} I_\lambda) \circ K = \sum\nolimits_{*\lambda} (I_\lambda \circ K)$$

for any $K \in J(A, R)$. This follows from Proposition 8(b) and the trivial observation that

$$(\sum\nolimits_\lambda I_\lambda)^* \circ K = (\sum\nolimits_\lambda I_\lambda) \circ K.$$

The star sum $+_*$ and the star product $\circ$ make $J_*(A, R)$ a commutative semiring (with 1), and $\tilde{D}_*(A, R)$ is its group of units.

Proposition 4.8. *If $\mathfrak{a}$ is R-invertible, then for every star module $I \in J(A, R)$ the A-module $\mathfrak{a}I$ is again a star module and $\mathfrak{a}I = \mathfrak{a} \circ I$. If $I \in \tilde{D}_*(A, R)$, then $\mathfrak{a}I \in \tilde{D}_*(A, R)$.*

Proof. This follows from the relation $\mathfrak{a}I^* = (\mathfrak{a}I)^*$, stated in Proposition 1(b). □

In classical multiplicative ideal theory, where A is a domain, $R = \mathrm{Quot}(A)$ and $A^* = A$, a central notion is that of a "fractional ideal". It is defined as follows.

A *fractional ideal* of A is an A-submodule $L \neq \{0\}$ of R such that there exists some $d \in A \setminus \{0\}$ with $L \subset d^{-1}A$.[10] One then studies the group of those fractional ideals which are star modules in present terminology. A comparison of this group with our group $\tilde{D}_*(A, R)$ reveals that $\tilde{D}_*(A, R)$ is "too big". Thus a definition of fractional ideals is also needed in our much more general setting. We propose the following.

Definition 5. An *R-fractional ideal* of A is an A-submodule I of R such that there exists an R-invertible ideal $\mathfrak{a}$ of A with $\mathfrak{a} \subset I \subset \mathfrak{a}^{-1}$. The set of all these modules will be denoted by $\text{Fract}(A, R)$.

As long as the extension $A \subset R$ is fixed we usually say "fractional ideal" instead of "R-fractional ideal".

Notice that, if A is a domain and R its quotient field, this definition of fractional ideal coincides with the classical one, since any invertible A-module is finitely generated ([Vol. I, Remarks II.1.10]), hence contained in $d^{-1}A$ for some $d \in A\setminus\{0\}$.

Proposition 4.9.

a) *If I and J are fractional ideals of A then $I + J$, $I \cap J$, IJ and $[I : J]$ are again fractional ideals.*

b) *If A^* is a fractional ideal of A, then, for any $I \in \text{Fract}(A, R)$, also I^* is a fractional ideal. In particular this holds if the star operation is strict.*

Proof. a) Assume that $\mathfrak{a} \subset I \subset \mathfrak{a}^{-1}$ and $\mathfrak{b} \subset J \subset \mathfrak{b}^{-1}$ with $\mathfrak{a}, \mathfrak{b} \in \text{Inv}(A, R)$. Then $\mathfrak{a}\mathfrak{b} \subset K \subset \mathfrak{a}^{-1}\mathfrak{b}^{-1}$, if K is any of the modules $I + J, I \cap J, IJ$. Also

$$[I : J] \subset [I : \mathfrak{b}] \subset [\mathfrak{a}^{-1} : \mathfrak{b}] = \mathfrak{a}^{-1}\mathfrak{b}^{-1},$$

and

$$[I : J] \supset [I : \mathfrak{b}^{-1}] \supset [\mathfrak{a} : \mathfrak{b}^{-1}] = \mathfrak{a}\mathfrak{b}.$$

b) We have $A \subset A^* \subset \mathfrak{b}^{-1}$ with some ideal $\mathfrak{b} \in \text{Inv}(A, R)$. Assume that $\mathfrak{a} \subset I \subset \mathfrak{a}^{-1}$ with $\mathfrak{a} \in \text{Inv}(A, R)$. Since $\mathfrak{a}^* = \mathfrak{a}A^*$ and $(\mathfrak{a}^{-1})^* = \mathfrak{a}^{-1}A^*$ by Proposition 1(b) we conclude that

$$\mathfrak{a} \subset \mathfrak{a}A^* \subset I^* \subset \mathfrak{a}^{-1}A^* \subset \mathfrak{a}^{-1}\mathfrak{b}^{-1},$$

hence a fortiori $\mathfrak{a}\mathfrak{b} \subset I^* \subset \mathfrak{a}^{-1}\mathfrak{b}^{-1}$. □

Scholium 4.10. The map $* : J(A, R) \to J(A, R)$ restricts to a map $I \mapsto I^*$ from $\text{Fract}(A, R)$ to itself. This map is completely determined by the values I^* of the ideals $I \in J(A) \cap \text{Fract}(A, R)$ (the fractional ideals which are "integral" in classical terminology). Indeed, if $I \subset \mathfrak{a}^{-1}$ with some $\mathfrak{a} \in \text{Inv}(A, R)$, then $\mathfrak{a}I \subset A$ and $I^* = \mathfrak{a}^{-1}(\mathfrak{a}I)^*$.

---

[10]Notice that L is **not** an ideal of A in the usual sense.

Definition 6. We define the following sets:

$\mathrm{Fract}_*(A,R) := \mathrm{Fract}(A,R) \cap J_*(A,R) = \{I \in \mathrm{Fract}(A,R) \mid I = I^*\}$,
$D_*(A,R) := \mathrm{Fract}_*(A,R) \cap \tilde{D}_*(A,R) = \{I \in \mathrm{Fract}_*(A,R) \mid I \text{ is star invertible}\}$,
$D_*^f(A,R) := D_*(A,R) \cap \tilde{D}_*^f(A,R) = \{I \in D_*(A,R) \mid I$ and $[A^* : I]$ are star finite$\}$.

$\mathrm{Fract}_*(A,R)$ is a subsemiring of $J_*(A,R)$, while $D_*(A,R)$ and $D_*^f(A,R)$ are subgroups of $\tilde{D}_*(A,R)$. In the classical setting, where A is a domain and R is its quotient field, the group $D_*(A,R)$ seems to be the central object of study if a star operation is present. In the case $* = \delta$ it is called the *"group of divisors"* of A, cf. [Bo, VII §1] and [Gi, §34].

In the classical setting, where A is an integral domain and R is its quotient field, there exists a very extended star-multiplicative "ideal theory" with impressive results, and it is an active area of research also now, cf. e.g. [AFZ, GP, Z].

There also exists an encouraging such literature for A a ring and R its total ring of quotients $\mathrm{Quot}(A)$ or, rather rarely, its complete ring of quotients $Q(A)$. But very little has been done beyond that.

While we believe that our definition of fractional ideals and the groups $D_*(A,R)$, $D_*^f(A,R)$ provides a good starting point for a general star multiplicative ideal theory, we refrain from delving seriously into it in the present chapter (and volume). It would be an all too big endeavour.

5 The Condition e.a.b.

The condition e.a.b. (= "endlich arithmetisch brauchbar") is a cancellation property for star products. Let $A \subset R$ be any ring extension.

Definition 1. Let $*$ be a star operation on $J(A,R)$ and $\mathcal{G}$ a multiplicative filter on R.

We say that $*$ is *e.a.b. for* $\mathcal{G}$, if the following holds: If $I, J, K \in J^f(A,R)$, $IR \in \mathcal{G}$ and $(IJ)^* = (IK)^*$, then $J^* = K^*$. We say that $*$ is e.a.b., if $*$ is e.a.b. for $\mathcal{G} = \{R\}$.

For A a domain, R its field of quotients, $\mathcal{G} = \{R\}$, a condition of this type has been discussed already by Gilmer [Gi, §32], but Gilmer restricts from the beginning to non zero fractional ideals. We leave a comparison of Gilmer's approach to our's (for A a domain and $R = \mathrm{Quot}(A)$) to the interested reader.

We can express the condition e.a.b. in other ways.

Proposition 5.1. *The following are equivalent:*

(1) $*$ *is e.a.b. for* $\mathcal{G}$.
(2) If $I, J, K \in J^f(A,R)$, $IR \in \mathcal{G}$, *and* $(IJ)^* \subset (IK)^*$, *then* $J^* \subset K^*$.
(3) If $I, K \in J^f(A,R)$, $IR \in \mathcal{G}$, *then* $[(IK)^* : I] = K^*$.

Proof. (1) ⇒ (2): Let $I, J, K \in J^f(A, R)$ be given with the conditions stated in (2). Then (cf. Proposition 4.7)

$$(IJ + IK)^* = ((IJ)^* + (IK)^*)^* = (IK)^{**} = (IK)^*.$$

It follows by assumption (1) that $(J + K)^* = K^*$, hence $J^* \subset K^*$.

(2) ⇒ (1): trivial.

(2) ⇒ (3): Let $I, K \in J^f(A, R)$ be given with $IR \in \mathcal{G}$. Without using (2) we know that $K^* \subset [(IK)^* : I]$. Let $x \in [(IK)^* : I]$. Then $I(Ax) \subset (IK)^*$, hence $(I(Ax))^* \subset (IK)^*$. By (2) we conclude that $(Ax)^* \subset K^*$, hence $x \in K^*$.

(3) ⇒ (2): Let I, J, K be given with the conditions stated in (2). From $IJ \subset (IJ)^* \subset (IK)^*$ and (3) we conclude that $J \subset [(IK)^* : I] = K^*$, hence $J^* \subset K^*$.□

We give an example of an e.a.b. star operation which will play a crucial role later on.

Let $A \subset R$ be a ring extension and $v : R \to \Gamma \cup \infty$ a valuation on R over A. For any $I \in J(A, R)$ let I^v denote the v-convex hull of I in R, i.e.

$$I^v = \{x \in R \mid \exists\, a \in I \text{ with } v(x) \geq v(a)\}.$$

We check the axioms St1–St3 and St4′ for the operation $I \mapsto I^v$. St1–St3 are evident. To prove St4′, let $I \in J(A, R)$ and $a \in R$ be given. If $x \in I^v$ there exists some $b \in I$ with $v(x) \geq v(b)$. Then $v(ax) \geq v(ab)$, hence $ax \in (aI)^v$. Thus indeed $aI^v \subset (aI)^v$. We have verified that $I \mapsto I^v$ is a star operation on $J(A, R)$. Notice that for $I = A$ we have $I^v = A_v$.

Theorem 5.2. *The star operation $I \mapsto I^v$ on $J(A, R)$ is e.a.b. for*

$$\mathcal{G}_v := \{I \in J(R) \mid I \not\subset \operatorname{supp} v\}.$$

To prove this we need some preparations.

Definition 2. We call an additive subgroup M of R *v-finite*, if

$$\exists \min_{x \in M} v(x) =: v(M) \in \Gamma \cup \infty.$$

Remarks 5.3.

i) If the additive group M is contained in supp v then M is v-finite, $v(M) = \infty$.
ii) If B is a subring of A_v and $M = Ba_1 + \ldots Ba_n$ is a finitely generated B-submodule of R, then M is v-finite and $v(M) = \min_i v(a_i)$. (N.B. We have used this fact already in Sect. 1.)
iii) If M is v-finite, the same holds for the v-convex hull M^v of M, and $v(M^v) = v(M)$. (The v-convex hull is defined as above for A-modules. We may take $A = \mathbb{Z}1_R$.)
iv) If M and N are v-finite then $M + N$ and MN are v-finite, and $v(M + N) = \min(v(M), v(N)), v(MN) = v(M) + v(N)$.

v) If M, N are v-finite additive subgroups of R, then

$$v(M) \geq v(N) \iff M^v \subset N^v.$$

Lemma 5.4. *Let L, M, N be v-finite additive subgroups of R. Assume that $(LM)^v \subset (LN)^v$ and $L \not\subset \operatorname{supp} v$. Then $M^v \subset N^v$.*

Proof. By the remarks we have $v(LM) = v((LM)^v) \geq v((LN)^v) = v(LN)$ hence $v(L) + v(M) \geq v(L) + v(N)$. Now $v(L) \neq \infty$. We conclude that $v(M) \geq v(N)$. This means that $M^v \subset N^v$. □

Proof of Theorem 5.2. Let I, J, K be finitely generated A-submodules of R and $IR \in \mathcal{G}_v$. Assume that $(IJ)^v \subset (IK)^v$. We have $IR \not\subset \operatorname{supp} v$. Thus $I \not\subset \operatorname{supp} v$. The lemma applies and yields $J^v \subset K^v$. This proves that the star operation $I \mapsto I^v$ is e.a.b. for $\mathcal{G}_v$. □

Lemma 5.5. *If $(\alpha_\lambda \mid \lambda \in \Lambda)$ is a family of star operations on $J(A, R)$ which all are e.a.b. for $\mathcal{G}$, then $\alpha := \Lambda_\lambda \alpha_\lambda$ (cf. 3.8) is e.a.b. for $\mathcal{G}$.*

Proof. Let $I, J, K \in J^f(A, R)$ be given with $IR \in \mathcal{G}$. Assume that $(IJ)^\alpha \subset (IK)^\alpha$. For every $\lambda \in \Lambda$ we have $IJ \subset (IK)^{\alpha_\lambda}$, hence $(IJ)^{\alpha_\lambda} \subset (IK)^{\alpha_\lambda}$. We conclude that $J^{\alpha_\lambda} \subset K^{\alpha_\lambda}$. Taking intersections we obtain $J^\alpha \subset K^\alpha$. □

By use of this lemma, Theorem 2 can be amplified as follows.

Theorem 5.6. *Let $A \subset R$ be a ring extension and Φ a set of valuations on R over A. We define*

$$\mathcal{G} := \bigcap_{v \in \Phi} \mathcal{G}_v = \{I \in J(R) \mid I \not\subset \operatorname{supp} v \textit{ for every } v \in \Phi\}.$$

This is a multiplicative filter on R. The star operation

$$I \mapsto I^* := \bigcap_{v \in \Phi} I^v$$

is e.a.b. for $\mathcal{G}$. □

Definition 3. We call $I^* = \bigcap_{v \in \Phi} I^v$ the *Φ-convex hull* of I in R, and we say that I is *Φ-convex* if $I = I^*$.[11] We denote the star operation $I \mapsto I^*$ by con_Φ. If $\Phi = \{v\}$ consists of a single valuation v, we speak of the v-convex hull (as above) and denote the v-convex hull operation by con_v.

Here is another class of examples of e.a.b star operations.

---

[11] In much of the literature the word "complete" is used instead of "convex". We are afraid of conflict with the topological meaning of "complete". (Any family Φ of valuations on R gives a uniform topology on R.)

Proposition 5.7. *Assume that $A \subset R$ is a Prüfer extension. Then every star operation on $J(A, R)$ is e.a.b. for $\{R\}$.*

Proof. Let $*$ be a star operation on $J(A, R)$, and assume that $I, J, K \in J^f(A, R), IR = R$, and $(IJ)^* = (KJ)^*$. Then I is R-regular, hence R-invertible (cf. [Vol. I, Theorem II.2.1]). We conclude by Proposition 4.1(b) that $IJ^* = IK^*$, then, multiplying by I^{-1}, that $J^* = K^*$. □

We indicate a way to obtain from a given e.a.b star operation a new one. Together with Proposition 3.7 this will provide us with very many e.a.b star operations.

Proposition 5.8. *Let $\varphi : (R, A) \to (T, B)$ be a morphism of ring extensions.*

a) Let γ be a star operation on $J(B, T)$. Then $\alpha : J(A, R) \to J(A, R)$, defined by $I^\alpha = R \cap (IB)^\gamma$,[12] is a star operation on $J(A, R)$.
b) Let $\mathscr{H}$ be a multiplicative filter on T and $\mathscr{G} := \{I \in J(R) \mid IT \in \mathscr{H}\}$, which is a multiplicative filter on R. If γ is e.a.b for $\mathscr{H}$, then α is e.a.b for $\mathscr{G}$.

Proof. a): We check the axioms St1–St4 for α. St1 and St2 are obvious.
St3: For any $I \in J(A, R)$ we have

$$IB \subset I^\alpha B = [R \cap (IB)^\gamma]B \subset (IB)^\gamma,$$

hence

$$(*) \qquad (IB)^\gamma = (I^\alpha B)^\gamma.$$

We conclude that

$$(I^\alpha)^\alpha = R \cap (I^\alpha B)^\gamma = R \cap (IB)^\gamma = I^\alpha.$$

St4: For $I, J \in J(A, R)$ we have

$$IJ^\alpha = I[R \cap (JB)^\gamma] \subset IR \cap [I(JB)^\gamma] \subset R \cap (IJB)^\gamma = (IJ)^\alpha.$$

We now know that α is a star operation.

b): Let $I, J, K \in J^f(A, R)$ be given with $IR \in \mathscr{G}$ and $(IJ)^\alpha = (IK)^\alpha$. Using the relation $(*)$ above we obtain

$$(IJB)^\gamma = ((IJ)^\alpha B)^\gamma = ((IK)^\alpha B)^\gamma = (IKB)^\gamma.$$

Furthermore $IT = (IR)T \in \mathscr{H}$. Since γ is e.a.b for $\mathscr{H}$ we conclude that $(JB)^\gamma = (KB)^\gamma$, hence $J^\alpha = K^\alpha$. □

---

[12] Recall that for $H \in J(B, T)$ we use $R \cap H$ as an abbreviation of $\varphi^{-1}(H)$, and that IB means $\varphi(I)B$.

Definition 4. We call the star operation α given in Proposition 8 the *pull back of* γ *by the morphism* φ, and we call $\mathscr{G}$ the *pull back of the filter* $\mathscr{H}$ *by* φ.

If we choose for γ the identity operation $H \mapsto H$ on $J(B,T)$ we obtain as pull back the star operation induced by φ, as described in Example 3.3. Thus Propositions 7 and 8 together give us the following fact, worth to be stated separately.

Theorem 5.9. *Let* $\varphi : (R,A) \to (T,B)$ *be a morphism of ring extensions. Let* $\mathscr{G} := \{I \in J(R) \mid IT = T\}$*, which is a multiplicative filter on* R*. Assume that* B *is Prüfer in* T*. Let* γ *be any star operation on* $J(B,T)$*. The pull back of* γ *by* φ *is a star operation on* $J(A,R)$ *which is e.a.b for* $\mathscr{G}$*.*

Example 5.10. We analyse the pull backs of two star operations by a morphism $\varphi : (R,A) \to (T,B)$ in the case that $B \subset T$ is a PM-extension. We are given a PM-valuation w on T with $A_w = B$. Let $v := w|R := w \circ \varphi$.

a) First we look at the pull back α of the w-convex hull operation con_w on $J(B,T)$. For $I \in J(A,R)$ we have $I^\alpha = R \cap (IB)^w$. Obviously this is I^v. Thus we have obtained the operation con_v on $J(A,R)$. The filter

$$\mathscr{H} := \{H \in J(T) \mid H \not\subset \operatorname{supp} w\} = \mathscr{G}_w$$

has the pull back

$$\mathscr{G} := \{I \in J(R) \mid I \not\subset \operatorname{supp} v\} = \mathscr{G}_v.$$

By Theorem 2 we know that con_w is e.a.b for $\mathscr{G}_w$ and con_v is e.a.b for $\mathscr{G}_v$. Without invoking Theorem 2 we obtain from Propositions 7 and 8 the weaker result, that α is e.a.b for the smaller filter

$$\mathscr{G}' := \{I \in J(R) \mid IT = T\}.$$

b) Now we take the pull back β of the identity operation id on $J(B,T)$ by φ. For any $I \in J(A,R)$ we have $I^\beta = R \cap (IB) = R \cap (IA_w)$, and we know by Propositions 7 and 8 that β is e.a.b for $\mathscr{G}'$. Since $\mathrm{id} \le \mathrm{con}_w$ in $\mathrm{Star}(T/B)$, we have $\beta \le \alpha$ in $\mathrm{Star}(R/A)$. Of course, this can also be seen by direct inspection.

It is interesting to realize that both operations α and β coincide on the set of all $I \in J(A,R)$ with $IR \in \mathscr{G}'$. Indeed, if $IR \in \mathscr{G}'$, then IB is T-regular, and we conclude by [Vol. I, Theorem III.2.2] that IB is w-convex, $IB = (IB)^w$. Intersecting with R we obtain $I^\beta = I^\alpha$. There seems to be no reason in general that α and β coincide on the bigger set

$$\{I \in J(A,R) \mid IR \in \mathscr{G}_v\} = \{I \in J(A,R) \mid I \not\subset \operatorname{supp} v\}.$$

□

6 Star Operations of Finite Type

We introduce more terminology which will serve us to round off several corners left in previous sections.

Definition 1. Let $*$ be a star operation on $J(A, R)$. We say that $*$ is of *finite type* if for every $I \in J(A, R)$ the module I^* is the union of the modules K^* with $K \in J^f(A, R)$ and $K \subset I$.

Notice that this condition may be stronger than just saying that every star module is a union of star finite star modules.

Example 6.1. Let $*$ be the star operation induced by a morphism of ring extensions $\varphi : (R, A) \to (T, B)$ (cf. Example 3.3), i.e. $I^* = R \cap (IB)$ for $I \in J(A, R)$. This star operation is of finite type. Indeed, if K runs through the finitely generated A-submodules of I, then $IB = \bigcup_K KB$, hence $R \cap (IB) = \bigcup_K R \cap KB$.

Example 6.2. Let $A \subset R$ be any ring extension and Φ a finite set of valuations on R over A. Then the Φ-convex hull operation on $J(A, R)$,

$$\mathrm{con}_\Phi(I) := \bigcap_{v \in \Phi} I^v,$$

is of finite type. Indeed, if $I \in J(A, R)$ and $x \in \mathrm{con}_\Phi(I)$ there exists for every $v \in \Phi$ some $y_v \in I$ with $v(x) \geq v(y_v)$. The module $K := \sum_v Ay_v$ is finitely generated, and $x \in \mathrm{con}_\Phi(K)$.

Not all relevant star operations are of finite type. In particular $\delta : I \mapsto [A : [A : I]]$ (cf. Proposition 3.6.) often is not of finite type. Nevertheless star operations of finite type abound.

Proposition 6.3. *Let $* : J(A, R) \to J(A, R)$ be any star operation. We define a map $\alpha : J(A, R) \to J(A, R)$ by $I^\alpha := \bigcup_K K^*$ with K running through the finitely generated A-submodules of $I \in J(A, R)$. Then α is a star operation of finite type, and $I^\alpha \subset I^*$ for every $I \in J(A, R)$.*

Proof. a) We check the validity of the axioms St1–St3 and St4′ for α (cf. Sect. 3). St1 and St2 hold obviously.

St3: Let $I \in J(A, R)$ be given and let L be a finitely generated A-submodule of I^α. Now I^α is the union of the modules K^* with K running through the finitely generated submodules of I. Since L is finitely generated we have $L \subset K^*$ for some K. This implies $L^* \subset K^* \subset I^\alpha$. We conclude that $(I^\alpha)^\alpha \subset I^\alpha$, hence $(I^\alpha)^\alpha = I^\alpha$.

St4′: Let $a \in R$ and $I \in J(A, R)$ be given. We have $I^\alpha = \bigcup_K K^*$ with K running through the finitely generated submodules of I. We conclude that

$$aI^\alpha = \bigcup_K aK^* \subset \bigcup_K (aK)^* \subset (aI)^\alpha,$$

since each aK is again finitely generated.

b) If $K \in J(A, R)$ is finitely generated, then clearly $K^{\alpha} = K^{*}$. Thus for any $I \in J(A, R)$ we have

$$I^{\alpha} = \bigcup_{K} K^{*} = \bigcup_{K} K^{\alpha}$$

with K running through the finitely generated submodules of I. Thus α is of finite type. Since $K^{*} \subset I^{*}$ for every K we conclude that $I^{\alpha} \subset I^{*}$. □

Definition 2. We call α the *companion of* $*$ *of finite type* (or: *the finite type companion of* $*$). We denote this operation α by $*_f$.

Remark 6.4. $*_f$ is the coarsest star operation α of finite type with $\alpha \le *$ in the natural partial ordering on Star(A, R) (Sect. 3, Definition 2) since we have $\alpha = \alpha_f \le *_f \le *$ for any such α.

In the following $A \subset R$ is any ring extension and $*$ is a star operation on $J(A, R)$. The transit from $*$ to $*_f$ is compatible with taking pullbacks.

Proposition 6.5. *Let $\varphi : (R, A) \to (T, B)$ be a morphism of ring extensions. Let γ be a star operation on $J(B, T)$ and α its pullback to $J(A, R)$ by φ. Then α_f is the pullback of γ_f by φ.*

Proof. Let β denote the pull back of γ_f by φ. For $I \in J(A, R)$ we have

$$(IB)^{\gamma_f} = \bigcup_{K} (KB)^{\gamma},$$

with K running through the finitely generated A-submodules of I, since every finitely generated B-submodule L of IB is contained in KB for some K, and KB is again a finitely generated B-module. Intersecting with R we obtain

$$I^{\beta} = \bigcup_{K} (R \cap (KB)^{\gamma}) = \bigcup_{K} K^{\alpha} = I^{\alpha_f}.$$

Thus $\beta = \alpha_f$. □

If γ is the identity operation on $J(B, T)$ the pull back α is the star operation induced by φ. We see by Proposition 5 that this operation is of finite type, as already observed in Example 1.

The star operations of finite type are easier to be handled than the other ones, as is witnessed by

Proposition 6.6. *If $*$ is of finite type then every star invertible A-star module in R is star finite. Thus $\tilde{D}_*(A, R) = \tilde{D}_*^f(A, R)$ and $D_*(A, R) = D_*^f(A, R)$ (cf. Sect. 4, Definitions 3 and 6).*

Proof. Let I be a star invertible star module, and let J denote its star inverse. Then $(IJ)^{*} = A^{*}$. Since $*$ is of finite type there exist finitely generated A-modules $K \subset I$

and $L \subset J$ with $1 \in (KL)^*$, hence $(KL)^* = A^*$. A fortiori $(KJ)^* = A^*$. From $I \circ J = A^*$ and $K \circ J = A^*$ we conclude that I and K are star equivalent, hence $I = K^*$. □

If we use star operations to build Kronecker extensions, we can always retreat to star operations of finite type. The same holds for the condition e.a.b from Sect. 5.

Remarks 6.7. Let $*$ be a star operation on $J(A, R)$ and $\mathcal{G}$ a multiplicative filter on R.

i) Then $A(X, \mathcal{G}, *) = A(X, \mathcal{G}, *_f)$. This is evident from the definition of $A(X, \mathcal{G}, *)$ (Sect. 3, Definition 3). Notice also that $A(X, \mathcal{G}, *) = A(X, \mathcal{G}_f, *)$.
ii) The operation $*$ is e.a.b for $\mathcal{G}$ iff $*_f$ is e.a.b for $\mathcal{G}$ iff $*_f$ is e.a.b for $\mathcal{G}_f$, as is evident from the definition of this property (Sect. 5, Definition 1).

7 Partial Star Operations, the Kronecker Operations

Let $A \subset R$ be a ring extension and $\mathcal{G}$ a multiplicative filter on R. Given a star operation $*$ on $J(A, R)$ we have introduced in Sect. 3 the Kronecker subring $A(X, \mathcal{G}, *)$ of $R(X, \mathcal{G})$. It will be in the focus of much of our study later on. But observe that, due to its definition (Sect. 3, Definition 3), $A(X, \mathcal{G}, *)$ is already determined by the values of $*$ on the subset $\Phi(\mathcal{G}/A)$ of $J(A, R)$, consisting of the modules $I \in J(A, R)$ with $IR \in \mathcal{G}$. (N.B. It is even determined by the values on $\Phi^f(\mathcal{G}/A)$.) Thus, mostly for systematic reasons, we now introduce "partial star operations", which are only defined on such sets $\Phi(\mathcal{G}/A)$ instead of $J(A, R)$.

Definition 1.

a) A *star operation on* $\Phi(\mathcal{G}/A)$ is a map $* : \Phi(\mathcal{G}/A) \to J(A, R), I \mapsto I^*$, which fulfills the axioms St1–St4 from Sect. 3, Definition 1, where now, of course the modules I, J there have to be taken in $\Phi(\mathcal{G}/A)$.
b) If we do not specify the filter $\mathcal{G}$, we say that such a map $*$ is a *partial star operation on* $J(A, R)$. We say that a partial star operation $*$ is *defined on* $\Phi(\mathcal{G}/A)$ if it is a star operation on $\Phi(\mathcal{H}/A)$ for some multiplicative filter $\mathcal{H}$ on R with $\mathcal{H} \supset \mathcal{G}$. (Notice that $\Phi(\mathcal{G}/A) \subset \Phi(\mathcal{H}/A)$ iff $\mathcal{G} \subset \mathcal{H}$.)
c) We call a partial star operation $*$ on $J(A, R)$ *strict*, if $A^* = A$.

If $*$ is a star operation on $\Phi(\mathcal{G}/A)$ then $I^* \in \Phi(\mathcal{G}/A)$ for every $I \in \Phi(\mathcal{G}/A)$. Thus the map $*$ goes in fact from $\Phi(\mathcal{G}/A)$ to $\Phi(\mathcal{G}/A)$.

If we take for $\mathcal{G}$ the biggest multiplicative filter $J(R)$ on R then $\Phi(\mathcal{G}/A) = J(A, R)$, and now a star operation on $\Phi(\mathcal{G}/A)$ is a star operation on $J(A, R)$ as defined in Sect. 3.

Comment. In the classical literature on multiplicative ideal theory, where A is a domain and R its quotient field, partial star operations in the present sense are no issue, since on R there exist only two multiplicative filters, $\{R\}$ and $\{R, \{0\}\}$.

But classically star operations are only defined on the subset Fract(A, R) of fractional ideals of $\Phi(\{R\}/A) = \{I \in J(A, R) \mid I \neq (0)\}$, cf. [Gi, §32]. In our much more general setting, where arbitrary ring extensions $A \subset R$ are admitted, it seems to be clumsy to develop the theory of star operations only for fractional ideals, as defined in Sect. 4. In general there seem not to be enough R-invertible A-modules at hand to make this agreeable, at least in the beginning of the theory. Notice also that for any $I \in$ Fract(A, R) we have $IR = R$. Thus to catch relevant A-modules I with $IR \in \mathcal{G}$ in case $\mathcal{G} \neq \{R\}$ (which is important for us, cf. Sect. 2), we first would need to develop a notion of "$\mathcal{G}$-fractional ideal". We refrain from doing this here.

Let Star$(\mathcal{G}/A)$ denote the set of star operations on $\Phi(\mathcal{G}/A)$ and let Star$_0(\mathcal{G}/A)$ denote the subset of strict star operations on $\Phi(\mathcal{G}/A)$. As in the special case $\mathcal{G} = J(R)$ we have a partial ordering on Star$(\mathcal{G}/A)$ decreeing for star operations α, β on $\Phi(\mathcal{G}/A)$ that $\alpha \leq \beta$ iff $I^\alpha \subset I^\beta$ for every $I \in \Phi(\mathcal{G}/A)$, and as in Sect. 3 we see that $\Phi(\mathcal{G}/A)$ with this ordering is a complete lattice. If α and β are partial star operations on $J(A, R)$, both defined on $\Phi(\mathcal{G}/A)$ we say that α is *finer than* β *on* $\Phi(\mathcal{G}/A)$, and β is *coarser then* α *on* $\Phi(\mathcal{G}/A)$ if $\alpha' \leq \beta'$ for the restrictions α' and β' of α and β to $\Phi(\mathcal{G}/A)$.

Almost all definitions in Sects. 3–6 related to star operations can be extended to partial star operations in an obvious way.[13] We mention two of them.

Definition 2. Let $*$ be a star operation on $\Phi(\mathcal{H}/A)$ for some multiplicative filter $\mathcal{H} \supset \mathcal{G}$ on R. We call $*$ *e.a.b for* $\mathcal{G}$, if for any three A-modules $I \in \Phi^f(\mathcal{G}/A)$, $J \in \Phi^f(\mathcal{H}/A)$, $K \in \Phi^f(\mathcal{H}/A)$ with $(IJ)^* = (IK)^*$ it follows that $J^* = K^*$.

Definition 3. Assume that $\mathcal{G}$ has finite type, i.e. $\mathcal{G} = \mathcal{G}_f$. A star operation $*$ on $\Phi(\mathcal{G}/A)$ is of *finite type*, if for $I \in \Phi(\mathcal{G}/A)$ the A-module I^* is the union of the modules K^* with K running through the finitely generated $\mathcal{G}$-regular A-submodules of I.

Notice that Definition 3 makes good sense due to the following easy lemma.

Lemma 7.1. *Assume again that* $\mathcal{G}$ *is of finite type. For any* $I \in \Phi(\mathcal{G}/A)$ *the set* $\{K \in \Phi^f(\mathcal{G}/A) \mid K \subset I\}$ *is cofinal in the set of all finitely generated* A-*submodules of* I.

Proof. It suffices to verify that there exists some $K_0 \in \Phi^f(\mathcal{G}/A)$ with $K_0 \subset I$. Then, for any A-submodule K of I the module $K_0 + K$ is contained in I and is an element of $\Phi^f(\mathcal{G}/A)$.

Since $RI \in \mathcal{G}$ and $\mathcal{G}$ has finite type, there exist finitely many elements $y_1, \ldots, y_n$ of RI such that $L := Ry_1 + \cdots + Ry_n \in \mathcal{G}$. Write

[13] A notable exception is the axiom St4′ which may loose its sense, since in general $\mathcal{G}$ may not contain enough principal ideals.

$$y_i = \sum_{j=1}^{N_i} \lambda_{ij} z_{ij}$$

with $z_{ij} \in I$ and $\lambda_{ij} \in R$. Then $K_0 := \sum_{ij} A z_{ij}$ is a finitely generated A-submodule of I and $RK_0 \supset L \in \mathscr{G}$, hence $RK_0 \in \mathscr{G}$. □

Also almost all statements involving star operations proved in Sects. 3–6 can be generalized to partial star operations in an obvious way, and will be used later without much comment. We only mention the following.

Assume that $\mathscr{G}$ is of finite type. Given any star operation $*$ on $\Phi(\mathscr{G}/A)$, we have an associated star operation $*_f$ of finite type on $\Phi(\mathscr{G}/A)$, defined by $I^{*_f} = \bigcup_K K^*$ with K running through all finitely generated A-submodules of I with $RK \in \mathscr{G}$, and we see as before that $*_f$ is the finest star operation α of finite type on $\Phi(\mathscr{G}/A)$ with $\alpha \leq *$. Again we call $*_f$ the *finite type companion* of $*$.

We now define Kronecker operations. They will be star operations on the whole of $J(A, R)$.

Let $*$ be a star operation on $J(A, R)$. We have the natural morphism of ring extensions $j = j_{\mathscr{G},*} : (R, A) \to (R(X, \mathscr{G}), A(X, \mathscr{G}, *))$, as explained in Sect. 3. Let α denote the star operation on $J(A, R)$ induced by j, i.e., for $I \in J(A, R)$, $I^\alpha = R \cap (IB)$ with $B := A(X, \mathscr{G}, *)$.

Definition 4. We call $*$ a *Kronecker operation on* $J(A, R)$ *for* $\mathscr{G}$, if $* = \alpha$.

Remarks 7.2.

i) If $*$ is a Kronecker operation for $\mathscr{G}$, then all values I^*, $I \in J(A, R)$, are determined by the values I^* for $I \in \Phi^f(\mathscr{G}/A)$, since only these values matter for the definition of $A(X, \mathscr{G}, *)$.
ii) We have observed in Sect. 6 that any star operation induced by a morphism of ring extensions is of finite type (Example 6.1). Thus every Kronecker operation is of finite type.
iii) A star operation on $J(A, R)$ is a Kronecker operation for $\mathscr{G}$ iff it is a Kronecker operation for $\mathscr{G}_f$, since both rings $R(X, \mathscr{G})$ and $A(X, \mathscr{G}, *)$ do not change if we replace $\mathscr{G}$ by $\mathscr{G}_f$.

Proposition 7.3. *Every Kronecker operation on* $J(A, R)$ *for* $\mathscr{G}$ *is e.a.b for* $\mathscr{G}$.

Proof. Let $*$ be such a Kronecker operation. Then $*$ is induced by $j = j_{\mathscr{G},*}$: $(R, A) \to (T, B)$ with $T = R(X, \mathscr{G})$ and $B = A(X, \mathscr{G}, *)$. Since B is Prüfer in T we know by Proposition 5.8 that $*$ is e.a.b for the filter

$$\mathscr{G}' := \{\mathfrak{a} \in J(R) \mid \mathfrak{a}T = T\}.$$

(Apply Proposition 5.8 with γ there the identity operation or $J(B, T)$.) Now it is rather obvious that $\mathscr{G}'$ contains the filter $\mathscr{G}_f$, hence $*$ is also e.a.b for $\mathscr{G}$. Indeed, if $\mathfrak{a} \in \mathscr{G}$ is given, and $\mathfrak{a}$ is finitely generated, we may choose a polynomial $f \in R[X]$

with $c_R(f) = \mathfrak{a}$. Then $f \in S_{\mathscr{G}}$, hence f is a unit in T. Trivially $\mathfrak{a}T \supset fT = T$, and thus $\mathfrak{a}T = T$. □

Remark 7.4. In fact $\mathscr{G}' = \mathscr{G}_f$. The inclusion $\mathscr{G}' \subset \mathscr{G}_f$ can be seen as follows. We have $\mathscr{G}' = \mathscr{G}'_f$. Let $\mathfrak{a} = Ra_1 + \cdots + Ra_n \in \mathscr{G}'$ be given. Then $\sum_1^n a_i T = T$. Thus there exist polynomials $u_1, \ldots, u_r \in R[X]$ and $g \in S_{\mathscr{G}}$ such that $\sum_1^n a_i u_i = g$. This implies that

$$\mathfrak{a} = \sum_1^n a_i R \supset \sum_1^n a_i c_R(u_i) \supset c_R(g),$$

hence $\mathfrak{a} \in \mathscr{G}$, since $c_R(g) \in \mathscr{G}$.

We describe a procedure which gives us all Kronecker operations on $J(A, R)$ for $\mathscr{G}$. If $*$ is a partial star operation defined on $\Phi(\mathscr{G}/A)$, we define the subring $A(X, \mathscr{G}, *)$ exactly as in Sect. 3, Definition 3. Again $A(X, \mathscr{G}, *)$ is a Kronecker subring of $R(X, \mathscr{G})$. We have $R(X, \mathscr{G}) = R(X, \mathscr{G}_f)$ and

$$A(X, \mathscr{G}, *) = A(X, \mathscr{G}_f, *) = A(X, \mathscr{G}_f, *_f)$$

as already stated in Sect. 3 for $*$ a star operation on $J(A, R)$, and again we have a natural morphism of ring extensions $j := j_{\mathscr{G},*} : (R, A) \to (R(X, \mathscr{G}), A(X, \mathscr{G}, *))$.

Theorem 7.5. *Let α denote the operation on $J(A, R)$ induced by this morphism j.*

a) $A(X, \mathscr{G}, \alpha) = A(X, \mathscr{G}, *)$.
b) α is a Kronecker operation for $\mathscr{G}$.

Proof. a): This follows from Theorem 3.11.
b): Now obvious. □

Definition 5. We call this Kronecker operation α the *Kronecker companion for $\mathscr{G}$* of the partial star operation $*$, and we write $\alpha = \operatorname{kro}(*, \mathscr{G})$.

We stress the fact that $\operatorname{kro}(*, \mathscr{G})$ is a star operation on $J(A, R)$ while $*$ may be a star operation on $\Phi(\mathscr{H}/R)$ for some multiplicative filter $\mathscr{H} \supset \mathscr{G}$. Also $\operatorname{kro}(*, \mathscr{G})$ is completely determined by the values of $*$ on $\Phi^f(\mathscr{G}/A)$. In particular $\operatorname{kro}(*, \mathscr{G}) = \operatorname{kro}(*_f, \mathscr{G}_f)$.

We compute the values of $\alpha = \operatorname{kro}(*, \mathscr{G})$ on $\Phi^f(\mathscr{G}/A)$ in terms of the values of $*$ on $\Phi^f(\mathscr{G}/A)$.

Theorem 7.6. *Let $*$ be a partial star operation defined on $\Phi^f(\mathscr{G}/A)$, and let α be its Kronecker companion for $\mathscr{G}$. For any $I \in \Phi^f(\mathscr{G}/A)$ we have*

$$I^\alpha = \bigcup_H [(IH)^* : H]$$

with H running through $\Phi^f(\mathscr{G}/A)$.

Proof. We abbreviate $T := R(X, \mathcal{G}), B := A(X, \mathcal{G}, *)$. Let $I \in \Phi^f(\mathcal{G}/A)$ be given. We choose a polynomial $f \in R[X]$ with $c_A(f) = I$. We have $c_A(f)R \in \mathcal{G}$, hence $f \in S_{\mathcal{G}}$.

a) Let $a \in [(IH)^* : H]$ for some $H \in \Phi^f(\mathcal{G}/A)$, hence $aH \subset (IH)^*$. The constant polynomial a has content $c_A(a) = aA$. Thus $c_A(a)H \subset (c_A(f)H)^*$, which says that $a/f \in B$. It follows that $a/1 \in fB$. Since $f \in S_{\mathcal{G}}$ we have $fB = c_A(f)B = IB$ (Proposition 2.6), and we conclude that $a/1 \in IB$, which means that $a \in R \cap (IB) = I^\alpha$. This proves that $[(IH)^* : H] \subset I^\alpha$ for every $H \in \Phi^f(\mathcal{G}/A)$.
b) Let $a \in I^\alpha = R \cap (IB)$. This means that $a \in R$ and $a/1 \in IB = fB$. Thus $a/1 = fg/h$ with $g \in R[X], h \in S_{\mathcal{G}}$, and $g/h \in B$, hence

(1) $ahh' = fgh'$ with some $h' \in S_{\mathcal{G}}$ and
(2) $c_A(g)U \subset (c_A(h)U)^*$ for some $U \in \Phi^f(\mathcal{G}/A)$.

Taking contents in (1) and applying the Dedekind–Mertens formula we see that there exists some $V \in \Phi^f(\mathcal{G}/A)$ with

$$ac_A(h)c_A(h')V = c_A(f)c_A(g)c_A(h')V.$$

Using (2) we conclude that

$$\begin{aligned} ac_A(h)c_A(h')(UV) &= c_A(f)c_A(g)c_A(h')UV \\ &\subset c_A(f)c_A(h')V(c_A(h)U)^* \\ &\subset (c_A(f)c_A(h')Vc_A(h)U)^*. \end{aligned}$$

Thus, with $\tilde{H} := c_A(h)c_A(h')(UV) \in \Phi^f(\mathcal{G}/A)$, we have $a\tilde{H} \subset (I\tilde{H})^*$, hence $a \in [(I\tilde{H})^* : \tilde{H}]$. This finishes the proof of the theorem. □

Corollary 7.7. *Assume that $*$ is a star operation on $\Phi^f(\mathcal{G}/A)$ which is e.a.b for $\mathcal{G}$. Let α be its Kronecker companion for $\mathcal{G}$ of $*$. Then $I^* = I^\alpha$ for every $I \in \Phi^f(\mathcal{G}/A)$. Thus α extends $*$ to a star operation on $J(A, R)$ which is again e.a.b for $\mathcal{G}$.*

Proof. If $I, H \in \Phi^f(\mathcal{G}/A)$ then $[(IH)^* : H] = I^*$, cf. Proposition 5.1. By Theorem 6 we conclude that $I^\alpha = I^*$. □

Proposition 7.8. *Let $*$ be a partial star operation defined on $\Phi(\mathcal{G}/A)$. Then $*_f \le \mathrm{kro}(*, \mathcal{G})$ on $\Phi(\mathcal{G}/A)$. If γ is a Kronecker operation for $\mathcal{G}$ on $J(A, R)$ with $*_f \le \gamma$ on $\Phi(\mathcal{G}/A)$ then $\mathrm{kro}(*, \mathcal{G}) \le \gamma$.*

Proof. Let $\alpha := \mathrm{kro}(*, \mathcal{G})$. If $I \in \Phi^f(\mathcal{G}/A)$ then we learn from Theorem 6 that $I^\alpha \supset [IA^* : A] = I^*$. Since α is of finite type it follows that $\alpha \ge *_f$ on $\Phi(\mathcal{G}/A)$.

Let now γ be a Kronecker operation on $J(A,R)$ with $\gamma \geq *_f$ on $\Phi(\mathcal{G}/A)$. By definition of $A(X,\mathcal{G},\gamma)$ and $A(X,\mathcal{G},*_f)$, we have

$$A(X,\mathcal{G},\gamma) \supset A(X,\mathcal{G},*_f) = A(X,\mathcal{G},*).$$

Thus the star operation induced by the natural morphism from (R,A) to $(R(X,\mathcal{G}),A(X,\mathcal{G},\gamma))$, which is γ, is coarser than the operation induced by the natural morphism from (R,A) to $(R(X,\mathcal{G}),A(X,\mathcal{G},*))$, which is α. □

Example 7.9. Let v be a $\mathcal{G}$-regular valuation on R over A, and let $* = \mathrm{con}_v$. This is the star operation $I \to I^v$ on $J(A,R)$. What can be said about the Kronecker companion $\alpha := \mathrm{kro}(*,\mathcal{G})$ of $*$? The ring $A(X,\mathcal{G},*)$ consists of all fractions $f/g \in R(X,\mathcal{G})$ with $f \in R[X], g \in S_{\mathcal{G}}$ and $c_A(f) \subset c_A(g)^v$. This is just the valuation ring A_{v^*} of the Gauß extension v^* of v to $R(X,\mathcal{G})$. Indeed, for $f \in R[X], g \in S_{\mathcal{G}}$ we have $v^*(f/g) = v(c_A(f)) - v(c_A(g))$, and this is ≥ 0 iff $c_A(f) \subset c_A(g)^v$. Thus α is the star operation induced by the natural morphism from (R,A) to $(R(X,\mathcal{G}),A_{v^*})$.

For $I \in J(A,\mathcal{G})$ we have $I^\alpha = R \cap (IA_{v^*})$. If $IR \in \mathcal{G}$, then $IR(X,\mathcal{G}) = R(X,\mathcal{G})$. Let I^{v^*} denote the v^*-convex hull of the additive subgroup $\{a/1 \mid a \in I\}$ of $R(X,\mathcal{G})$. If $IR \in \mathcal{G}$ then $IR(X,\mathcal{G}) = R(X,\mathcal{G})$, and it follows from [Vol. I, Corollary III.2.3] that $IA_{v^*} = I^{v^*}$, hence $I^\alpha = R \cap (I^{v^*}) = I^v$. This can also be read off from Corollary 7. In general, if $I \in J(A,R)$, we can only say that $IA_v^* \subset I_v^*$, hence

$$IA_v \subset R \cap (IA_v^*) = I^\alpha \subset I^v.$$

Thus α lies in between the star operations $I \mapsto IA_v$ and con_v.

8 Star Regular Valuations

As before, let $A \subset R$ be a ring extension, $\mathcal{G}$ a multiplicative filter on R and $*$ a star operation on $\Phi(\mathcal{H}/A)$ for some multiplicative filter $\mathcal{H} \supset \mathcal{G}$ on R. We then have the Kronecker extension $A(X,\mathcal{G},*) \subset R(X,\mathcal{G})$ at our disposal.

We now study the following problem. Let $v : R \to \Gamma \cup \infty$ be a valuation over A which is $\mathcal{G}$-regular (cf. Sect. 2, Definition 4) and let $v^* : R(X,\mathcal{G}) \to \Gamma \cup \infty$ denote the Gauß extension of v to $R(X,\mathcal{G})$, as defined in Sect. 2. We know that

$$A_{v^*} \supset AR(X,\mathcal{G})_{\mathrm{kr}} = A(X,\mathcal{G},\mathrm{id})$$

(cf. Corollary 3.12). When does it happen that $A_{v^*} \supset A(X,\mathcal{G},*)$?

Theorem 8.1. *If v is a $\mathcal{G}$-regular valuation on R over A, the following are equivalent.*

(1) $A_{v^*} \supset A(X, \mathscr{G}, *)$.
(2) If $I \in J^f(A, R)$, $J \in \Phi^f(\mathscr{G}/A)$ *and* $I \subset J^*$ *then* $v(I) \geq v(J)$.
(3) If $I \in \Phi^f(\mathscr{G}/A)$*, then* I^* *is v-finite (cf. Sect. 5, Definition 2) and* $v(I^*) = v(I)$.
(4) $I^* \subset I^v$ *for every* $I \in \Phi^f(\mathscr{G}/A)$.

Proof. Condition (1) means the following: If $f \in R[X], g \in S_{\mathscr{G}}$, and $c_A(f)H \subset (c_A(g)H)^*$ for some $H \in \Phi^f(\mathscr{G}/A)$ then $v^*(f/g) \geq 0$. Let $c_A(f) = I, c_A(g) = J$. Then $v^*(f/g) = v(I) - v(J)$. Thus (1) is clearly equivalent to the following condition.
(2′) If $I \in J^f(A, R), J, H \in \Phi^f(\mathscr{G}/A)$, and $IH \subset (JH)^*$, then $v(I) \geq v(J)$.
Of course, (2′) implies (2); take $H = A$. But also (2) implies (2′): Indeed under the premises of (2′) we conclude by (2) that

$$v(I) + v(H) \geq v(J) + v(H),$$

and then, since $H \not\subset \mathrm{supp} v$, i.e. $v(H) \neq \infty$, that $v(I) \geq v(J)$. We have verified that (1) and (2) are equivalent.
(2) ⇒ (3): Let $x \in I^*$. We have $I^* = (I + ax)^*$ and $IR \in \mathscr{G}, (I + ax)R \in \mathscr{G}$. By (2) it follows that $v(I) = v(I + ax)$. Thus $v(x) \geq v(I)$. This proves that $\min_{x \in I^*} v(x)$ exists and equals $v(I)$.
(3) ⇒ (4): Let $\gamma := v(I)$. Then $I^v = \{x \in R \mid v(x) \geq \gamma\}$. By (3) we know that $v(x) \geq \gamma$ for every $x \in I^*$. Thus $I^* \subset I^v$.
(4) ⇒ (2): Clear by applying condition (4) to J. □

Definition 1. If v is a valuation on R over A, we say that v is *star regular for* $\mathscr{G}$ *and* $*$, or, in short, v is $(\mathscr{G}, *)$*-regular*, if v is $\mathscr{G}$-regular and fulfills the equivalent conditions (1)–(4) in Theorem 1. In the case $\mathscr{G} = \{R\}$ we often simply say that v is $*$*-regular*, or star regular with respect to $*$.

Remarks 8.2.

a) In all these conditions we may replace $\mathscr{G}$ by $\mathscr{G}_f$ and $*$ by $*_f$ without changing their meaning. Thus v is $(\mathscr{G}, *)$-regular iff v is $(\mathscr{G}_f, *_f)$-regular.
b) Condition (1) does not change if we replace $*$ by its Kronecker companion $\mathrm{kro}(*, \mathscr{G})$, since the ring $A(X, \mathscr{G}, *)$ remains the same. Thus v is $(\mathscr{G}, *)$-regular iff v is $(\mathscr{G}, \mathrm{kro}(*, \mathscr{G}))$ regular.

Example 8.3. Let v be a valuation on R over A and w a $\mathscr{G}$-regular valuation on R over A. What does it mean that w is $(\mathscr{G}, *)$-regular for $* = \mathrm{con}_v$, i.e. $I^* = I^v$ for every $I \in J(A, R)$? By Theorem 1, condition (4), w is $(\mathscr{G}, *)$-regular iff $I^v \subset I^w$ for every $I \in \Phi^f(\mathscr{G}/A)$, i.e. $\mathrm{con}_v \leq \mathrm{con}_w$ on $\Phi^f(\mathscr{G}/A)$, and hence on $\Phi(\mathscr{G}/A)$. □

Let again $v : R \to \Gamma \cup \infty$ be a $\mathscr{G}$-regular valuation over R and let $*$ be a star operation on $\Phi(\mathscr{H}/A)$ for some multiplicative filter $\mathscr{H} \supset \mathscr{G}$. If v is $(\mathscr{G}, *)$-regular, we know by Theorem 1 that I^* is v-finite and $v(I^*) = v(I)$ for $I \in \Phi^f(\mathscr{G}/A)$. We want to extend this result to more general modules $I \in J^f(A, R)$.

Let $\mathfrak{q} := \operatorname{supp} v$. We introduce the set

$$\Delta := \Delta_{v,\mathscr{G}} = \{v(I) \mid I \in \Phi^f(\mathscr{G}/A)\}.$$

Clearly Δ is a submonoid of the abelian monoid $v(R \setminus \mathfrak{q}) \subset \Gamma_v$, i.e., $0 \in \Delta$, and Δ is closed under addition.

Lemma 8.4. *$\Delta = \Delta_{v,\mathscr{G}}$ is a lower set of the totally ordered abelian monoid $v(R)$. In other terms, if $\alpha \in \Delta, \gamma \in v(R)$, and $\gamma \leq \alpha$, then $\gamma \in \Delta$.*

Proof. Choose some $I \in \Phi^f(\mathscr{G}/A)$ with $v(I) = \alpha$ and some $x \in R$ with $v(x) = \gamma$. Then $I + Ax \in \Phi^f(\mathscr{G}/A)$ and $v(I + Ax) = \gamma$. □

In Sect. 2 we have introduced the group $H_{v,\mathscr{G}}$. It is the convex hull of the group

$$\{v(I) - v(J) \mid I, J \in \Phi^f(\mathscr{G}/A)\}$$

in Γ_v. Thus $H_{v,\mathscr{G}}$ is the smallest convex subgroup of Γ_v containing $\Delta_{v,\mathscr{G}}$.

Lemma 8.5. $v(R) \cap H_{v,\mathscr{G}} = \Delta_{v,\mathscr{G}}$.

Proof. Of course $\Delta_{v,\mathscr{G}} \subset v(R) \cap H_{v,\mathscr{G}}$. Conversely, let $x \in R$ be given with $v(x) \in H_{v,\mathscr{G}}$. We have modules $K, L \in \Phi^f(\mathscr{G}/A)$ with $v(x) \leq v(K) - v(L)$. Then $v(xL) \leq v(K)$. The A-module $U := K + xL$ is $\mathscr{G}$-regular, finitely generated, and $v(U) = v(x)$. Thus $v(x) \in \Delta_{v,\mathscr{G}}$. □

We arrive at the following remarkable result:

Theorem 8.6. *Assume, as before, that $*$ is a star operation on $\Phi(\mathscr{H}/A)$ for some multiplicative filter $\mathscr{H} \supset \mathscr{G}$. Assume further that v is a $(\mathscr{G}, *)$-regular valuation on R over A. Then, for every $I \in \Phi^f(\mathscr{H}/A)$ with $v(I) \in H_{v,\mathscr{G}}$ the module I^* is v-finite and $v(I^*) = v(I)$.*

Proof. Let $\gamma := v(I) \in H_{v,\mathscr{G}}$. We have to verify that $v(x) \geq \gamma$ for every $x \in I^*$. By Lemma 5 we know that $\gamma \in \Delta_{v,\mathscr{G}}$. We choose some $J \in \Phi^f(\mathscr{G}/A)$ with $v(J) = \gamma$. Then $I + J \in \Phi^f(\mathscr{G}/A)$ and $v(I + J) = \gamma$. By Theorem 1, condition (3), we have $v(x) \geq \gamma$ for every $x \in (I + J)^*$. In particular $v(x) \geq \gamma$ for $x \in I^*$. □

We add to Theorem 1 still another characterization of $(\mathscr{G}, *)$-regularity, employing the A_v-modules $I_\gamma := I_{\gamma,v} := \{x \in R \mid v(x) \geq \gamma\}$, with $\gamma \in v(R)$. (N.B. We used these modules much in [Vol. I, Chap. III].)

Proposition 8.7. *Assume that $*$ is a star operation of finite type on $\Phi(\mathscr{H}/A)$ for some $\mathscr{H} \supset \mathscr{G}$. Assume further that v is a $\mathscr{G}$-regular valuation on R over A.*

a) Let $\gamma \in v(R)$. Then I_γ is $\mathscr{G}$-regular iff $\gamma \in \Delta_{v,\mathscr{G}}$.
b) v is $(\mathscr{G}, *)$-regular iff $I_\gamma = I_\gamma^*$ for every $\gamma \in \Delta_{v,\mathscr{G}}$.

Proof. a) If $\gamma \in \Delta_{v,\mathscr{G}}$ there exists some $K \in \Phi^f(\mathscr{G}/A)$ with $v(K) = \gamma$. We have $K \subset I_\gamma$ and $KR \in \mathscr{G}$, hence $I_\gamma R \in \mathscr{G}$. Thus I_γ is $\mathscr{G}$-regular.

Conversely, let $I_\gamma R \in \mathcal{G}$. We choose some $x \in R$ with $v(x) = \gamma$ and some finitely generated submodule K of I_γ with $KR \in \mathcal{G}$. Then $K + ax \in \Phi^f(\mathcal{G}/A)$ and $v(K + ax) = \gamma$. Thus $\gamma \in \Delta_{v,\mathcal{G}}$.

b) Assume first that v is $(\mathcal{G}, *)$-regular. Let $\gamma \in \Delta_{v,\mathcal{G}}$ be given. Then I_γ is $\mathcal{G}$-regular, as just proved. Thus I_γ is the union of all finitely generated $\mathcal{G}$-regular submodules K of I. By Theorem 1, condition (3), we have $v(K^*) = v(K) \geq \gamma$ for each such K, i.e. $K^* \subset I_\gamma$. Since $*$ is of finite type, we conclude that $I_\gamma^* = I_\gamma$.

Conversely assume that $I_\gamma^* = I_\gamma$ for every $\gamma \in \Delta_{v,\mathcal{G}}$. We verify condition (3) in Theorem 1 and then will know that v is $(\mathcal{G}, *)$-regular. Thus, let $I \in \Phi^f(\mathcal{G}/A)$ be given. Then $\gamma := v(I) \in \Delta_{v,\mathcal{G}}$ and $I \subset I_\gamma$. If follows that $I \subset I^* \subset I_\gamma^* = I_\gamma$. Thus I^* is v-finite and $v(I^*) = \gamma$. □

Asking for $(\mathcal{G}, *)$-regularity of a given valuation v we can always retreat to the primary specialization of v which is $\mathcal{G}$-special (cf. Sect. 2).

Theorem 8.8. *Assume that $*$ is a partial star operation defined on $\Phi(\mathcal{G}/A)$ and $v : R \to \Gamma \cup \infty$ is a valuation on R over A which is epimorphic, i.e. $\Gamma = \Gamma_v$. Let $w := v|H_{v,\mathcal{G}}$. Then v is $(\mathcal{G}, *)$-regular iff w is $(\mathcal{G}, *)$-regular.*

Proof. We know already from Sect. 2, Proposition 2.5., that v is $\mathcal{G}$-regular iff w is $\mathcal{G}$-regular, and thus may assume now that both v and w are $\mathcal{G}$-regular.

Let $I \in \Phi^f(\mathcal{G}/A)$ be given. Then $v(I) = w(I)$ by definition of $w = v|H_{v,\mathcal{G}}$. If $x \in R$ then also $v(I + Ax) = w(I + Ax)$. Let $v(I) = \gamma$. Then $v(x) \geq \gamma$ iff $v(I + Ax = \gamma$ iff $w(I + Ax) = \gamma$ iff $w(x) \geq \gamma$. We conclude that $I^* \subset I_{\gamma,v} = I^v$ iff $I^* \subset I_{\gamma,w} = I^w$. Looking at condition (4) in Theorem 1 we see that v is $(\mathcal{G}, *)$-regular iff w is $(\mathcal{G}, *)$-regular. □

We now analyse star-regularity for $\mathcal{G}$-special valuations. For simplicity we assume that $*$ is defined on all of $J(A, R)$.

Theorem 8.9. *Let v be a $\mathcal{G}$-special valuation on R over A. Then v is $(\mathcal{G}, *)$-regular iff $I^* \subset I^v$ for every $I \in J^f(A, R)$ not contained in* supp(v).

Proof. Now $H_{v,\mathcal{G}} = \Gamma_v$. By Theorems 1 and 4 we know that v is $(\mathcal{G}, *)$-regular iff $I^* \subset I^v$ for every $I \in J^f(A, R)$ with $v(I) \in H_{v,\mathcal{G}}$. In the present situation $v(I) \in H_{v,\mathcal{G}}$ means that $v(I) \neq \infty$. □

The result can be simplified under a mild condition on v.

Corollary 8.10. *Assume that there exists some $\gamma \in v(R)$ with $0 < \gamma < \infty$, and that v is $\mathcal{G}$-special. Then v is $(\mathcal{G}, *)$-regular iff $*_f \leq \mathrm{con}_v$.*

Proof. Since v is $(\mathcal{G}, *)$-regular iff v is $(\mathcal{G}, *_f)$-regular, we may assume that $*$ is of finite type. The condition $* \leq \mathrm{con}_v$ means that $I^* \subset I^v$ for every (finitely generated) A-module I in R. Let $\mathfrak{q} := \mathrm{supp}\, v$. In view of Theorem 9 we may assume that v is $(\mathcal{G}, *)$-regular. It only remains to verify that $I^* \subset I^v$ for every (finitely generated) A-module $I \subset \mathfrak{q}$.

By Proposition 7 we know that $I_\gamma = I_\gamma^*$ for every $\gamma \in v(R \setminus \mathfrak{q})$. Due to our extra assumption on v the ideal $\mathfrak{q}$ of R is the intersection of these modules I_γ. Thus

$$\mathfrak{q}^* = (\bigcap_\gamma I_\gamma)^* \subset \bigcap_\gamma I_\gamma^* = \bigcap_\gamma I_\gamma = \mathfrak{q}$$

i.e. $\mathfrak{q}^* = \mathfrak{q}$. If $I \subset \mathfrak{q}$ then $I^* \subset \mathfrak{q}$ and $I^v = \mathfrak{q}$, hence $I^* \subset I^v$. □

We return to partial star operations and now relate $(\mathcal{G}, *)$-regularity to properties of the Kronecker companion $\alpha = \mathrm{kro}(x, \mathcal{G})$ (cf. Sect. 7) of $*$.

Proposition 8.11. *Let $*$ be a partial star operation defined on $\Phi(\mathcal{G}/A)$ and $\alpha = \mathrm{kro}(*, \mathcal{G})$. Assume that v is a $(\mathcal{G}, *)$-regular valuation of R over A. Then $I^\alpha \subset I^v$ for every $I \in J(A, R)$, in short, $\alpha \leq \mathrm{con}_v$.*

Proof. We have $I^\alpha = R \cap (IB)$ with $B := A(X, \mathcal{G}, *)$, the "product" IB being taken in $R(X, \mathcal{G})$. Assume first that I is finitely generated. Let $\gamma := v(I)$. Since $B \subset A_{v^*}$ it follows that $\gamma = v^*(IB)$, and we conclude that $v(x) \geq \gamma$ for every $x \in I^\alpha$. This means that $I^\alpha \subset I^v$.

If I is not finitely generated and $x \in I^\alpha$, then $x \in K^\alpha$ for some finitely generated A-submodule K of I, since α is of finite type. We conclude that $x \in K^v \subset I^v$. Thus $I^\alpha \subset I^v$ for every $I \in J(A, R)$. □

If I is $\mathcal{G}$-regular we will obtain a description of I^α in terms of $(\mathcal{G}, *)$-regular valuations. Recall from Sect. 2 (Theorem 2.15) that the $\mathcal{G}$-special valuations v on R over A correspond bijectively with the special valuations w on $R(X, \mathcal{G})$ over $A(X, \mathcal{G}, *)$ via $w = v^*$, since $A(X, \mathcal{G}, *)$ is a Kronecker subring of $R(X, \mathcal{G})$ over A.

We have to introduce some more notation. Abusively we will denote a valuation and its equivalence class by the same letter (as we did often previously).

Definition 2.

a) Let $\sigma(\mathcal{G}/A, *)$ denote the set of equivalence classes of non trivial $\mathcal{G}$-special valuations on R over A which are also $(\mathcal{G}, *)$-regular.
b) We define on $\sigma(\mathcal{G}/A, *)$ a partial ordering by decreeing that $v \leq w$ iff w is coarser then v. If (w.l.o.g) $v : R \to \Gamma \cup \infty$ is epimorphic, this means that w is equivalent to v/U for some unique convex subgroup U of $\Gamma_v = \Gamma$, cf. [Vol. I, Definition 9 in I §1].
c) Let $\tau(\mathcal{G}/A, *)$ denote the set of minimal elements of $\sigma(\mathcal{G}/A, *)$.

In the case that $A \subset R$ is Prüfer, $\mathcal{G} = \{R\}$ and $*$ is the identity operation id, we have met these sets $\sigma(\mathcal{G}/A, *)$ already in Chap. 1. Then $\sigma(\mathcal{G}/A, *)$ is the restricted PM-spectrum $S(R/A)$ of R over A, as defined in Chap. 1, Sect. 3 and $\tau(\mathcal{G}/A, *)$ is the set $\omega(R/A)$ of minimal elements of $S(R/A)$ (cf. Definition 6 in Chap. 1, Sect. 3).

We arrive at a central result of this section, namely a *description of the restricted PM-spectrum of $R(X, \mathcal{G})$ over $A(X, \mathcal{G}, *)$* for a $(\mathcal{G}, *)$-regular partial star operation. It is a sharpening of Theorem 2.15.

Theorem 8.12. *The Gauß extension map $v \mapsto v^*$ on $\sigma(\mathcal{G}/A, *)$ is a well defined bijection from $\sigma(\mathcal{G}/A, *)$ to $S(R(X,\mathcal{G})/A(X,\mathcal{G},*))$, which preserves the ordering: $v \le w$ iff $v^* \le w^*$. Thus this map restricts to a bijection from $\tau(\mathcal{G}/A, *)$ to $\omega(R(X,\mathcal{G})/A(X,\mathcal{G},*))$.*

Proof. This follows from Theorem 3.15 and the following two facts, which are obvious.

1) If $v : R \to \Gamma \cup \infty$ is a $\mathcal{G}$-special and $(\mathcal{G}, *)$-regular valuation on R over A, then, for any convex subgroup U of Γ, the coarsening $v/U : R \to \Gamma/U \cup \infty$ has the same properties and $(v/U)^* = v^*/U$.
2) If w_1,w_2 are *PM*-valuations (= special valuations) of $R(X,\mathcal{G})$ over $A(X,\mathcal{G},*)$ with $w_1 \le w_2$, then $w_1|R \le w_2|R$. Here $w_i|R$ is an abbreviation for $w_i \circ j$ with j the natural map from R to $R(X,\mathcal{G})$. □

We now are ready to determine the values of a Kronecker operation for $\mathcal{G}$ on $\Phi(\mathcal{G}/A)$ in terms of valuations.

Theorem 8.13. *Let $*$ be a partial star operation defined on $\Phi(\mathcal{G}/A)$, and let α denote its Kronecker companion. For any $I \in \Phi(\mathcal{G}/A)$ we have*

$$I^\alpha = \bigcap_{v\in\tau} I^v$$

with $\tau := \tau(\mathcal{G}/A, *)$, the set of non trivial minimal $\mathcal{G}$-special and $(\mathcal{G}, *)$-regular valuations on R over A.

Proof. Let $T := R(X,\mathcal{G})$ and $B := A(X,\mathcal{G},*)$. From $IR \in \mathcal{G}$ it follows that $IT = B$. Thus the B-module IB is T-regular. Since $B \subset T$ is Prüfer, we know by [Vol. I, Proposition III.1.10 & Corollary III.2.3]) that

$$IB = \bigcap_{w\in\omega} (IB)A_w = \bigcap_{w\in\omega} (IB)^w$$

with $\omega := \omega(T/B)$ the set of minimal non trivial *PM*-valuations on T over B. We know by Theorem 12 that $\omega = \{v^* \mid v \in \tau\}$. For any $v \in \tau$ the set $(IB)^{v^*}$ coincides with the v^*-convex hull I^{v^*} of $\{a/1 \mid a \in I\}$ in T, since $B \subset A_{v^*}$. Thus $IB = \bigcap_{v\in\tau} I^{v^*}$. Intersecting with R we obtain $I^\alpha = \bigcap_{v\in\tau}(I^{v^*} \cap R) = \bigcap_{v\in\tau} I^v$.□

9 A Generalization: Weak Stars and Semistars

Much of what has been done in Sect. 8 can be seen as a contribution to a concern which we pose in slightly vague terms as follows.

Problem 9.1. Let R be a ring and $v : R \to \Gamma \cup \infty$ a valuation on R. Let $x_1, x_2, \ldots, x_n \in R$ be given, and let γ denote the minimum of the values

$v(x_1), v(x_2), \ldots, v(x_n)$. Find in a systematic manner a subset J of R, as big as possible, containing $x_1, \ldots, x_n$ such that $v(x) \geq \gamma$ for every $x \in J$!

Of course, if I_0 denotes the additive subgroup of R generated by $x_1, \ldots, x_n$, then the v-convex hull I_0^v is the biggest such set J, but we are looking for an "accessible" set $J \subset I_0^v$ which depends on $x_1, \ldots, x_n, \gamma$ in a uniform way, if v varies in a—not too small—family of valuations on R.

Assume that a subring A of R is known which is contained in A_v. (In the worst case take $A = \mathbb{Z} \cdot 1_R$.) Then our answer to this problem has been as follows (cf. Theorem 8.1): Let $I := Ax_1 + \cdots + Ax_n$, hence $I_0^v = I^v$. We choose a multiplicative filter $\mathscr{G}$ on R with $IR \in \mathscr{G}$ and a partial star operation $*$ of $\Phi(\mathscr{G}/A)$ such that v is $(\mathscr{G}, *)$-regular. {N.B. This is possible in various ways.} Then $I \subset I^* \subset I^w$ for every $(\mathscr{G}, *)$-regular valuation w on R over A.

The value of this answer depends on the flexibility to use a sufficiently rich stock of star operations, and to understand the nature of $(\mathscr{G}, *)$-regularity for valuations. Here our method can be strengthened by modifying the notion of a star operation.

Reviewing the steps leading to the results in Sect. 8 you may observe that we did not need all axioms St1–St4 valid for a star operation. While St1, St2, St4 seem to be indispensable we nowhere used the idempotency axiom St3, which says the $(I^*)^* = I^*$.

We hasten to admit that St3 is certainly indispensable for building a multiplicative ideal theory related to star products, the beginning of which had been displayed in Sect. 4, and in particular for investigating star invertibility. There also the axiom St5, saying that $A^* = A$, is important.

Historically such a multiplicative ideal theory has been the main motivation for introducing star operations, beginning with the double-inverse operation δ (cf. [vdW], [Bo, Chap. VII], [Gi]). But for our approach to Problem 1 we can weaken the notion of a star operation, and then will have stronger results than those in Sect. 8 without extra costs.

One possibility is simply to omit the axiom St3; another one is, to omit St3 but to strengthen St4 somewhat. Both have their merits, as we will see.

Definition 1. Let $A \subset R$ be a ring extension and $\mathscr{H}$ a multiplicative filter on R. Let $*$ be a map from $\Phi(\mathscr{H}/A)$ to $\Phi(\mathscr{H}/A)$.

a) We call $*$ a *weak star operation* on $\Phi(\mathscr{H}/A)$, if the following holds for any $I, J \in \Phi(\mathscr{H}/A)$:

   (St1) $I \subset I^*$,
   (St2) $I \subset J \Rightarrow I^* \subset J^*$,
   (St4) $IJ^* \subset (IJ)^*$.

b) We call $*$ a *semistar operation* on $\Phi(\mathscr{H}/A)$ if St1, St2 hold together with the following strengthening of St4.

   (St4$^+$) $I^*J^* \subset (IJ)^*$.

c) We call a weak star operation $*$ *strict* if

(St5) $A^* = A$.

In order to ease language we very often will abbreviate the term "weak star operation" (resp. "semistar operation", "star operation") to "*weak star*" (resp. "*semistar*", "*star*").

Remark 9.2. If $*$ is a semistar on $\Phi(\mathscr{H}/A)$, then A^* is an overring of A in R. But, if $*$ is a weak star, then in general A^* will only be an A-submodule of R containing A.

Example 9.3. Let $\alpha : I \mapsto I^\alpha$ and $\beta : I \mapsto I^\beta$ be "operations" on $\Phi(\mathscr{H}/A)$, i.e., maps from $\Phi(\mathscr{H}/A)$ to itself. We define a product $\alpha\beta$ of these operations by putting $I^{\alpha\beta} := (I^\alpha)^\beta$. {Thus $\alpha\beta = \beta \circ \alpha$.} If α, β are semistars, then $\alpha\beta$ is a semistar, and if α, β are weak stars, then $\alpha\beta$ is a weak star. But if α and β are stars on $\Phi(\mathscr{H}/A)$ then most often $\alpha\beta$ is not a star, only a semistar. If α and β are strict, then $\alpha\beta$ is strict.

Example 9.4. Let α and β be weak stars on $\Phi(\mathscr{H}/A)$. Then the operation $I^\gamma := I^\alpha + I^\beta$ is again a weak star on $\Phi(\mathscr{H}/A)$, as is easily checked. But if α and β are semistars, there is no reason in general, why γ should be again a semistar. Usually it will be just a weak star. If α and β are strict, then γ is strict. □

Example 9.5. Assume that α and β are weak star operations on $\Phi(\mathscr{H}/A)$ with $I^\beta \subset I^\alpha$ for every $I \in \Phi(\mathscr{H}/A)$. We fix a module $H \in \Phi(\mathscr{H}/A)$, and then define a third operation γ on $\Phi(\mathscr{H}/A)$ by $I^\gamma := [(IH)^\alpha : H^\beta]$. We claim that γ is again a weak star on $\Phi(\mathscr{H}/A)$. Indeed, let $I, J \in \Phi(\mathscr{H}/A)$. Then $IH^\beta \subset (IH)^\beta \subset (IH)^\alpha$, hence $I \subset I^\gamma$. If $I \subset J$, then $(IH)^\alpha \subset (JH)^\alpha$, hence

$$I^\gamma = [(IH)^\alpha : H^\beta] \subset [(JH)^\alpha : H^\beta] = J^\gamma.$$

Finally

$$IJ^\gamma = I \cdot [(JH)^\alpha : H^\beta] \subset [I(JH)^\alpha : H^\beta] \subset [(IJH)^\alpha : H^\beta] = (IJ)^\gamma.$$

□

Example 9.6. Assume that $A \subset R$ is a ring extension and $\mathscr{F}$ is a subset of $J(A)$ closed under multiplication. Then it is immediate, that the map $* : J(A, R) \to J(A, R)$,

$$I^* := I^R_{[\mathscr{F}]} := \{x \in R \mid \exists H \in \mathscr{F} \text{ with } Hx \in I\}$$

is a semistar operation on $J(A, R)$. As in Example 3.2, where $\mathscr{F}$ has been a multiplicative filter on A of finite type, we call $I^R_{[\mathscr{F}]}$ the *$\mathscr{F}$-hull of I in R*. In Example 3.2 it has been important for proving St3, that $\mathscr{F}$ is of finite type. □

Definition 2. Given a ring extension $A \subset R$ and a multiplicative filter $\mathscr{H}$ on R, we denote the set of weak stars on $\Phi(\mathscr{H}/A)$ by $\mathrm{WStar}(\mathscr{H}/A)$ and the subset of semistars on $\Phi(\mathscr{H}/A)$ by $\mathrm{SStar}(\mathscr{H}/A)$, finally the subset of stars on $\Phi(\mathscr{H}/A)$ by $\mathrm{Star}(\mathscr{H}/A)$. In the special case $\mathscr{H} = J(R)$ we also write $\mathrm{WStar}(R/A)$, $\mathrm{SStar}(R/A)$, $\mathrm{Star}(R/A)$ for these sets respectively. We define a partial ordering on the set $\mathrm{WStar}(\mathscr{H}/A)$ by decreeing for two weak stars α, β on $\Phi(\mathscr{H}/A)$, that $\alpha \le \beta$ iff $I^\alpha \subset I^\beta$ for every $I \in \Phi(\mathscr{H}/A)$, and then we say that α is *finer* than β, or that β is *coarser* than α. □

Proposition 9.7. *Assume that $(\alpha_\lambda \mid \lambda \in \Lambda)$ is a family in $\mathrm{WStar}(\mathscr{H}/A)$. We define operations γ and η on $\Phi(\mathscr{H}/A)$ by putting*

$$I^\gamma := \bigcap_{\lambda\in\Lambda} I^{\alpha_\lambda}, \ I^\eta := \sum_{\lambda\in\Lambda} I^{\alpha_\lambda}$$

for $I \in \Phi(\mathscr{H}/A)$.

i) Both γ and η are weak stars on $\Phi(\mathscr{H}/A)$. They are the infimum and supremum respectively of the family (α_λ) in $\mathrm{WStar}(\mathscr{H}/A)$.
ii) If every α_λ is a semistar, then γ is a semistar.
iii) If every α_λ is a star, then γ is a star.

Proof. i): The verification of St1, St2, St4 for γ and η is an easy straight forward matter. After that it is obvious that γ is the infimum and η is the supremum of the family (α_λ) in $\mathrm{WStar}(\mathscr{H}/A)$.

ii): Given $I, J \in \Phi(\mathscr{H}/A)$, we have $I^\gamma J^\gamma \subset I^{\alpha_\lambda} J^{\alpha_\lambda} \subset (IJ)^{\alpha_\lambda}$ for each $\lambda \in \Lambda$, hence $I^\gamma \cdot J^\gamma \subset (IJ)^\gamma$.

iii): Given $I \in \Phi(\mathscr{H}/A)$, we have $(I^\gamma)^\gamma \subset (I^{\alpha_\lambda})^{\alpha_\lambda} = I^{\alpha_\lambda}$ for each $\lambda \in \Lambda$, hence $(I^\gamma)^\gamma \subset I^\gamma$, hence $(I^\gamma)^\gamma = I^\gamma$. □

By part (i) of the proposition we know that the poset $\mathrm{WStar}(\mathscr{H}/A)$ is a complete lattice.

Definition 3. If $(\alpha_\lambda \mid \lambda \in \Lambda)$ is a family in $\mathrm{WStar}(\mathscr{H}/A)$, we denote the weak stars γ and η obtained in Proposition 7 by $\bigwedge_\lambda \alpha_\lambda$ and $\bigvee_\lambda \alpha_\lambda$ respectively. If Λ is a finite index set, say $\Lambda = \{1, \dots, n\}$, then we also write $\alpha_1 \wedge \dots \wedge \alpha_n$ and $\alpha_1 \vee \dots \vee \alpha_n$ for these weak stars. Thus, for any $I \in \Phi(\mathscr{H}/A)$,

$$I^{\alpha_1\wedge\dots\wedge\alpha_n} = I^{\alpha_1} \cap \dots \cap I^{\alpha_n}, I^{\alpha_1\vee\dots\vee\alpha_n} = I^{\alpha_1} + \dots + I^{\alpha_n}.$$

□

This is common notation for partially ordered sets, applied to the poset $\mathrm{WStar}(\mathscr{H}/A)$.

Scholium 9.8. If X is one of the posets $\mathrm{SStar}(\mathscr{H}/A)$, $\mathrm{Star}(\mathscr{H}/A)$, then by general abstract nonsense every family $(\alpha_\lambda \mid \lambda \in \Lambda)$ in X has a supremum in X, namely the infimum of the set $\{\beta \in X \mid \alpha_\lambda \le \beta \text{ for every } \lambda \in \Lambda\}$. {Notice that this set is not empty, since X has a maximal element, the trivial star $I \mapsto R$ $(I \in \mathscr{H})$.} But the supremum may be bigger than $\bigvee_\lambda \alpha_\lambda$. We can only say that

$$\bigvee_{\lambda} \alpha_\lambda \leq \sup_{\mathrm{SStar}}(\alpha_\lambda) \leq \sup_{\mathrm{Star}}(\alpha_\lambda).$$

Thus X is again a complete lattice, but most often not a sublattice of $\mathrm{WStar}(\mathcal{H}/A)$. □

Definition 4. Assume that $\mathcal{H}$ has finite type. Following the terminology in Sects. 6 and 7 we say that a *weak star* $*$ on $\Phi(\mathcal{H}/A)$ has *finite type*, if for any $I \in \Phi(\mathcal{H}/A)$ the module I^* is the union of the modules K^*, where K runs through the finitely generated A-submodules of I with $RI \in \mathcal{H}$. □

Example 9.9. We see, as in Sects. 6 and 7 for stars, that every weak star $*$ on $\Phi(\mathcal{H}/A)$ has a *finite type companion* $*_f$, defined by $I^*_f = \bigcup_K K^*$, where again K runs through the finitely generated A-submodules of I with $RI \in \mathcal{H}$. This companion $*_f$ is a weak star of finite type.

More precisely $*_f$ is the coarsest weak star of finite type on $\Phi(\mathcal{H}/A)$, which is finer than $*$. If $*$ is a semistar, then also $*_f$ is a semistar. □

We are ready to discuss an observation which has been our main motivation for including weak stars into the present volume: The definition of the subring $A(X, \mathcal{G}, *)$ of $R(X, \mathcal{G})$ in Sect. 3, Definition 3 makes completely sense if $*$ is only a weak star, and the subsequent proofs of Lemma 3.9 and Theorem 3.10 remain valid for $*$ a weak star, and, of course, for all that it suffices that $*$ is defined on some $\Phi(\mathcal{H}/A)$ for some multiplicative filter $\mathcal{H} \supset \mathcal{G}$ instead of $J(A, R)$. In particular the ring $A(X, \mathcal{G}, *)$ is Kronecker in $R(X, \mathcal{G})$.

The path is open to relate weak stars to Kronecker star operations as done in Sect. 7 for stars. We extend Theorem 7.5 as follows.

Theorem 9.10. *Assume that $\mathcal{G}$ is a multiplicative filter on R. Assume further that $*$ is a weak star on $\Phi(\mathcal{H}/A)$ for some multiplicative filter $\mathcal{H} \supset \mathcal{G}$ on R. Let α denote the star (!) on $J(A, R)$ induced by the natural morphism $j : (R, A) \to (R(X, \mathcal{G}), A(X, \mathcal{G}, *))$ of ring extensions. Then $A(X, \mathcal{G}, *) = A(X, \mathcal{G}, \alpha)$, and α is a Kronecker operation for $\mathcal{G}$.*

Proof. The arguments in Sect. 7 for proving this if $*$ is a star operation remain valid if $*$ is a weak star. □

Theorem 9.11. *In the situation of Theorem 10 we have $I^\alpha = \bigcup_H [(IH)^* : H]$, with H running through $\Phi^f(\mathcal{G}/A)$, for every $I \in \Phi^f(\mathcal{G}/A)$.*

Proof. Verbatim as the proof of Theorem 7.6. □

Again we call the star on $J(A, R)$ the *Kronecker companion* of $*$ for $\mathcal{G}$, and we write $\alpha = \mathrm{kro}(*, \mathcal{G})$. As in the case of $*$ a star we have $\mathrm{kro}(*, \mathcal{G}) = \mathrm{kro}(*, \mathcal{G}_f) = \mathrm{kro}(*_f, \mathcal{G}_f)$.

We turn to valuations. Assume that a valuation v on R over A is given. If $*$ is a weak star on $\Phi(\mathcal{H}/A)$ and $\mathcal{G}$ is a multiplicative filter on R contained in $\mathcal{H}$, we define $(\mathcal{G}, *)$-regularity of v exactly as in Sect. 8 for $*$ a star: We call v $(\mathcal{G}, *)$-regular if v is $\mathscr{G}$-regular and A_{v^*} contains $A(X, \mathscr{G}, *)$. Theorem 8.1 and its proof remain in force. In particular

Scholium 9.12. v is $(\mathscr{G}, *)$-regular iff v is $\mathscr{G}$-regular and $I^* \subset I^v$ for every $I \in \Phi^f(\mathscr{G}/A)$. $\square$

It is now easily checked that all results on star-regularity in Sect. 8 extend from stars to weak stars.

So weak stars indeed contribute to Problem 1 posed above. But they also do other services. They are of help in creating new stars, as we will see in the next section.

10 Upper Stars

Continuing the basic study of weak star operations begun in Sect. 9 we exhibit a way to build stars from sets of weak stars.

In the whole section $A \subset R$ is a fixed ring extension and $\mathscr{H}$ *is a multiplicative filter on R of finite type*. In Sect. 9 we introduced the partially ordered sets, in fact complete lattices

$$\mathrm{WStar}(\mathscr{H}/A) \supset \mathrm{SStar}(\mathscr{H}/A) \supset \mathrm{Star}(\mathscr{H}/A)$$

consisting of the weak stars, semistars, and stars on $\Phi(\mathscr{H}/A)$ respectively. Some of the later constructions will work only for weak stars *of finite type*. So we now introduce the partially ordered sets

$$\mathrm{WStar}_f(\mathscr{H}/A) \supset \mathrm{SStar}_f(\mathscr{H}/A) \supset \mathrm{Star}_f(\mathscr{H}/A)$$

consisting of the weak stars, semistars, and stars on $\Phi(\mathscr{H}/A)$ of finite type. Notice that this chain of sets is obtained from the chain of sets above by intersecting it with $\mathrm{WStar}_f(\mathscr{H}/A)$.

Remark 10.1. These three partially ordered sets are again complete lattices: If S is a non empty subset of any of these sets, then in all three cases the infimum of S is the finite type companion $(\bigwedge S)_f$ of the infimum $\bigwedge S$ in $\mathrm{WStar}(\mathscr{H}/A)$ (cf. Proposition 9.7 and Sect. 9, Definition 3). $\square$

It is a useful fact that the inclusion mapping from $\mathrm{WStar}_f(\mathscr{H}/A)$ to $\mathrm{WStar}(\mathscr{H}/A)$ retains suprema.

Proposition 10.2. *For any nonempty subset S of* $\mathrm{WStar}_f(\mathscr{H}/A)$ *the supremum $\bigvee S$ in* $\mathrm{WStar}(\mathscr{H}/A)$ *has again finite type.*

Proof. Let $\gamma := \bigvee S$. Let $I \in \Phi(\mathscr{H}/A)$ and $x \in I^\gamma = \sum_{\alpha \in S} I^\alpha$ be given. There exist finitely many elements $\alpha_1, \dots, \alpha_n \in S$ and elements $x_i \in I^{\alpha_i}$ such that $x = x_1 + \cdots + x_n$. Since the weak stars α_i are of finite type, there exist finitely generated A-submodules K_i of I such that $RK_i \in \mathscr{H}$ and $x_i \in K_i^{\alpha_i}$.

Then $K := K_1 + \ldots + K_n$ is again a finitely generated submodule of I with $RK \in \mathcal{H}$, and $x \in \sum_{i=1}^{n} K^{\alpha_i} \subset K^{\gamma}$. □

We address a new problem: Given a family $(\alpha_\lambda \mid \lambda \in \Lambda)$ of weak stars on $\Phi(\mathcal{H}/A)$, the subset M of Star$(\mathcal{H}/A)$ consisting of all stars on $\Phi(\mathcal{H}/A)$, which are coarser than every α_λ, is certainly not empty, since it contains the trivial star operation $I \mapsto R, \Phi(\mathcal{H}/R) \to \Phi(\mathcal{H}/R)$. Thus this set has an infimum γ in Star$(\mathcal{H}/A)$, and the star γ coincides with the infimum $\bigwedge M$ of M in WStar$(\mathcal{H}/A)$ (cf. Proposition 9.7 and Sect. 9, Definition 3). γ is the finest star operation on $\Phi(\mathcal{H}/A)$ which is coarser than every α_λ.

Definition 1. We call γ the *upper star of the family* $(\alpha_\lambda \mid \lambda \in \Lambda)$, and we write $\gamma = \mathrm{st}(\alpha_\lambda \mid \lambda \in \Lambda)$. If the family is finite, say $\Lambda = \{1, 2, \ldots, n\}$, we also write $\gamma = \mathrm{st}(\alpha_1, \ldots, \alpha_n)$. □

Our problem now is to find a somewhat explicit description of upper stars. There is one thing which we can say in advance.

Remark 10.3. For any family $(\alpha_\lambda \mid \lambda \in \Lambda)$ in WStar$(\mathcal{H}/A)$ we have $\mathrm{st}(\alpha_\lambda \mid \lambda \in \Lambda) = \mathrm{st}(\bigvee_\lambda \alpha_\lambda)$. □

Thus in principle it suffices to find a description of st(α) for a single weak star α. On the other hand, given a specific family $(\alpha_\lambda \mid \lambda \in \Lambda)$ of weak stars it may be advantageous not to take a route via the formula in Remark 3.

We will be able to give a good answer to the problem only in the case that the α_λ are of finite type. To prepare for this we need lemmas about "products" in WStar$(\mathcal{H}/A)$.

Definition 2. Let $\alpha_1, \ldots, \alpha_n$ be weak stars on $\Phi(\mathcal{H}/A)$. The product $\alpha_1\alpha_2 \cdot \ldots \cdot \alpha_n$ is the composite $\alpha_n \circ \ldots \circ \alpha_1$ of the maps $\alpha_i : \Phi(\mathcal{H}/A) \to \Phi(\mathcal{H}/A)$, (as already considered in Example 9.3 for n = 2). □

$\alpha_1\alpha_2 \cdot \ldots \cdot \alpha_n$ is again a weak star. If the α_i are semistars (or even stars) then $\alpha_1\alpha_2 \cdot \ldots \cdot \alpha_n$ is a semistar (cf. Example 9.3).

Lemma 10.4. *If α, β, γ are weak stars on $\Phi(\mathcal{H}/A)$ and $\alpha \leq \beta$, then $\alpha\gamma \leq \beta\gamma$.*

Proof. For any $I \in \Phi(\mathcal{H}/A)$ we have $I^\alpha \subset I^\beta$, hence $(I^\alpha)^\gamma \subset (I^\beta)^\gamma$. □

Lemma 10.5. *Let $\alpha_1, \ldots, \alpha_n$ be weak stars on $\Phi(\mathcal{H}/A)$. For any sequence of indices $1 \leq i_1 < i_2 \ldots < i_r \leq n$ we have $\alpha_{1_1}\alpha_{i_2} \cdot \ldots \cdot \alpha_{i_r} \leq \alpha_1\alpha_2 \cdot \ldots \cdot \alpha_n$.*

Proof. We define weak stars $\beta_1, \ldots, \beta_n$ on $\Phi(\mathcal{H}/A)$ by putting $\beta_j = \alpha_{i_k}$ if $j = i_k$ for some $k \in \{1, \ldots, r\}$ and $\beta_i = \mathrm{id}$ otherwise. {id is the identity operation $I \mapsto I$} Then $\beta_j \leq \alpha_j$ for every j. It follows by Lemma 4 that

$$\alpha_{i_1} \cdot \ldots \cdot \alpha_{i_r} = \beta_1 \cdot \ldots \cdot \beta_n \leq \alpha_1 \cdot \ldots \cdot \alpha_n.$$

□

Recall that *we always assume that the filter $\mathcal{H}$ has finite type.*

Lemma 10.6. *Assume that α and β are weak stars on $\Phi(\mathscr{H}/A)$ of finite type. Then the product $\alpha\beta$ is again of finite type.*

Proof. Let $x \in I^{\alpha\beta}$ be given. There exists a finitely generated A-submodule K of I^α with $RK \in \mathscr{H}$ and $x \in K^\beta$. Since K is finitely generated and α is of finite type there exists a finitely generated submodule H of I with $RH \in \mathscr{H}$ and $K \subset H^\alpha$. It follows that $x \in K^\beta \subset H^{\alpha\beta}$. □

Lemma 10.7. *Assume that S is a nonempty set of weak stars on $\Phi(\mathscr{H}/A)$ and β is a weak star on $\Phi(\mathscr{H}/A)$. Then the following holds.*

a) $\beta(\bigvee S) = \bigvee_{\alpha\in S}(\beta\alpha)$.
b) $(\bigvee S)\beta \geq \bigvee_{\alpha\in S}(\alpha\beta)$.
c) If β is of finite type and S is upward directed (i.e., for $\alpha, \beta \in S$ there exists $\gamma \in S$ with $\alpha \leq \gamma, \beta \leq \gamma$), then $(\bigvee S)\beta = \bigvee_{\alpha\in S}(\alpha\beta)$.

Proof. Let $\eta := \bigvee S$.

a): For any $I \in \Phi(\mathscr{H}/A)$

$$(I^\beta)^\eta = \sum_{\alpha\in S}(I^\beta)^\alpha = \sum_{\alpha\in S} I^{\beta\alpha}.$$

b): This is evident form Lemma 4.

c): Assume now that β is of finite type and S is upward directed. Then also the set $\{\alpha\beta \mid \alpha \in S\}$ is upward directed. Thus, for $\eta := \bigvee S, \zeta := \bigvee_{\alpha\in S}(\alpha\beta)$ we have

$$I^\eta = \sum_{\alpha\in S} I^\alpha = \bigcup_{\alpha\in S} I^\alpha, \quad I^\zeta = \sum_{\alpha\in S} I^{\alpha\beta} = \bigcup_{\alpha\in S} I^{\alpha\beta}.$$

Let $x \in I^{\eta\beta}$ be given. Since β is of finite type, we know by Lemma 7.1 that there exists a finitely generated submodule K of I^η with $RK \in \mathscr{H}$ and $x \in K^\beta$. Since K is finitely generated and S is upward directed, there exists some $\alpha \in S$ with $K \subset I^\alpha$. If follows that $x \in K^\beta \subset I^{\alpha\beta} \subset I^\zeta$. Thus $I^{\eta\beta} \subset I^\zeta$. We have proved that $\eta\beta \leq \zeta$, and we conclude by b) that $\eta\beta = \zeta$. □

Theorem 10.8. *Assume that S is a non empty set of weak stars of finite type on $\Phi(\mathscr{H}/A)$.*

i) The upper star st(S) *has again finite type.*
ii) st$(S) = \bigvee M$ *with* $M := \{\alpha_1\alpha_2 \cdot \ldots \cdot \alpha_r \mid r \in \mathbb{N}, \alpha_1 \in S, \ldots, \alpha_r \in S\}$. *This set of weak stars is upward directed.*

Proof. We know by Lemma 6 that the elements of M are weak stars of finite type, and by Lemma 5 that M is upward directed. Let $\gamma := \bigvee M$. This weak star has again finite type by Proposition 2. We fix some $\alpha \in M$. For any $\beta \in M$ the product $\beta\alpha$ is again in M. By Lemma 7(a) we conclude that $\gamma\alpha \leq \gamma$. Since $\gamma\alpha \geq \gamma$ holds anyway, we conclude that $\gamma\alpha = \gamma$ for every $\alpha \in M$. Now it follows by Lemma 7(c) that $\gamma\gamma = \gamma$, i.e., γ is a star.

Clearly $\alpha \le \gamma$ for each $\alpha \in S$. If η is any star on $\Phi(\mathscr{H}/A)$ which is coarser than each $\alpha \in S$, we know by Lemma 4 that $\alpha_1 \cdot \ldots \cdot \alpha_r \le \eta^r = \eta$ for any $\alpha_1, \ldots, \alpha_r \in S$, and we conclude that $\gamma \le \eta$. Thus $\gamma = \mathrm{st}(S)$. □

Remark 10.9. That $\mathrm{st}(S)$ has finite type can be seen in a more conceptual way as follows. For any $\alpha \in S$ we have $\alpha \le \mathrm{st}(S)$, hence $\alpha = \alpha_f \le \mathrm{st}(S)_f$. It follows that $\mathrm{st}(S) \le \mathrm{st}(S)_f$, hence $\mathrm{st}(S) = \mathrm{st}(S)_f$.

Scholium 10.10.

a) In the case that S is a singleton $\{\alpha\}$ the theorem reads as follows: If α is a weak star of finite type on $\Phi(\mathscr{H}/A)$ then $\mathrm{st}(\alpha)$ has again finite type and $\mathrm{st}(\alpha) = \bigvee_n \alpha^n$. More explicitly, $I^{\mathrm{st}(\alpha)} = \bigcup_{n\in\mathbb{N}} I^{\alpha^n}$ for any $I \in \Phi(\mathscr{H}/A)$.
b) If $\alpha_1, \ldots, \alpha_r$ are weak stars on $\Phi(\mathscr{H}/A)$, then

$$\mathrm{st}(\alpha_1, \ldots, \alpha_r) = \mathrm{st}(\alpha_1 \vee \ldots \vee \alpha_r) = \mathrm{st}(\alpha_1\alpha_2 \cdot \ldots \cdot \alpha_r).$$

Indeed, let η be a star on $\Phi(\mathscr{H}/A)$ which is coarser than all α_j. By Lemmas 4 and 5

$$\alpha_1 \vee \ldots \vee \alpha_r \le \alpha_1 \cdot \ldots \cdot \alpha_r \le \eta^r = \eta.$$

This gives the claim (cf. also Remark 3).
c) If $\alpha_1, \ldots, \alpha_r$ are weak stars of finite type on $\Phi(\mathscr{H}/A)$ then

$$\mathrm{st}(\alpha_1, \ldots, \alpha_r) = \bigvee_{n\in\mathbb{N}} (\alpha_1 \cdot \ldots \cdot \alpha_r)^n.$$

□

Theorem 8 has an interesting consequence for star-regularity of valuations (Theorem 12 below). Let $\mathscr{G}$ be a multiplicative filter of finite type on R contained in $\mathscr{H}$. For any weak star α on $\Phi\mathscr{H}/A)$ let $\alpha|\mathscr{G}$ denote the restriction of α to $\Phi(\mathscr{G}/A)$. Similarly, for any subset S of $\mathrm{WStar}(\mathscr{H}/A)$ let $S|\mathscr{G}$ denote the set of restrictions $\alpha|\mathscr{G}$ of the weak stars $\alpha \in S$.

Proposition 10.11. *If S is a set of weak stars of finite type on $\Phi\mathscr{H}/A)$, then $\mathrm{st}(S|\mathscr{G}) = \mathrm{st}(S)|\mathscr{G})$.*

Proof. This follows from the explicit description of $\mathrm{st}(S)$ and $\mathrm{st}(S|\mathscr{G})$ by Theorem 8. □

Theorem 10.12. *Let v be a valuation on R over A. Let S be a set of weak stars of finite type on $\Phi\mathscr{H}/A)$ and γ is the upper star of S, $\gamma = \mathrm{st}(S)$. Assume that v is $(\mathscr{G},\alpha)$-regular for every $\alpha \in S$. Then v is $(\mathscr{G},\gamma)$-regular.*

Proof. If α is a weak star on $\Phi\mathscr{H}/A)$, then regularity of v for $\mathscr{G}$ and α depends only on $\mathscr{G}$ and $\alpha|\mathscr{G}$. Since $\gamma|\mathscr{G} = \mathrm{st}(S|\mathscr{G})$ (Proposition 10), we may replace S by $(S|\mathscr{G})$ in advance, hence may assume that $\mathscr{H} = \mathscr{G}$. As before (Sect. 5, Definition 3), let con_v denote the v-convex hull operation on $J(A, R)$. If α is any weak star of finite type on $\Phi(\mathscr{G}/A)$ then Scholium 9.12 tells us that v is $(\mathscr{G}, \alpha)$-regular iff $\alpha \leq \mathrm{con}_v|\mathscr{G}$. Since $\mathrm{con}_v|\mathscr{G}$ is a star on $\Phi(\mathscr{G}/A)$, the claim now follows directly from the definition of $\mathrm{st}(S)$. □

Notice that our assumption, that the $\alpha \in S$ are of finite type, is not an essential restriction of generality in Theorem 12 (in contrast to Theorem 8), since for any weak star α the valuation is $(\mathscr{G}, \alpha)$-regular iff v is $(\mathscr{G}, \alpha_f)$-regular.

Appendix: Errata Volume I

- p. 25, line +10:
  "*of* Γ" instead of "*of* v"
- p. 48, line +14:
  "*R-overring of A*" instead of "*overring B of R*"
- p. 60, line +6:
  "*semireal*" instead of "*real*"
- p. 64, line +10:
  "$F(0) = 0$ *and* $\deg(F) \geq 2$" instead of "$y \in R$"
- p. 64, line +14:
  better "$v_i(G(F(x))) \leq 2v_i(x)$" instead of "$v_i(G(F(x))) < v_i(x)$"
- p. 92, line −1:
  "*submodule of R with* $A \subset L$" instead of "*submodule of R*"
- p. 93, line +4:
  "$y \in A$" instead of "$y \in R$"
- p. 140, line +13:
  "*implications*" instead of "*implication*"
- p. 148, line +2:
  "$I_{-\gamma}$" instead of "I_γ"
- p. 149, line −9:
  "*R-regular*" instead of "*regular*"
- p. 149, line −8:
  Delete the equation "$1 = \sum f_i g_i$"
- p. 152, Proposition 10.16:
  "*If* $A \subset R$ *is a Bezout extension*" instead of "*If A is a Bezout extension*"
- p. 165, Definition 3, line +3:
  The meaning of the symbol $\Omega(R/A)$ in II, §11 deviates from the meaning in the rest of the book, since here only valuations with value group $\mathbb{Z}$ are under consideration.
- p. 178, end of Summary:
  Everywhere "*§10, §11*" instead of "*§9, §10*"

M. Knebusch and T. Kaiser, *Manis Valuations and Prüfer Extensions II*,
Lecture Notes in Mathematics 2103, DOI 10.1007/978-3-319-03212-2,

- p. 187, end of proof of Theorem 3.3:
  Add the sentence *"Now Proposition I.5.1 tells us that $A \subset R$ is PM."*
- p. 206, line +9:
  "$x' \in B$" instead of "$x' \in \mathfrak{p}$"
- p. 267, Symbol Index, $\Omega(R/A)$:
  Refer to p. 180 instead of p. 163.

References

[Al] J. Alajbegović, Approximation theorems for Manis valuations with the inverse property. Commun. Algebra **12**, 1399–1417 (1984)

[Al_1] J. Alajbegović, R-Prüfer rings and approximation theorems, in *Methods in Ring Theory*, ed. by F. van Ostayen (D. Reidel Publisher, Dordrecht/Boston/London, 1984), pp. 1–36

[Al-M] J. Alajbegović, J. Močkoř, *Approximation Theorems in Commutative Algebra* (Kluwer Academic, Dordrecht, 1992)

[Al-O] J. Alajbegovic, E. Osmanagić, Essential valuations of Krull rings with zero divisors. Commun. Algebra **18**, 2007–2020 (1990)

[AFZ] D. Anderson, M. Fontana, M. Zafrullah, Some remarks on Prüfer $*$-multiplication domains and class groups. J. Algebra **319**, 272–295 (2008)

[Ar] M. Arapović, Approximation theorems for Manis valuations. Can. Math. Bull. **28**(2), 184–189 (1985)

[Bo] N. Bourbaki, *Algebrè Commutative*, Chap. 1–7 (Hermann, Paris, 1961–1965)

[E] O. Endler, *Valuation Theory* (Springer, Berlin, 1972)

[FL] M. Fontana, K.A. Loper, Kronecker function rings: a general approach, in *Ideal Theoretic Methods in Commutative Algebra*, ed. by D.D. Anderson, I.J. Papick. Lecture Notes in Pure and Applied Mathematics, vol. 220 (Marcel Decker, New York, 2001), pp. 189–205

[GP] S. Gabelli, G. Picozza, Star stable domains. J. Pure Appl. Algebra **208**, 853–866 (2007)

[GJ] L. Gillman, J. Jerison, *Rings of Continuous Functions* (D. Van Nostrand, Princeton, NJ, 1960). Reprint Springer 1976

[Gi] R. Gilmer, *Multiplicative Ideal Theory* (Marcel Dekker, New York, 1972)

[Gr] J. Gräter, Der allgemeine Approximationssatz für Manisbewertungen. Mh. Math. **93**, 277–288 (1982)

[Gr_1] J. Gräter, Der Approximationssatz für Manisbewertungen. Arch. Math. **37**, 335–340 (1981)

[Gr_2] J. Gräter, R-Prüferringe und Bewertungsfamilien. Arch. Math. **41**, 319–327 (1983)

[Gr_3] J. Gräter, Über die Distributivität des Idealverbandes eines kommutativen Ringes. Mh. Math **99**, 267–278 (1985)

[G_2] M. Griffin, Valuations and Prüfer rings. Can. J. Math. **26**, 412–429 (1074)

[G_3] M. Griffin, Rings of Krull type. J. Reine Angew. Math. **229**, 1–27 (1968)

[G_4] M. Griffin, Families of finite character and essential valuations. Trans. Am. Math. Soc. **130**, 75–85 (1968)

[Ha-K] F. Halter-Koch, Kronecker function rings and generalized integral closures. Commun. Algebra **31**, 45–59 (2003)

[Ha-K_1] F. Halter-Koch, *Ideal Systems; An Introduction to Multiplicative Ideal Theory* (Marcel Dekker, New York, 1996)

M. Knebusch and T. Kaiser, *Manis Valuations and Prüfer Extensions II*, Lecture Notes in Mathematics 2103, DOI 10.1007/978-3-319-03212-2,

[Ho] M. Hochster, Prime ideal structure in commutative rings. Trans. Am. Math. Soc. **142**, 264–292 (1969)

[HK] R. Huber, M. Knebusch, On valuation spectra. Contemp. Math. **155**, 167–206 (1994)

[Huc] J.A. Huckaba, *Commutative Rings with Zero Divisors* (Marcel Dekker, New York, 1988)

[KZ] M. Knebusch, D. Zhang, *Manis Valuations and Prüfer Extensions I*. Lecture Notes in Mathematics, vol. 1791 (Springer, Berlin, 2002)

[KZ_1] M. Knebusch, D. Zhang, Convexity, valuations and Prüfer extensions in real algebra. Doc. Math. **10**, 1–109 (2005)

[Kr] L. Kronecker, Grundzüge einer arithmetischen Theorie der algebraischen Größen. J. Reine Angew. Math. **92**, 1–122 (1988)

[LM] M. Larsen, P. McCarthy, *Multiplicative Theory of Ideals* (Academic, New York, 1971)

[M] M.E. Manis, Valuations on a commutative ring. Proc. Am. Math. Soc. **20**, 193–198 (1969)

[Me] F. Mertens, Über einen algebraischen Satz. S.-B. Akad. Wiss. Wien (2a) **101**, 1560–1566 (1892)

[N] M. Nagata, *Local Rings* (Interscience Publ./Wiley, New York, 1962)

[Rib] P. Ribenboim, *The Theory of Classical Valuations* (Springer, Berlin, 1999)

[vdW] B. van der Waerden, *Algebra, Zweiter Teil* (Springer, Berlin, 1959)

[W] H. Weyl, *Algebraic Theory of Numbers* (Princeton University Press, Princeton, 1940)

[Z] M. Zafrullah, t-invertibility and Bazzoni-like statements. J. Pure Appl. Algebra **214**, 654–657 (2010)

Subject Index

M. Knebusch and T. Kaiser, *Manis Valuations and Prüfer Extensions II*, Lecture Notes in Mathematics 2103, DOI 10.1007/978-3-319-03212-2,

Symbol Index

M. Knebusch and T. Kaiser, *Manis Valuations and Prüfer Extensions II*, Lecture Notes in Mathematics 2103, DOI 10.1007/978-3-319-03212-2,

LECTURE NOTES IN MATHEMATICS

Springer

Edited by J.-M. Morel, B. Teissier; P.K. Maini

Editorial Policy (for the publication of monographs)

1. Lecture Notes aim to report new developments in all areas of mathematics and their applications - quickly, informally and at a high level. Mathematical texts analysing new developments in modelling and numerical simulation are welcome.
   Monograph manuscripts should be reasonably self-contained and rounded off. Thus they may, and often will, present not only results of the author but also related work by other people. They may be based on specialised lecture courses. Furthermore, the manuscripts should provide sufficient motivation, examples and applications. This clearly distinguishes Lecture Notes from journal articles or technical reports which normally are very concise. Articles intended for a journal but too long to be accepted by most journals, usually do not have this "lecture notes" character. For similar reasons it is unusual for doctoral theses to be accepted for the Lecture Notes series, though habilitation theses may be appropriate.

2. Manuscripts should be submitted either online at www.editorialmanager.com/lnm to Springer's mathematics editorial in Heidelberg, or to one of the series editors. In general, manuscripts will be sent out to 2 external referees for evaluation. If a decision cannot yet be reached on the basis of the first 2 reports, further referees may be contacted: The author will be informed of this. A final decision to publish can be made only on the basis of the complete manuscript, however a refereeing process leading to a preliminary decision can be based on a pre-final or incomplete manuscript. The strict minimum amount of material that will be considered should include a detailed outline describing the planned contents of each chapter, a bibliography and several sample chapters.
   Authors should be aware that incomplete or insufficiently close to final manuscripts almost always result in longer refereeing times and nevertheless unclear referees' recommendations, making further refereeing of a final draft necessary.
   Authors should also be aware that parallel submission of their manuscript to another publisher while under consideration for LNM will in general lead to immediate rejection.

3. Manuscripts should in general be submitted in English. Final manuscripts should contain at least 100 pages of mathematical text and should always include
   - a table of contents;
   - an informative introduction, with adequate motivation and perhaps some historical remarks: it should be accessible to a reader not intimately familiar with the topic treated;
   - a subject index: as a rule this is genuinely helpful for the reader.

   For evaluation purposes, manuscripts may be submitted in print or electronic form (print form is still preferred by most referees), in the latter case preferably as pdf- or zipped ps-files. Lecture Notes volumes are, as a rule, printed digitally from the authors' files. To ensure best results, authors are asked to use the LaTeX2e style files available from Springer's web-server at:

   ftp://ftp.springer.de/pub/tex/latex/svmonot1/ (for monographs) and
   ftp://ftp.springer.de/pub/tex/latex/svmultt1/ (for summer schools/tutorials).

Additional technical instructions, if necessary, are available on request from lnm@springer.com.

4. Careful preparation of the manuscripts will help keep production time short besides ensuring satisfactory appearance of the finished book in print and online. After acceptance of the manuscript authors will be asked to prepare the final LaTeX source files and also the corresponding dvi-, pdf- or zipped ps-file. The LaTeX source files are essential for producing the full-text online version of the book (see http://www.springerlink.com/openurl.asp?genre=journal&issn=0075-8434 for the existing online volumes of LNM). The actual production of a Lecture Notes volume takes approximately 12 weeks.

5. Authors receive a total of 50 free copies of their volume, but no royalties. They are entitled to a discount of 33.3 % on the price of Springer books purchased for their personal use, if ordering directly from Springer.

6. Commitment to publish is made by letter of intent rather than by signing a formal contract. Springer-Verlag secures the copyright for each volume. Authors are free to reuse material contained in their LNM volumes in later publications: a brief written (or e-mail) request for formal permission is sufficient.

Addresses:

Professor J.-M. Morel, CMLA,
École Normale Supérieure de Cachan,
61 Avenue du Président Wilson, 94235 Cachan Cedex, France
E-mail: morel@cmla.ens-cachan.fr

Professor B. Teissier, Institut Mathématique de Jussieu,
UMR 7586 du CNRS, Équipe "Géométrie et Dynamique",
175 rue du Chevaleret
75013 Paris, France
E-mail: teissier@math.jussieu.fr

For the "Mathematical Biosciences Subseries" of LNM:

Professor P. K. Maini, Center for Mathematical Biology,
Mathematical Institute, 24-29 St Giles,
Oxford OX1 3LP, UK
E-mail: maini@maths.ox.ac.uk

Springer, Mathematics Editorial, Tiergartenstr. 17,
69121 Heidelberg, Germany,
Tel.: +49 (6221) 4876-8259

Fax: +49 (6221) 4876-8259
E-mail: lnm@springer.com